HUMAN ACTIVITY AND
ENVIRONMENTAL PROCESSES

Contributors

B. W. Atkinson	Professor of Geography, Queen Mary College, University of London, London E1 4NS, UK
E. C. F. Bird	Reader in Geography, University of Melbourne, Australia
D. R. Coates	Professor of Geology, State University of New York, Binghamton, New York, USA
H. M. French	Associate Professor of Geography, University of Ottawa, Canada
D. Greenland	Professor of Geography, University of Colorado, Boulder, Colorado, USA
K. J. Gregory	Professor of Geography, University of Southampton, Southampton SO9 5NH, UK
C. M. Harrison	Lecturer in Geography, University College, 26 Bedford Way, London WC1E 6BT, UK
A. C. Imeson	Lecturer, Laboratory of Physical Geography and Soil Science, University of Amsterdam, The Netherlands
C. A. M. King	Emeritus Professor of Geography, University of Nottingham, Nottingham NT7 2RD, UK
A. Rapp	Professor of Physical Geography, University of Lund, Sweden
M. J. Selby	Professor of Earth Sciences, University of Waikato, Hamilton, New Zealand
I. G. Simmons	Professor of Geography, University of Durham, Science Laboratories, Durham, DH1 3LE, UK
P. G. Sly	National Water Research Institute, Canada Centre for Inland Waters, Burlington, Ontario, Canada
S. T. Trudgill	Senior Lecturer in Geography, University of Sheffield, Sheffield, S10 2TN, UK
D. E. Walling	Professor of Physical Geography, University of Exeter, Exeter, UK
R. Whitlow	Senior Lecturer in Geography, University of Zimbabwe, Harare, Zimbabwe

Contents

Part I ATMOSPHERE

Introduction

K. J. Gregory and D. E. Walling

Part II HYDROSPHERE

D. E. Walling and K. J. Gregory

D. E. Walling

E. C. F. Bird

Part III   GEOSPHERE

Part IV   PEDOSPHERE

Part V   BIOSPHERE

Part VI CONCLUSION

D. E. Walling and K. J. Gregory

Preface

As explained in the introduction, this book was first published by W. M. Dawson & Sons in 1980 and was entitled *Man and Environmental Processes*. Publication subsequently passed to Butterworths who later decided that they were unable to continue with the series of which the book formed a part. In view of the encouraging comments that we had received about the original book we felt that it was desirable to produce a further edition and John Wiley & Sons agreed to publish this later edition. The present book differs from the original in that most of the original chapters have been retained but have been updated to an extent which is appropriate for their subject. In addition we have included an extra four chapters which are intended to widen the scope to provide useful introductory coverage for the theme of human activity and environmental processes as demonstrated by oceans, lakes, and African vegetation and in relation to desertification.

We hope therefore that the present volume will be even more useful than its predecessor, will indicate something of the interesting analyses that have been undertaken on the theme of human activity and environmental processes, and will also indicate the usefulness of these analyses in relation to environmental management.

In completing this volume we have been very grateful to the contributors, and to the assistance we have received from Mrs June Gandhi, Mrs Rosemary Dempster, and to Mr C. T. Hill who assisted with preparation of the index and final preparation of the manuscript.

January 1986   Ken J. GREGORY, Des E. WALLING

Human Activity and Environmental Processes
Edited by K. J. Gregory and D. E. Walling
©1987 John Wiley & Sons Ltd.

1

Introduction

K. J. GREGORY and D. E. WALLING

> The object of the present volume is: to indicate the character and, approximately, the extent of the changes produced by human action in the physical condition of the globe we inhabit; to point out the dangers of imprudence and the necessity of caution in all operations which, on a large scale, interfere with the spontaneous arrangements of the organic or the inorganic world; to suggest the possibility and the importance of the restoration of disturbed harmonies and the material movement of waste and exhausted regions; and, incidentally, to illustrate the doctrine, that man is, in both kind and degree, a power of a higher order than any of the other forms of animated life, which, like him, are nourished at the table of bounteous nature. . . .
>
> G. P. Marsh, 1864

With these words George Perkins Marsh began the preface of his volume *Man and Nature* (1864). The book was conceived as a 'little volume showing that whereas others think that the earth made man, man in fact made the earth', it proved to have a great impact on the way in which men visualize and use the land (Lowenthal, 1965), and became a fundamental basis for the conservation movement (Mumford, 1931). Although this significant book did include reference to the effects of human activity upon environmental processes its primary aim was to demonstrate the magnitude of the changes wrought by man. This is shown by the way in which chapters were devoted to vegetable and animal species (Chapter 2), to the woods (3), to the waters (4), to the sands (5), and to the projected or possible geographical changes by man (6). The aim of this present volume is to review the effects of human activity on physical environment processes, and this is justified, not only as a complement to the approach taken by G. P. Marsh, but also as a sequel to the work produced since 1864. Contributions since the mid-nineteenth century to the study of the significance of human activity may be visualized in four categories: those produced up to 1960 (Section 1.1), those produced between 1960 and 1970 (1.2), those produced since 1970 (1.3) and the emphasis upon processes which may be detected since 1980 (1.4).

An earlier version of this book entitled *Man and Environmental Processes* (Gregory and Walling, 1980, 1981) was first published in 1980. However, despite the favourable reception which that book received, decisions by the two publishers about the extent of their publishing programme meant that the book is no longer available. We believe that despite the books published since 1980 (see pp. 7) there remains a need for a book which is focused on human activity in relation to process. All the material originally prepared for 1979 has been updated, and in some cases completely rewritten, and in addition new chapters on desertification, lakes and oceans have been included. In this way it is hoped that the content now does broader justice to the theme of human activity and environmental processes.

1.1   A CENTURY OF MILESTONES

Although the importance of the study of human activity is now acknowledged as an integral part of physical geography and of environmental sciences, it is salutary that in the century after the publication of *Man and Nature* there were comparatively few major and comprehensive reviews by geographers of the subject afforded by this theme. Explanation of the paucity of contributions may be found in the fragmentation of physical geography and in the way in which progress in the subject was dominated conceptually by evolution. Whereas physical geography was viewed as an integral discipline in the nineteenth century, so that Huxley embraced air, land surface, vegetation, soil and organisms within the compass of his *Physiography* (1877), in the twentieth century physical geography increasingly separated into geomorphology, climatology, and biogeography. Emphasis on evolutionary concepts provided an intellectual environment of physical geography in which human activity was not accorded prime consideration. Evolutionary ideas which may have originated from the influence of Darwin's *Origin of Species* (1859) became incorporated in conceptual approaches throughout physical geography (Stoddart, 1966a). The cycle of erosion came to dominate geomorphology, and parallels can be found in the concepts of plant succession, in the development of zonal soils and in air mass climatology that pervaded other parts of physical geography.

Despite the evolutionary focus which dominated the branches of physical geography and of environmental science in the first half of the twentieth century there were several contributions which may be regarded as milestones in the elucidation of the impact of human activity upon physical landscape. One of these was provided by R. L. Sherlock in *Man as a Geological Agent*, published in 1922. In this book, and in a related article (Sherlock, 1923), Sherlock developed the theme of the significance of man as an agent in geographical change. This was achieved by distinguishing geological and biological aspects. The latter, which included the effects on plant and animal species, was not included in his book which concentrated on geological aspects with particular attention devoted to denudation by excavation and attrition, to subsidence, to accumulation, to alterations of the sea coast, to the circulation of water and to climate and scenery. Sherlock thus introduced the study of man, emphasized the contrasts between natural and human denudation, proposed that there were indications that the doctrine of uniformitarianism had been carried too far and concluded that, in a densely populated country like England, 'Man is many more times more powerful, as an agent of denudation, than all the atmospheric denuding forces combined' (Sherlock, 1922, p. 333).

Whereas this milestone was general in character, more specific ones became evident in the succeeding twenty years. These were inspired particularly by contemporary problems and, although not devoted exclusively to the influence of human activity, books on soil erosion appeared, reflecting growing awareness of a worldwide problem which necessitated conservation measures. Thus the book by G. V. Jacks and R. O Whyte (1939), *The Rape of the Earth*, provided a world survey of soil erosion as a basis for consideration of control measures, conservation and the political and social consequences. Perhaps most significant was an international symposium, 'Man's Role in Changing the Face of the Earth', an interdisciplinary symposium with international participation, which was organized by the Wenner-Gren Foundation for Anthropological Research and held at Princeton, New Jersey, in 1955. This meeting prompted an 1193-page volume with the same title (Thomas, 1956), which has subsequently been republished in parts. The intellectual stimulation for this impressive venture was acknowledged to be the work of G. P. Marsh and that of the Russian geographer, A. I. Woeikof (1842–1914), who had also developed, independently, a utilitarian approach to the study of the earth's surface which acknowledged the activities of man. This volume embraced more than fifty-two chapters collected in three parts. The first part was retrospective, elaborating the ways in which man has changed the face of the earth; the second reviewed many ways in which processes of seas and land waters, atmosphere, slope and soils, biotic communities, ecology of wastes, and urban and industrial demands upon land have been modified; and the third drew attention to the prospect raised by the limits of man and the earth and the role of man. This monumental achievement certainly provided a milestone in material available to earth scientists by collating research from a broad spectrum of earth science disciplines and by providing an exhaustive compendium of reviews of the state of knowledge in 1956. In the shadow of this large volume the present book must be conceived as a small contribution which focuses on process and particularly on developments during the last two decades.

1.2 A DECADE OF PAPERS

Increasing attention accorded to the significance of human activity in the physical environment was subsequently apparent in general review papers and in more specific research contributions that were published in the decade beginning in 1960. A series of general review papers appeared and a sample of these reflect the general trends which were becoming established. In a view of 'Man and the natural environment', Wilkinson (1963) illustrated the transmogrification of the face of the earth under the impact of the destructive, the conservative and the creative agency of man by illustrating the scale and extent of human activity in relation to the atmosphere, the lithosphere and hydrosphere, the soil, and the biosphere. The previous deficiency of studies of the form-creating activity of man and of the influence of man on natural phenomena was regretted by Fels (1965) who advocated the necessity of study of anthropogenous geomorphological processes. J. N. Jennings in a presidential address in Australia (Jennings, 1966) revived the title used by Sherlock, of 'Man as a geological agent', for his review in which he stressed that the greater significance accorded to man arises because of more studies of process and also because measurements of contemporary processes are nearly always heavily biased by anthropogenic effects. The geomorphological significance of man was also reviewed in a significant paper by Brown (1970) in which he characterized man as a geomorphological process in relation to his direct, purposeful modifications of landform and to indirect effects, and this involved reference to the influence of human activity upon geomorphological processes. General statements also occurred in the USSR and for some this has involved the designation of the influence of man as generating a new geological epoch, the Noosphere. This concept involves the active role played by the conscious mind and science in the purposefully directed, and not simply haphazard, development of the man–nature relationship (Trusov, 1969). It is a concept which is capable of wider use in geography (Bird, 1963).

Perhaps most significantly at the end of this decade Chorley explored the role and relations of physical geography (Chorley, 1971) and this was succeeded by affirmation of the possibilities of a systems approach (Chorley and Kennedy, 1971). One of the systems proposed was the control system, which is concerned particularly with man and the way human activity operates as a regulator in natural systems. In the ecosystem state control merely involves the manipulation of the negative-feedback loops in order to stabilize the system operation at some optimum state (Chorley, 1973). In an evaluation of this approach in relation to others Chorley (1973) noted, 'It is clear, however, that social man is, for better or worse, seizing control of his terrestrial environment and any geographical methodology which does not acknowledge this fact is doomed to inbuilt obsolescence'.

This sentiment was therefore emerging in a number of general conceptual papers published in the 1960s and the ideas have continued to develop subsequently as well. Thus the problem of Quaternary relief development and man was reviewed by Demek (1973) against the fact that 55 per cent of the world's dry land surface is intensively used by man, that 30 per cent is partly modified by man, and the remaining 15 per cent is only slightly modified if at all.

Such general movements in the focus of geographic enquiry were accompanied in this decade by at least four other trends within and beyond the canvas of writings by geographers. Developments within the subject included firstly an increasing awareness of the need to document and to study processes, and secondly growth of interest in the hazards presented by the natural environment pertinent to human activity. More widely across the spectrum of the earth sciences there was, thirdly, the initiation of international programmes which incorporated the effects of man and, fourthly, the emergence of the worldwide concern for the exhaustibility of earth resources in relation to human activity on spaceship earth.

The first of these themes, the emphasis upon environmental processes, occurred as a necessary requirement to permit the further development of models of the development of the physical environment. Therefore as rates of erosion were deduced and compared for various world areas it was demonstrated, for example, that erosion rates near Rome, Italy, are now between 100 and 1000 m^3 km^{-2} per year whereas they were between 20 and 30 m^3 km^{-2} per year prior to man's influence (Judson, 1968). A number of studies investigated the effects of man's influence expressed through land use pressure as reflected, for example, in the activity of gullying in the southwest USA (Denevan, 1967) and in fires in relation to floods in the Bow Valley of Alberta (Nelson and Byrne, 1966). In some areas, study of processes, of the influence of man upon these processes and of the environmental effects of man's role as a regulator of these processes proceeded to introduce greater concern for an evaluation of the engineering strategies available to minimize man's influence. Thus in areas of permafrost greater knowledge of permafrost equilibrium in relation to human activity is a prelude to an understanding of the solutions whereby disturbance of the permafrost by man can be minimized (e.g. Ferrians *et al.*, 1969).

In the field of climatology the effects of man upon atmospheric processes and in the creation of man-induced climates attracted studies including the analysis of urban climates epitomized by an examination of the climate of London (Chandler, 1965). Studies of man's effect on soils had been available for two decades in the form of works on soil erosion but these were complemented by investigations of the less obvious ways whereby soil processes and characteristics had been affected by man. Thus in the area adjacent to the Silesian industrial region, Gilewska (1964) identified the changes in the geographic environment brought about by industrialization and urbanization, including the incidence of metal ions in soils derived from emission into the atmosphere from industrial sources. The influence of man has always been acknowledged in the biogeographic field, but this also received renewed attention, as for example from Fosberg (1963) who reviewed man's place in the island ecosystem. Greater use of the ecosystem concept thus allowed evaluation of the significance of man in particular situations such as the effects of catastrophic human interference with coral island ecosystems (Stoddart,

1966b) and comparison of the productivity of agricultural with the natural systems which man replaces. A number of studies therefore embraced consideration of the significance of recent plant invasions (Harris, 1966).

A second theme, which often stemmed from consideration of process extremes, was introduced by investigations of natural hazards. Because a physical event becomes a hazard only when the mechanics of physical environment are treated in relation to human activity, this approach provided an additional impetus for the growing interest in the significance of human activity. The initial interest was devoted to the flood hazard but this was soon complemented by consideration of the range of environmental hazards (Burton *et al.*, 1968). The way in which such an approach provided an integrated theme especially appropriate in geography is exemplified by the appearance of books like *The Value of the Weather* (Maunder, 1970) and *Water, Earth and Man* (Chorley, 1969).

Two additional influences which encompassed and affected, but did not originate within, geography were provided by international programmes and by environmental concern. A number of international programmes were inaugurated. They included the International Hydrological Decade (1965–74), which dealt with man's influence as one of the salient themes, and the later Man and the Biosphere programme (1970) which patently devolves upon the magnitude of human influence.

Environmental concern was generated during the decade and stemmed from salutary warnings such as *Silent Spring* (Carson, 1962), from consideration of *Man and Environment* (Arvill, 1967) and subsequently from debates about population and the limits to growth (Ehrlich and Ehrlich, 1970). Subsequently in a work entitled *The Environmental Revolution*, Nicholson (1972) provided a chart of human impacts on the countryside, of the areas affected and of the consequent effects.

1.3   A DECADE OF READINGS

One consequence of the 1960s may therefore be seen as the development of specific studies of man's influence on particular processes and sections of environment, accompanied by general conceptual attention accorded to human activity. These developments internal to geography were achieved within an intellectual environment which embraced growing concern for the effects of man in the past and concern for his future, and this provided one of the motivating reasons for the initiation of international research programmes.

The culmination of these trends and their continuation into the decade of the 1970s found the available geographic literature not entirely appropriate. This deficiency therefore was remedied by a number of books which were collections of papers or readings. Some of these were specially commissioned. The Association of American Geographers Commission on College Geography in

1971 conceived a collection of essays devoted to geographic research on environmental problems (Manners and Mikesell, 1974). Although the ensuing volume was not specifically devoted to human activity, this theme was prominent in each of its twelve chapters. An alternative variant in books produced in the 1970s was demonstrated by collections of previously published papers often derived from a wide spectrum of journals. Thus Detwyler (1971) collated previously published papers to provide *Man's Impact on Environment* and Coates (1972, 1973) provided volumes of readings on *Environmental Geomorphology and Landscape Conservation*. These volumes drew together papers from a wide range of environmental publications and also showed how papers of significance existed prior to 1970. In a volume devoted to non-urban areas (Coates, 1973) the papers were organized into sections dealing with man-induced terrain degradation, soil conservation and landscape management, and of the twenty-six papers reprinted in the volume, half were originally published before 1960, indicating that a number of early significant papers existed even if they were sporadically disseminated throughout the literature.

Subsequently a number of volumes composed of edited collections or of readings appeared which were devoted to particular fields germane to the theme of man and environment. Most numerous amongst these were a number of volumes devoted to urban physical geography. This theme was covered broadly by *Urbanization and Environment* (Detwyler and Marcus, 1972) and more specifically by *Urban Geomorphology* (Coates, 1976).

The collection of papers previously published as readings or as edited collections was a convenient way of juxtaposing much useful and potentially related material. Thus, in the context of cities, *Cities and Geology* (Legget, 1973) was one of the books which advocated that attention be paid to environmental geology. Such an environmental geology has been defined in several ways but provided a very fertile way of collating the aspects of geology pertinent to human activity (Keller, 1976; Coates 1981).

1.4 EMPHASIS UPON PROCESSES

Increasing attention devoted to the study of the impact of human activity upon environment was often justified because, if we know the magnitude and effects of human activity in the past, then it may be feasible to make use of that knowledge in environmental management and decision-making in the future. However, whereas the initial contributions had focused on the nature and extent of the human impact (section 1.2) and, subsequently, the precise nature of the impact were explored (1.3), it was realized that only by a more complete understanding of processes would the optimum application of research units be achieved. Therefore, although the human impact was explored in a broad way in a textbook (Goudie, 1981) structured around the effect of human activity on environment and on processes in the conventional subdivisions of physical geography, namely vegetation, animals, soil, water, geomorphology and atmosphere, and although the theme has continued to be popular in environmental geology as exemplified in Janet Watson's *Geology and Man* (1983) and in a more environmental physical geography (e.g. Drew, 1983; Wilcock, 1983; White *et al.*, 1984), there is still a need for a focus on process. Emphasis upon processes is included in volumes recently published to deal with specific branches of physical geography, including climatology (Bach *et al.*, 1979), geomorphology (Nir, 1983) and hydrology (Hollis, 1979), but this book endeavours to cover all branches of physical geography.

Such a focus can be vindicated in view of four trends which may be detected in, or which are reflected by, recent research. *First* has been the continuation of concern for environment and this has been acknowledged in general reviews in particular disciplines as exemplified by Jones's (1983) article on *environments of concern*. More generally there has been greater acknowledgement of the changes which may be assuming large-scale and sometimes world proportions. Such changes include desertification, particularly as expressed in Africa, deforestation, which in an area as large as Amazonia could have a feedback effect on global climate, and world climatic change influenced by major changes including atmospheric CO_2 concentrations. Whereas in the 1970s the extent to which CO_2 concentration could have an influence upon world temperature was debated, by the 1980s the consequent temperature increase was established but debate continued about the magnitude and the possible consequences. Thus the ICSU–WMO–UNEP conference at Villach, Austria, in 1980 issued a statement about the possible effects of atmospheric CO_2 and concluded that a major international and interdisciplinary research effort is needed to develop a set of impact scenarios — scenarios that affect developing as well as developed countries — and called for a special partnership of effort (*Nature and Resources*, 1981).

It is still not always appreciated exactly what the effects of human activity may be, so *secondly* there have been studies undertaken to show what results may arise from particular scenarios of human activity. One of the largest excavations on the earth's surface at present is the magnetite working at Zhelenogorsk in the USSR, where a depression $5 \, km \times 3 \, km \times 175 \, m$ in dimensions had been excavated by the late 1970s and where excavation to a depth of 500 m is expected by the next century. The local effects on atmospheric dustiness and on groundwater quality have been the subject of careful monitoring.

To use the results from such empirical investigations it is necessary *thirdly* to devise new conceptual methods or approaches, that are extensions of previous ones, to facilitate the optimum use of empirical data. There are at least three aspects of such advance in conceptualization. First there has been the emphasis upon particular environments and Douglas (1983), for example, has focused attention upon the integrity of the urban system and has advocated an emphasis upon the energy flows in that system. Second there have been conceptual advances which could be valuable inputs to model construction; whereas linear relationships are not always appropriate for modelling environmental change due to human activity, techniques centred on step functions (Dury, 1982) or on catastrophe theory (Wilson, 1981) all afford considerable potential. Some models have now been constructed which allow more precise modelling of human impact and general circulation models (GCMs) can be employed to indicate the pattern and magnitude of global climatic change consequent upon specific types of modification. This enables quantitative values to be placed upon world patterns and supplements the use of analogues as suggested by Butzer (1980) when he used the world distribution of climates 5000 to 8000 years ago as a potential analogue to indicate the way in which global climates could be manifested in 2050. Further refined global models should usher in a future decade of more reliable prediction and of more precise environmental impact statements.

Such modelling is a useful adjunct to a *fourth* trend which may be detected in the 1980s and this is a concern with environmental management. Although not all authors have agreed that the potential benefits have been fully realized (e.g. Johnston, 1983), it is inevitable that a greater understanding of the ways in which processes are modified and an enhanced use of predictive models should allow the development of a greater contribution to environmental management. However, whereas in the past it may have been assumed that applicable research may contribute to management, such contributions are only of implicit relevance to questions of management. For research to be really applied and explicitly relevant to management it is necessary for contributions to be directed towards impact, prediction and design (Gregory, 1985).

1.5   A BOOK FOR PROCESSES

It was against this background of a shift in emphasis within physical geography, which crystallized in the 1960s and which was supported in the 1970s by the publication of a number of edited books and volumes of readings, that this book was conceived. It seemed timely to envisage a volume specially written to review the significance of human activity to the range of environmental processes studied by the physical geographer. The processes considered are those which have traditionally been studied by the physical geographer, so pollution and global element cycles are not included. However, the contents, although composed in the light of recent foci in physical geography should be of interest to other disciplines in the earth and environmental sciences.

Such an endeavour cannot achieve completeness, but the use of contributions from authors from several continents was designed to provide as international a perspective as possible. Although each author reviews the significance of human activity for physical landscape processes, the approach adopted necessarily varies, partly because the emphasis always has and still does differ between the component branches of the subject. This volume is directed primarily towards land system processes with chapters grouped under the major headings of atmosphere, hydrosphere, geosphere, pedosphere and biosphere. Certain important ingredients of these spheres are given less attention, partly inevitably and partly as a reflection of the research activity of physical geographers and their tendency to ignore the oceans, which are represented by a single chapter.

The Chateau de Val, reconstructed in the fifteenth century, is now virtually surrounded by a reservoir created in 1951 when the dam was built at Bort-les-Orgues to impound the headwaters of the Dordogne for the generation of hydroelectric power. Today the Chateau figures within what is obviously a man-made lake and such direct influences of man are rapidly deduced. The less obvious and indirect effects of the man-impounded lake on the downstream flows and river channel, on the local water balance through evaporation and on plant communities are less immediately apparent. It is therefore to such indirect effects on process that this book is dedicated. In *Man and Nature* (1864) George Perkins Marsh titled the final chapter 'Projected or Possible Geographical Changes by Man'. The range of future changes which could be induced by man is now much greater than could be imagined in 1864, but a knowledge of the possible consequences for landscape processes is just as significant! There are implications for physical geography because in the past we may, according to Hewitt and Hare (1973), have been captives of tradition. These authors argued that '. . . the geographer, when he analyses the properties of man–environment systems, must base himself on the central functions of that system, rather than on the traditional divisions of physical geography' (Hewitt and Hare, 1973). By focusing more on human influence on physical landscape processes we may achieve a vital and relevant physical geography. As Hare (1985) has recently noted, the public demand is for prediction. Progress towards more effective prediction depends upon knowledge of the way in which human activity has influenced environmental processes and this may provide an important key to a more vital and relevant physical geography.

REFERENCES

Arvill, R., 1967, *Man and Environment*, Penguin, Harmondsworth, 332 pp.
Bach, W., Pankrath, J., and Kellog, C. E. (Eds), 1979, *Man's Impact on Climate*, Elsevier, Amsterdam.
Bird, J. H., 1963, 'The noosphere: a concept possibly useful to geographers', *Scottish Geographical Magazine*, **79**, 54–56.
Brown, E. H., 1970, 'Man shapes the earth', *Geographical Journal*, **136**, 74–85.
Burton, J., Kates, R. W., and White, G. G., 1968, 'The human ecology of extreme geophysical events', *Natural Hazard Research Working Paper No. 1*, Department of Geography, University of Toronto.
Butzer, K. W., 1980, 'Adaptation to global environmental change. *Professional Geographer*', **32** 269–278.
Carson, R. L., 1962, *Silent Spring*, Penguin, Harmondsworth.

Chandler, T. J., 1965, *The Climate of London*, Hutchinson, London.

Chorley, R. J. (Ed.), 1969, *Water, Earth and Man*, Methuen, London, 588 pp.

Chorley, R. J., 1971, The role and relations of physical geography, *Progress in Geography*, **3**, 87–109.

Chorley, R. J., 1973, 'Geography as human ecology', in Chorley, R. J. (Ed.), *Directions in Geography*, Methuen, London, 155–169.

Chorley, R. J., and Kennedy, B. A., 1971, *Physical Geography: A Systems Approach*, Prentice-Hall, London, 370 pp.

Coates, D. R. (Ed.), 1972, *Environmental Geomorphology and Landscape Conservation, Volume I, Prior to 1900*, Benchmark papers in Geology, Dowden, Hutchinson & Ross, Stroudsburg.

Coates, D. R. (Ed.), 1973, *Environmental Geomorphology and Landscape Conservation, Volume III, Non-urban regions*. Benchmark Papers in Geology, Dowden, Hutchinson & Ross, Stroudsburg, 483 pp.

Coates, D. R. (Ed.), 1976, *Urban Geomorphology*, Geological Society of America, Special Paper 174.

Coates, D. R., 1981, *Environmental Geology*, Wiley, Chichester, 701 pp.

Demek, J., 1973, 'Quaternary relief development and man', *Geoforum*, **15**, 68–71.

Denevan, W. M., 1967, 'Livestock numbers in nineteenth-century New Mexico and the problem of gullying in the Southwest', *Annals Association America Geographers*, **57**, 691–703.

Detwyler, T. R., 1971, *Man's Impact on Environment*, McGraw-Hill, New York, 731 pp.

Detwyler, T. R., and Marcus, M. G. (Eds), 1972, *Urbanization and Environment: The Physical Geography of the City*, Duxbury Press, Belmont, California.

Douglas, I., 1983, *The Urban Environment*, Edward Arnold, London, 229 pp.

Drew, D. P., 1983, *Man–Environment Processes*, George Allen & Unwin, London, 135 pp.

Dury, G. H., 1982, 'Step-functional analysis of long records of streamflow', *Catena*, **9**, 379–96.

Ehrlich, P. R., and Ehrlich, A. H., 1970, *Population, Resource, Environment: Issues in Human Ecology*, W. H. Freeman, San Francisco.

Fels, E., 1965, 'Nochmals: Anthropogen Geomorphologie', *Petermanns Geographische Mitteilungen*, **109**, 9–15.

Ferrians, O. J., Kachadoorian, R., and Greene, G. W., 1969, Permafrost and related engineering problems in Alaska, *US Geological Survey Professional Paper 678*.

Fosberg, P. R., 1963, *Man's Place in the Island Ecosystem*, Symposium Tenth Pacific Science Congress, Honolulu, Bishop Museum Press, Bishop.

Gilewska, A., 1964, 'Changes in the geographic environment brought about by industrialization and urbanization', *Problems of Applied Geography*, **2**, 201–210.

Goudie, A. S., 1981, *The Human Impact, Man's role in Environmental Change*, Basil Blackwell, Oxford, 316 pp.

Gregory, K. J., 1985, *The Nature of Physical Geography*, Edward Arnold, London, 262 pp.

Gregory, K. J., and Walling, D. E. (Eds), 1980, *Man and Environmental Processes*, Dawson, Folkestone, 276 pp.

Gregory, K. J., 1981, *Man and Environmental Processes*, Butterworths, London, 276 pp.

Hare, F. K., 1985, 'Future environments; can they be predicted?', *Transactions Institute of British Geographers*, **10**, 131–137.

Harris, D. R., 1966, 'Recent plant invasions in the arid and semi-arid south-west of the United States', *Annals Association American Geographers*, **56**, 408–422.

Hewitt, K., and Hare, F. K., 1973, 'Man and environment. Conceptual frameworks', *Association of American Geographers Commission on College Geography Resource Paper no. 20*.

Hollis, G. E. (Ed.), 1979, *Man's Impact on the Hydrological Cycle in the United Kingdom*, Geobooks, Norwich.

Huxley, T. H., 1877, *Physiography; an Introduction to the Study of Nature*. Macmillan, London.

Jacks, G. V., and Whyte, R. O., 1939, *The Rape of the Earth*, Faber, London.

Jennings, J. N., 1966, 'Man as a geological agent', *Australian Journal of Science*, **28**, 150–156.

Johnston, R. J., 1983, 'Resource analysis, resource management, and the integration of physical and human geography', *Progress in Physical Geography*, **7**, 127–146.

Jones, D. K. C., 1983, 'Environments of concern', *Transactions Institute of British Geographers*, **N38**, 429–457.

Judson, S., 1968, 'Erosion rates near Rome, Italy', *Science*, **160**, 144–146.

Keller, E. A., 1976, *Environmental Geology*, C. E. Merrill, Columbus.

Legget, C. F., 1973, *Cities and Geology*, McGraw-Hill, New York.

Lowenthal, D. (Ed.), 1965, *Man and Nature by George Perkins Marsh*, Harvard University Press, Cambridge, Massachusetts.

Manners, I. R., and Mikesell, M. W., 1974, Perspectives on environment, *Association American Geographers Publication no. 13*.

Marsh, G. P., 1864, *Man and Nature or Physical Geography as Modified by Human Action*, Charles Scribner, New York.

Maunder, W. H., 1970, *The Value of the Weather*, Methuen, London, 388 pp.

Mumford, L., 1931, *The Brown Decades: a Study of the Arts in America 1865–1895*, Dover, New York.

Nature and Resources, 1981, 'CO_2 and the effects of human activities on climate: a global issue', *Nature and Resources*, **17** (3), 2–5.

Nelson, J. G., and Byrne, A. B., 1966, 'Man as an instrument of landscape change. Fires, floods, and national parks in the Bow Valley, Alberta', *Geographical Review*, **56**, 226–238.

Nicholson, M., 1972, *The Environmental Revolution*, Penguin, Harmondsworth.

Nir, D., 1983, *Man as a Geomorphological Agent*, Reidel, Dordrecht, 165 pp.

Sherlock, R. L., 1922, *Man as a Geological Agent*, Witherby, London, 372 pp.

Sherlock, R. L., 1923, 'The influence of man as an agent in geographical change', *Geographical Journal*, **61**, 258–273.

Stoddart, D. R., 1966a, 'Darwin's impact on geography', *Annals Association American Geographers*, **56**, 683–698.

Stoddart, D. R., 1966b, 'Catastrophic human interference with coral island ecosystems', *Geography,* **53**, 25–40.

Thomas, W. L., 1956, *Man's Role in Changing the Face of the Earth*, University of Chicago Press, Chicago, 1193 pp.

Trusov, Yuili, 1969, 'The concept of the noosphere', *Soviet Geography*, **10**, 220–236.

Watson, J., 1983, *Geology and Man: an Introduction to Applied Earth Science*, George Allen & Unwin, London.

White, I. D., Mottershead, D. N., and Harrison, S. J., 1984, *Environmental Systems: An Introductory Text*, George Allen & Unwin, London, 495 pp.

Wilcock, D. N., 1983, *Physical Geography*, Blackie, London, 218 pp.

Wilkinson, H., 1963, 'Man and the natural environment', Department of Geography, University of Hull, Occasional Papers in Geography No. 1.

Wilson, A. G., 1981, *Geography and the Environment. Systems Analytical Methods*, Wiley, Chichester, 297 pp.

PART I
ATMOSPHERE

INTRODUCTION

The role of human activity in relation to atmospheric processes has been studied extensively for many years and yet this subject is also the focus of much contemporary research interest. The longstanding interest in the effect of human activity in relation to atmospheric processes has devolved upon the deliberate change of meso-climates by the construction, for example of shelter belts or wind-breaks. In such ways human action has been responsible for the moderation of climates in relation to human activity. On a larger scale, interest in the deliberate modification of climate has occurred as there have been attempts at direct intervention, for example by cloud seeding or by hurricane modification. Of a similar nature is the way in which the climatology of human comfort has been developed by the use of clothing of various kinds and by the development of housing technology specifically to create a favourable and artificial climatic environment. Interest in such methods of climatic process modification have existed for several decades. Much more recently, however, there has been a greatly increased interest in problems which have assumed a national, international or sometimes global dimension. Such problems embrace the problem of acid rain, the increased carbon dioxide content of the atmosphere that has been associated with atmospheric warming, and the impact of high-flying supersonic aircraft which have been suggested to be responsible for modification of the ozone layer.

Modification of atmospheric processes as a result of human activity has therefore been studied over several decades and continues at the present time with attention to such global problems. These approaches illustrate two major reasons for interest in the effect of human activity on the physical environment. The first reason arises from the need to know how physical processes and the physical environment have deliberately been modified, and this is exemplified by study of shelter belts or subjects in the field of micro-climatology. Secondly, however, it is necessary to know the consequences of inadvertent modification of the earth's atmosphere and associated atmospheric processes. Therefore, it is necessary that we begin to estimate the significance of increased carbon dioxide content of the atmosphere and other global changes. Between these two extremes, of a longstanding interest in deliberate modification of atmospheric processes and more recently of concern for the consequences of inadvertent modification of climatic processes on a world scale, there has been a development of interest in climatology at different scales. Particularly at the meso-scale the physical geographer has been much concerned with the field of boundary layer climatology (Oke, 1978). In his text Oke analysed the boundary layer climates of the lowest kilometre of the atmosphere and dealt with natural climates of non-vegetated and vegetated surfaces, topographically affected climates and the climates of animals and then he proceeded to characterize man-modified environments embracing conscious and non-conscious modification of boundary layer climates, particularly including those of urban areas. Also at the meso-scale there are developments in the field of agroclimatology.

We now have the situation, therefore, where human activity can be analysed in relation to climatological processes at a series of spatial scales. It is possible to focus upon the effects of human action at the micro-climatological, the meso-climatological and the macro-climatological scale. In the latter case, which is concerned with world atmospheric circulation, it is now possible to use general circulation models (GCM) to indicate how the effects of human activity through forcing functions may cause macro-climatological changes to occur. Thus GCMs can be employed to indicate how increasing carbon dioxide content of the atmosphere in the future may lead to different climatic scenarios.

In the two succeeding chapters it is inevitable that the extremely large subject area cannot be completely covered. However, the subjects of the radiation balance covered by D. Greenland and the significance of human activity at the meso-scale of climatology as investigated by B. W. Atkinson are two essential subject areas of considerable prominence.

REFERENCE

Oke, T. R., 1978, *Boundary Layer Climates*, London, Methuen.

Human Activity and Environmental Processes
Edited by K. J. Gregory and D. E. Walling
©1987 John Wiley & Sons Ltd.

2

The Radiation Balance

D. GREENLAND

2.1 INTRODUCTION

Throughout the period of human presence on earth people have stood in awe of the sun. Every single sunrise (Fig. 2.1) moves the human spirit. The intuitive feeling of the importance of the sun to life on earth has often been expressed in a religious manner, as in the time of the fourth dynasty of ancient Egyptian civilization when the sun god was known as Re. Only in the last one hundred years or so have we really begun to realize the full importance of the sun. Radiation from the sun is the starting point for most physical systems on the earth's surface and in its atmosphere. It is the major energy source for moving great currents of air and ocean water around the globe. These currents make the heat from radiation more equitably distributed over the earth so that a large part of the planet has temperatures in a range habitable for human beings. As an energy source for evaporating water, radiation drives the hydrologic cycle of the earth thus providing, through precipitation, a continually renewed source of fresh water for the world's agriculture and life. Through the process of photosynthesis solar energy is changed into the chemical energy and material of plants that are later consumed by humans and other animals to sustain their life. Without solar energy all ecosystems and many physical systems on the planet would come to a halt. Whereas in the past both ancient and modern civilizations have tended to take the presence of solar energy for granted, in the last few years we have gradually realized that certain kinds of human activity can alter the values of its flow. On a small scale we can increase the amount of energy absorbed at the earth's surface and melt unwanted snow or inadvertently reduce incoming radiation by polluting the air of our cities. On the global scale it has become apparent that human activity through nuclear warfare could vastly reduce solar energy input to the earth and its ecosystems and conceivably even bring them to a stop. The continued existence of our species will be determined by our careful management of the inflow and outflow of the truly vital radiant energy to and from the planet which collectively is known as: the radiation balance.

FIGURE 2.1   Sunrise. The beginning of another cycle of shortwave radiation input to the earth's physical system

This chapter will first consider the laws governing the flows and balances of radiation in the earth–atmosphere system. Then attention will be turned to some of the ways that nature herself alters the balance. Dust from erupting volcanoes, for example, has a marked effect. Most space, however, will be devoted to the many ways in which humans can, and do, alter the radiation balance. This occurs on local, regional and global scales with interactions taking place between the scales. Considering the importance of radiation to life on earth, the implications of human influence on the radiation balance are profound. Just how profound is seen by speculation towards the end of this chapter of ways in which it seems possible that our activities could vastly alter the climate of the planet.

2.2   RADIATION LAWS AND THE RADIATION BALANCE

The sun acts as a huge nuclear reactor in which hydrogen–helium fusion takes place and mass is transformed into energy. The vast majority of this energy is beamed into space in what is called electromagnetic radiation. This radiation has a dual character. In some ways it behaves like particles. Atmospheric scientists, however, are concerned with its more common behaviour when it acts in the form of waves. Electromagnetic waves can pass through empty space and through materials that have any degree of transparency. Consequently, some of the radiation emitted by the sun can pass through space and be intercepted by the earth.

There are two basic physical laws which help our understanding of the flow and nature of radiation. The first, known as Wien's displacement law, states that the wavelength of radiation is inversely proportional to the absolute temperature of the emitting body. Thus the sun radiating at 6000°K has a radiation with wavelengths of between 0.2 and 10 micrometres (one micrometre is one millionth (or 10^{-6}) of a metre). This is called shortwave radiation in the meteorological context, and is short relative to the longwave radiation which is emitted from the earth's surface and the atmosphere. The latter radiate at about 300°K and the consequent radiation has wavelengths between 4 and 50 micrometres. A second law is the Stefan–Boltzmann law which tells us that the amount of flow, or flux, of radiation is proportional to the fourth power of the absolute temperature of the radiating body. Therefore, because of the temperatures given above, a far greater amount of radiant energy is emitted from the sun than from the earth.

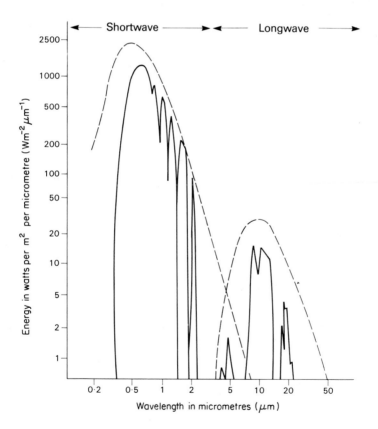

FIGURE 2.2 Distribution of radiation intensity with wavelength for atmospheric radiation (redrawn from Physical Climatology by W. D. Sellers, by permission of University of Chicago Press)

The results of the action of these two laws are seen in Fig. 2.2. The dashed lines show the amount of radiation coming from perfect emitters at 6000°K (left) and 300°K (right). The solid lines indicate, in the short wavelengths, the amount of direct-beam radiation arriving at the earth's surface, and, in the long wavelengths, the value of longwave radiation emitted from the earth's surface. The irregularity of the solid lines indicates the many wavebands where both short- and longwave radiation are absorbed by atmospheric constituents. Shortwave radiation with wavelengths less than 0.3 micrometres is absorbed by ozone and that with wavelengths above 0.7 micrometres is absorbed by water vapour and carbon dioxide. The main absorbers of longwave radiation are also water vapour (6.3 to 7.7 and greater than 20 micrometres), ozone (9.4 to 9.8 micrometres), carbon dioxide 13.1 to 16.9 micrometres) and clouds, which can absorb in all wavelengths. If the amount of any of these substances in the atmosphere is altered it is likely to affect the amount of radiant heat absorbed by the atmosphere.

As a practical aid to help in the understanding of the nature of the radiant flow and the various flows to and from the earth's surface, the method of measurement of the flows at the surface may be described. Fig. 2.3 illustrates the necessary measuring sensors. In this case they are located at 3490 m over alpine tundra near the continental divide in the Rocky Mountains of North America. The two glass-domed sensors on the left measure incoming shortwave

FIGURE 2.3   Platform with sensors for measuring various flows of radiant energy to and from the earth's surface

radiation. They measure both *direct radiation* from the sun and *diffuse radiation* that arrives at the earth's surface after being scattered from molecules of the gases of the atmosphere, clouds and any particulate matter in the atmosphere. It is important to note that the more clouds and particulate matter existing in the atmosphere the more scattering there will be and the higher will be the value of the ratio of diffuse to direct radiation. The direct and diffuse radiation together make up *incoming shortwave radiation*. These two sensors are not identical since one has a glass dome which lets through the visible part (0.40–0.74 micrometres) of shortwave radiation, and the other has a quartz dome permitting the passage of both visible and ultraviolet (0.15–0.4 micrometres) radiation. Subtraction of the recorded values of the latter from those of the former allows an estimation of the receipt of ultraviolet light. The shortwave sensor in the middle of the platform is surrounded by a shadow band which cuts out direct radiation from the sensor resulting in a measurement of only diffuse radiation. Some of the shortwave and longwave radiation from the atmosphere that reaches the ground is absorbed and can be used to heat the ground while another part of the radiant beams is reflected. Two sensors in the centre at the right-hand end of the platform measure the incoming shortwave radiation and the outgoing reflected shortwave radiation. The ratio of reflected radiation to incoming radiation is called the *albedo* of the surface. The value of the albedo is critical in determining the values of the upward and downward radiant flows and the amount of heat absorbed by a body. Finally, three sensors at the right-hand end of the platform measure various flows of long- and shortwave radiation together. The operative sensors on these three instruments are covered by polyethylene which permits the passage of both long- and shortwave radiation. One instrument measures the algebraic sum of the total upward and downward short- and longwave flows to which we give the name *net radiation*. Another measures the upward fluxes of short and longwave radiation while a third monitors the downward fluxes of short and longwave radiation. Appropriate subtraction of the values of the measured flows of upward and downward shortwave radiation yields estimates of values of upward and downward flows of longwave radiation.

A special form of the law of conservation of energy demands that the radiation balance (balance of radiation or net radiation) at the earth's surface will be equal to the sum of upward and downward flows of short- and longwave radiation. The concept may be applied not only to the earth's surface but to higher levels of the atmosphere, the complete atmosphere, or to the earth–atmosphere system as a whole.

There are many ways in which the values of the short- and longwave upward and downward flows may vary. Of particular concern are possible changes in the amount of shortwave radiation reaching the earth's surface, changes in the albedo of the surface or atmosphere, and changes in the composition of the atmosphere which might affect its albedo or its power to absorb radiant heat.

In order to put human-induced changes of the radiation balance into perspective it is useful to first examine some natural changes occurring in the system.

2.3 NATURAL ALTERATION OF THE RADIATION BALANCE

There are many short-term variations in the composition of the atmosphere due to changing humidity and cloud conditions and to the variation of impurity content, so that it is difficult to define a 'normal' atmospheric composition. Yet there are some states which are clearly abnormal relative to the time scale of the present century. Two examples of these are the possibility of variation of the amount of energy from the sun actually arriving at the outer part of the earth's atmosphere and the possibility of excessive quantities of dust being placed in the atmosphere by frequent and intense volcanic eruptions.

The amount of solar radiation of all wavelengths received per unit time and unit area at the outer part of the earth's atmosphere is called the *solar constant*. It is becoming increasingly clear that the value of the solar constant is not necessarily constant over time. Measurements from a satellite in 1980 for about a 200-day period indicated a mean value of the solar constant of $1368.31 \, \text{W m}^{-2}$ with mean variations around ± 0.05 per cent but with two significant decreases of 0.2 per cent which lasted about a week (Willson *et al.*, 1981). Earlier rocket flights had detected a variability of ± 0.3 per cent over a two-year period. There are a number of pieces of direct and proxy evidence suggesting an 11 (or some multiple of 11) -year periodicity in solar activity which might have an influence on terrestrial and atmospheric phenomena (Pittock, 1983). Furthermore, variations in the earth's orbital parameters which change the latitudinal and seasonal distribution of radiation received from the sun and have periodicities of about 20 000, 40 000, and 100 000 years are believed by many to account in part for some of the major climatic changes occurring on the earth over the last two million years or so (Kerr, 1978). At the very least it can be said that a change in the value of the solar constant or the distribution of radiation in time and space across the earth will have some effect on the climate of the earth.

In addition to the possibility of changing solar output, the transparency of the atmosphere may alter, especially when large amounts of dust are put into it. In the natural world, volcanic eruptions spew enormous masses of particulate matter and gases into the atmosphere. The Krakatoa eruption of 1883 is said to have injected $54 \, \text{km}^3$ of ash, lava and mud into the atmosphere. Following this eruption, shortwave radiation arriving at the earth's surface was reduced by about 10 per cent. Bryson and Murray (1977, p. 147) quote a report that, after the eruption, the Fire Department in Poughkeepsie, New York, were called out because the brilliant sunset, due to the scattering of red light by dust in the atmosphere, was mistaken for the glow of a fire in the western part of the sky. More recently, after the eruption of the El Chichon volcano in Mexico, measurements in Michigan showed a decrease of 25 per cent in incoming direct solar radiation while the diffuse radiation was about 85 per cent above normal (Baker *et al.*, 1984).

Traditionally the net effect of volcanic dust is believed to lead to a cooling of the earth's surface. Volcanic dust in the atmosphere tends to prevent shortwave radiation reaching the ground and to reflect it away from the top of the dust cloud, but the particles concerned are mainly small and cannot hold much heat. Lamb (1970, p 461) therefore suggests that periods of decreased earth temperatures follow times of major volcanic activity. More recently, attention has been given to the processes of absorption by the particles (aerosols) and the relative magnitudes of backscattering and absorption of radiation. Idso (1981) has suggested that aerosols may act to either heat or cool the earth, depending on the relative magnitudes of the backscattering and absorption characteristics of the particulates and the albedo of the underlying surface.

2.4 HUMAN INFLUENCE ON THE RADIATION BALANCE

Changes of solar output and volcanic dust are just two of a myriad of natural physical factors that can alter the flows that make up the radiation balance. Set against such a background is the relatively new presence of human influence on the balance. This influence also takes many diverse forms—some of which are certain and some of which also remain in the realm of speculation. Because of its diversity, human influence on the radiation balance is, perhaps, most appropriately illustrated with reference to examples acting on different spatial scales. Whilst clearest at the local scale, the effects of human activity at the global scale still have to be brought more clearly into focus but, at the same time, offer some of the gravest implications.

(a) The local scale

No surface on earth is artificially altered more from its natural state than the urban surface. The activities on this surface also have a marked effect on the air above it (Fig. 2.4). Alterations to the surface and the overlying air affect the radiation balance in a significant way. Measurements generally show that the polluted air over a city decreases the amount of shortwave radiation arriving at the surface and that the decrease is greatest during the winter months when the radiation has to pass through the greatest thickness of polluted air. The loss is sometimes somewhat offset by the lower albedo of the city compared to its surroundings thus allowing more radiation to be absorbed. City surfaces also usually have higher emitting temperatures than those of surrounding areas which gives rise to higher values of outgoing longwave radiation. The absorption of both long- and shortwave radiation in some of the components of the urban pollution dome may heat this air and return a compensating longwave flow to the city (Oke, 1982).

FIGURE 2.4   Polluted atmosphere over the town of Keswick, United Kingdom

The modified urban atmosphere represents an inadvertent modification to the radiation balance. There are many deliberate ways of altering the radiation balance for beneficial purposes. Perhaps the best known of these is the construction of a greenhouse. Here the glass permits the passage of incoming shortwave radiation but prohibits the passage of outgoing longwave radiation thus, together with the convective sensible heat from the surface, raising the temperature of the air inside the structure. The alteration of albedo on a small scale is also deliberately undertaken in some places. Coal dust, or other dark material, for example, is sometimes spread on snow surfaces at airports or on highways to promote absorption of incoming radiation and the consequent melting of the snow. Some scientists believe the use of dark materials has potential on larger scales, as is discussed in the next section.

(b)   The regional scale

Probably the most important alteration of the radiation balance at the regional scale is the adjustment of surface albedo associated with the development of agriculture. It is estimated that 18 to 20 per cent of the land surface of the globe has been significantly influenced by humans—primarily for agricultural purposes (SMIC), 1971, p. 63). All of these changes alter the radiation balance in some manner. For example, a representative albedo for arable land or grassland is 20 per cent, which will mean less radiant heat being absorbed compared to coniferous forests with an albedo of 12 per cent or deciduous forests with an albedo of 18 per cent. Alternatively, the irrigation of land in dry areas often gives rise to a green surface with an albedo of some five or more per cent less than a surrounding desert area (Fig. 2.5). Once more an increased amount of shortwave radiation would be absorbed and, in this case, much of it would be used in the evapotranspiration process of the plants and therefore indirectly aid plant growth.

FIGURE 2.5 Contrast in surface cover and colour between an irrigated area and surrounding desert. Sevier Valley, Utah

A different, but as yet untried, way of altering the radiation balance on a regional scale is once more with the use of dark materials. Existing observational and modelling information suggests that absorption of heat by carbon aerosol particles of diameter 0.1 micrometres or less can be best used on a regional scale for several purposes. These include precipitation enhancement, the alleviation of tropical storm destruction, and increasing the rapidity of fog burn-off and snowmelt. In most cases the carbon would be spread from aircraft (Gray *et al.*, 1974). Despite favourable environmental impact assessments of this technique (Gray *et al.*, 1976) some people remain sceptical because operation of the method would need continual repetition every time a heat source was required (Glantz, 1977, p. 317).

Another suggested method of radiation balance alteration which might lead to enhanced precipitation in arid areas is the asphalt island concept (Black, 1970).

Here the idea is to lower surface albedo by spraying a large area of land near a body of water with a thick layer of asphalt. Air passing over the surface would be heated and rise promoting cloud and possible precipitation formation. It was estimated that along the Mediterranean coast of North Africa annual rainfall could be increased from about 80 mm to between 500 mm and 800 mm per year near the mountains (Black and Popkin, 1967) and produce 2 or 3 acres of arable land for each acre of asphalt. Rather obvious practical and aesthetic difficulties have so far also prevented the application of this idea.

It is in the arid areas of the world that one of the more important real effects of human, as well as natural, alteration of the radiation balance occurs—the periodic increase of quantities of airborne dust. In the natural sense it is well documented that airborne dust may have large effects on the flows of the radiation balance. Brinkmann and McGregor (1983) report a drop of visibility at the surface to less than 500 m due to dust transported by the Harmattan wind in the Sahara desert and report reduction of surface shortwave radiation by as much as 90 per cent.

An increase of airborne dust loadings over arid areas due to human activities and its effect on the atmospheric radiation balance on a regional scale has been the focus of considerable discussion relating to possible climatic change. One of the best examples of this is the suggestion of Bryson and Murray (1977, p. 113) that the Rajasthan desert in western India results from over-intensive use of land in the area, which has given rise to a change from a relatively fertile land system to a self-perpetuating desert. An increase of dust in the atmosphere associated with over-intensive agricultural and pastoral practices is said to cause increased atmospheric subsidence which suppresses precipitation from air which otherwise has enough moisture content. According to Bryson and Murray the effect of the dust on the radiation balance is threefold. First, there is a reflection of incoming radiation leading to lower surface temperatures than otherwise would occur, thus retarding the development of atmospheric instability and rainfall production. Secondly, longwave radiation at night from the top of the dust cloud further results in the cooling and the subsidence of the air. Thirdly, the trapping of longwave radiation from the ground at night can give temperatures high enough to prevent dew formation. The fact that the Rajasthan desert is caused by human influence is disputed by the evidence of Singh *et al.* (1974) but the example shows quite plausibly a possible interrelation between human activity and the radiation balance. We will return to the argument in the next section because it is equally important on the global scale.

Whether or not the Rajasthan desert has been produced by human activity, there is no doubt that the development of large agglomerated urban and industrial areas can affect the radiation balance on a regional scale. Satellite photographs show the drift of haze from the eastern United States to the central Atlantic. Flowers *et al.* (1969) have demonstrated that turbidity values, a measure of the reduction of atmospheric transparency due to suspended particles, can be twice as large in the eastern United States as those that occur in the less densely populated western states. Similarly lower turbidity values are found for Goose Bay, in central Labrador, compared to those recorded at Montreal and near Toronto (Uboegbulam and Davies, 1983). Ball and Robinson (1982) examining the effect of haze estimated an average annual depletion of shortwave radiation arriving at the surface in the central United States of the order of 7.5 per cent. It has been suggested that increasing industrialization and urbanization, particularly in the northern hemisphere, is a feature that may have global effects and it to these that attention is now turned.

(c) The global scale

In treating the global scale three examples will be taken, all of which demonstrate the extreme importance of present or possible future human impacts on the radiation balance of the atmosphere and subsequently on the climate of the earth. Emphasis will be given to the problem of the effect of dust in the atmosphere, the possibility of an increase of carbon dioxide in the atmosphere, and some of the events that might follow a global nuclear war, especially the hypothesis of the development of a nuclear winter.

The extension of the theory of climatic cooling related to increased atmospheric dust from treating dust of mainly volcanic origin to dust of industrial and other human activity origin was primarily advocated by Bryson (1967) who was supported by the investigations of many other workers. An alternative view presented by Charlson and Pilat (1969) points out that attention has been focused only on scattering effects of dust and that additional consideration of absorption showed that either heating or cooling could occur, depending on the relative magnitudes of absorption and backscatter. In reviewing all this material Idso (1981) quotes further studies indicating that atmospheric cooling or warming in the presence of increased dust load might also be dependent on the albedo of the underlying surface. In addition, Idso presents some evidence to suggest that if longwave radiation is taken into account, once more warming or cooling can take place at the earth's surface under a variety of different conditions.

So most scientists believe that the radiation balance of the earth will be altered if atmospheric dust content is increased for either natural or anthropogenic reasons, but there is disagreement on the nature of the alteration and whether the final effect would be to heat or cool the earth's surface. There is greater consensus, although it is not complete, on the possible effect of increased carbon dioxide in the atmosphere.

Observations at the Mauna Loa observatory in Hawaii and from the South Pole and several other locations manifest a steady increase in the amount of carbon dioxide in the atmosphere over the last two decades at a rate of about 9 parts per million per year. Since carbon dioxide in the air has the ability to absorb longwave radiation its increase will directly affect the radiation balance of the atmosphere and earth. More longwave radiation that leaves the surface will be absorbed in the atmosphere which, all other factors being held constant, would be expected to raise the temperature of the atmosphere. The higher temperature of the atmosphere would give higher rates of longwave radiation transfer back to the earth. Therefore, because of the warmer atmosphere and an increased longwave radiant flow to the earth, earth surface temperatures might be expected to rise. There are some possible further indirect effects that might also amplify the surface temperature increase. The carbon dioxide induced temperature increase might increase evaporation rates thus placing more water vapour in the atmosphere and altering the radiation balance in the same way as carbon dioxide does. Furthermore, the rise in the surface temperature would tend to decrease the surface area covered by snow and ice leading to a decrease in the average albedo of the earth's surface in turn allowing more radiation to be absorbed by the surface and increasing further the surface temperatures. On the other hand, an increase in evaporation rates might increase cloud cover across the globe which then would increase the planetary albedo and allow less shortwave radiation to be absorbed by the system (Neiburger *et al.*, 1982, p. 363).

The increase of carbon dioxide in the atmosphere is believed to be associated with human activities in the increased burning of fossil fuel and the burning of biomass, especially in the area of tropical rainforests. Computer models suggest the possibility of a global temperature rise of between $2°$ and $5°K$ in the above scenario, assuming continued rates of carbon dioxide input into the atmosphere. This might lead to ice sheet melting and a sea level rise of as much as 5 m submerging many of our coastal areas and cities. Other possible effects have been mentioned, such as decreased yields of fish, adverse effects on agriculture and expansion of desert areas. Not all scientists agree with this kind of scenario (Idso, 1982) but all agree that because of the potential severity of this effect of human activity on the radiation balance and the global climate a large amount of research effort into the subject is clearly warranted and that this should be done quickly.

While the rise of atmospheric carbon dioxide content and its attendant possible climatic changes is a relatively slow-acting phenomenon in terms of the human lifespan, increasing attention is being given to the possible rapid changes in the radiation balance and global climate that could accompany the frightening prospect of a nuclear war. In the last decade, studies on the effects of nuclear war on the atmosphere (NAS, 1975) concentrated on the global depletion of stratospheric ozone which would be accompanied by an increase in the amount of harmful ultraviolet light arriving at the surface. More recently it has been pointed out that fires associated with a nuclear war might well alter the radiation balance of earth and atmosphere so as to plunge large areas of the earth into sub-freezing temperatures and long-lasting darkness—a concept that has become known as nuclear winter (Crutzen *et al.*, 1984).

FIGURE 2.6   Injection of pollutants into the atmosphere from agricultural land burning. South Island, New Zealand. Hundreds of such fires would follow a nuclear attack. Some of the fires would form their own clouds. Crutzen *et al.* (1984) show that rain coming from such clouds would not be very effective either in putting the fires out or in removing the pollutants from the atmosphere

Even today large forest, grassland or agricultural fires put great quantities of pollutants into the atmosphere (Fig. 2.6). During a large nuclear war it has been proposed that forest fires would destroy an area 55 times larger than that which today is annually lost to wildfires in the continental United States. In addition to this would be pollutants from fires in urban and industrial areas, cultivated lands and grasslands. Crutzen *et al.*, (1984) estimate that 10^{14} g of black smoke would be injected into the atmosphere. Penetration of shortwave radiation to the surface would be greatly reduced over large areas of the northern hemisphere and into the southern hemisphere. There would be darkness in large areas of the earth and photosynthetic activity of otherwise undamaged plants would cease for an undetermined length of time. The situation would last for at least several weeks. Because of the absorption of shortwave radiation in the upper part of the atmosphere the land areas of the earth would cool — probably well below freezing and even to 45°K below the present-day normal. The upper atmosphere being warm while the surface is cool would present conditions unfavourable for the dispersion of the pollutants thus prolonging the situation. The resulting atmospheric stability would greatly alter the way in which the general circulation of the atmosphere currently operates and cause considerable disruption to the current weather systems of the world.

Any survivors of a nuclear war would therefore be faced with the prospect of a dark and cold environment in which it was difficult, if not impossible, to grow any food. The additional probability of environmental contamination by radioactivity and later, further cancer-producing, increased ultraviolet radiative penetration is awesome indeed.

The implications of this and the other examples of possible alteration of the radiation balance of the earth and its atmosphere are clearly great and demand a consideration of the future of the radiation balance on our planet.

2.5   THE FUTURE

Clearly major changes to the radiation balance of the earth and its atmosphere are possible resulting from the activities of the human inhabitants of the planet. Equally clearly, some of the possible changes, if they take place, will be detrimental to human and other life on earth. Unfortunately, despite considerable research, there is still doubt surrounding the magnitudes and rates of change involved in the human processes. There is even greater uncertainty as to the resulting effects on other parts of the radiation balance and the earth–atmosphere system as a whole. All these statements hold good for the majority of radiation balance changes that humans have induced or could produce on the earth whether the alteration be at a local or broad geographic scale.

The radiation balance is the key to the climate of the planet and any alteration of the balance should be approached with extreme caution for fear of producing irreversible effects. Many people recommend the use of models to make predictions of future changes in the radiation balance and other factors. But it is possible that there is not time to construct adequate models. It is thus imperative that the development of new models of the radiation flow through the atmosphere, the changing radiation balance and its linkage with other aspects of the earth–atmosphere system, should be accompanied by a decrease in the rates of input of pollutant substances into the atmosphere and increased efforts to prevent the possibility of nuclear warfare.

In 1957, Sir Fred Hoyle wrote a captivating novel called *The Black Cloud*. In it he described in fictional terms what would become of the earth's climate if a black cloud were interposed between the sun and the earth (Hoyle, 1957). This is science fiction at its best; and some of Hoyle's predictions are uncannily close to those of the modern nuclear winter studies. It is very sobering to realize that it is within the power of human beings to alter the radiation balance of the planet to an extent that equals or supersedes natural alterations. We have good reason to fear the morning when the sun does not rise.

REFERENCES

Baker, C. B., Kuhn, W. R., and Ryznar, E., 1984, 'Effects of El Chichon volcanic cloud on direct and diffuse solar irradiances', *Journal of Climate and Applied Meteorology*, **23** (3), 449–452.

Ball, R. J., and Robinson, G. D., 1982, 'The origin of haze in the Central United States and its effects on solar irradiation', *Journal of Applied Meteorology*, **21** (2), 171–188.

Black, J. F., 1970, 'Asphalt island concept of weather modification', ESSO Memorandum, 4th June, Linden, New Jersey.

Black, J. F., and Popkin, A. H., 1967, 'New roles for asphalt in controlling man's environment', paper presented at the National Petroleum Refiners Association Annual Meeting, April 1967, at San Antonio, Texas.

Brinkman, A. W., and McGregor, J., 1983, 'Solar radiation in dense Saharan aerosol in Northern Nigeria', *Quarterly Journal of the Royal Meterological Society*, **109** (462), 831–848.

Bryson, R. A., 1967, 'Is man changing the climate of the earth?', *Saturday Review*, **50**, 52–55.

Bryson, R. A., and Murray, T. J., 1977, *Climates of Hunger*, University of Wisconsin Press, Madison, Wisconsin.

Charlson, R. J., and Pilat, M. J., 1969, 'Climate: the influence of aerosols', *Journal of Applied Meteorology*, **8**, 1001–1002.

Crutzen, P. J., Galbally, I. E., and Bruhl, C., 1984, 'Atmospheric effects from post-nuclear fires', *Climatic Change*, **6**, 323–364.

Flowers, E. C., McCormick, R. A., and Kurfis, K. R., 1969, 'Atmospheric turbidity over the United States', 1961–66, *Journal of Applied Meteorology*, **8**, 955–962.

Glantz, M. H., 1977, 'Climate and weather modification in and around arid lands in Africa', in Glantz, M. H. (Ed.), *Desertification*, Westview Pres, Boulder, Colorado, pp. 307–337.

Gray, W. M., Frank, W. M., Corrin, M. L., and Stokes, C. A., 1974, 'Weather modification by carbon dust absorption of solar energy', Atmospheric Science Paper No. 225, Department of Atmospheric Science, Colorado State University, Fort Collins, Colorado.

Gray, W. M., Frank, W. M., Covin, M. L., and Stokes, 1976, 'Weather modification by carbon dust absorption of solar energy', *Journal of Applied Meteorology*, **15** (4), 355–386.

Hoyle, Sir Fred, 1957, *The Black Cloud*, William Heinemann, London.

Idso, S. B., 1981, 'Climatic change: the role of atmospheric dust', in Peare, T. L., (Ed.), *Desert Dust: Origin, Characteristics, and Effect on Man*, Special paper 186, The Geological Society of America, pp. 207–215.

Idso, S. B., 1982, *Carbon Dioxide: Friend or Foe*, IBT Press, Tempe, Arizona, 92 pp.

Kerr, R. A., 1978, 'Climate control: How large a role for orbital variations', *Science*, **201**, 144–146.

Lamb, H. H., 1970, 'Volcanic dust in the atmosphere: with a chronology and an assessment of its meteorological significance', *Philosophical Transactions of the Royal Society of London: Mathematical and Physical Sciences*, **266** (1178), 425–533.

NAS, (National Academy of Sciences), 1975, *Long Term World Wide Effects of Multiple Nuclear Weapon Detonations*, Washington, D.C., 213 pp.

Neiburger, M., Edinger, J. G., and Bonner, W. D., 1982, *Understanding Our Atmospheric Environment*, 2nd edn, W. H. Freeman, San Francisco, California, 453 pp.

Oke, R. R., 1982, 'The energetic basis of the urban heat island', *Quarterly Journal of the Royal Meteorological Society*, **108**, (455), 1–24.

Pittock, A. B., 1983, 'Solar variability, weather and climate: an update', *Quarterly Journal of the Royal Meteorological Society*, **109** (459), 23–56.

Singh, G., Joshi, R. D., Chopra, S. K., and Singh, A. B., 1974, 'Late Quarternary history of vegetation and climate of the Rajasthan Desert, India', *Philosophical Transactions of the Royal Society*, B, **267**, 467–501.

SMIC; Report of the study of man's impact on climate, 1971, *Inadvertent Climate Modification*, Massachusetts Institute of Technology, Cambridge, Mass.

Uboegbulam, T. C., and Davies, J. A., 1983, Turbidity in eastern Canada. *Journal of Climate and Applied Meteorology*, **22** (8) 1384–1392.

Willson, R. C., Gulkis, S., Janssen, M., Hudson, H. S., and Chapman, G. A., 1981, Observation of solar irradiance variability, *Science*, **211**, 700–702.

Human Activity and Environmental Processes
Edited by K. J. Gregory and D. E. Walling
©1987 John Wiley & Sons Ltd.

3

Precipitation

B. W. ATKINSON

3.1 INTRODUCTION

Within the last two decades the atmosphere has increasingly been viewed as a resource (Chandler, 1970) and the relationships between man's activities and atmospheric behaviour have received renewed prominence in national scientific and technological programmes. Geographers may be tempted to argue that we are merely witnessing a delayed appreciation of some facets of climatic determinism! But sectarian interests and methodological squabbles apart, all scientists agree that in an increasingly crowded world, it is prudent, if not yet necessary, to reappraise the relationships between climate and man. Of the climatic elements, temperature and precipitation, particularly the former, have received the most attention as indicators of climatic change. The main reason for this emphasis is the availability of data. From the point of view of life on the planet, precipitation is arguably more important than temperature, as the droughts in the Sahel in the early 1970s and in the UK in 1976 graphically illustrated. The populations suffering those droughts were the latest in a historical succession where man has been essentially at the mercy of Nature's whims. Consequently any suggestion that man may be able to influence the quality, amounts and distribution of precipitation is eagerly examined. The results of this examination over a period of thirty years or so fall conveniently into two classes: first, man's conscious attempts to influence precipitation; and secondly, his inadvertent effects upon precipitation. This chapter covers these two topics.

3.2 PRECIPITATION MECHANISMS

Before considering man's effects upon precipitation in detail, a brief review of precipitation types and mechanisms may be of value. The main precipitation forms are rain, snow and hail. All these forms of precipitation ultimately owe their existence to the cooling of air, usually by adiabatic expansion in uplift, so that the water vapour condenses or turns into ice. But this apparently simple process involves a myriad of mechanisms which are far from fully understood, despite over two decades of intensive research (Mason, 1971). We can, however, paint a broad picture.

The condensation of vapour and freezing of water that may result from uplift and cooling require, respectively, condensation and ice nuclei. The resultant clouds may comprise any or all of the following: water droplets with temperatures greater than 0°C, supercooled water droplets with a temperature less than 0°C and ice crystals. A major problem of cloud physics was the explanation of the growth of these cloud particles to a size which would cause them to fall out of the cloud as precipitation. Two mechanisms were suggested over thirty years ago and, in essence, are still accepted.

The more obvious of the two mechanisms is that involving coalescence and growth of particles. The term 'coalescence' in fact covers three types of 'joining together'; first, in a more restricted sense, coalescence means the joining of two liquid water drops; secondly, if ice crystals collide and join they are said to 'aggregate'; and thirdly, if an ice crystal collects a water drop the process is known as 'accretion'. Whichever of these processes operates, the resultant particle will be larger and, as such, its ability to 'catch' more cloud droplets or ice crystals dramatically increases. Of the three main types of precipitation, rainfall results largely from coalescence, snowfall from aggregation and hailfall from accretion.

The second, and far less obvious, way in which cloud particles may grow to precipitation size is known as the Bergeron or ice-crystal mechanism. This process depends upon the fact that saturated vapour pressure is greater over liquid than over ice surfaces. Consequently, if ice-crystals and supercooled water co-exist in a cloud (as they frequently do in extra-tropical latitudes), then the liquid particles tend to evaporate and the ice crystals to grow. Since ice particles are relatively few in number, each one draws on a comparatively large water supply and may grow into a crystal large enough to have an appreciable fall velocity. The process of aggregation then proceeds rapidly, resulting in snowflakes that collect more cloud droplets and fall to lower, warmer altitudes where they may melt and continue descent as raindrops.

Both mechanisms outlined above are on the micro-scale, and can operate successfully only if cloud size, duration and internal airflow are favourable. Indeed the type, amount and intensity of precipitation from any cloud are largely determined by cloud dynamics. In the broadest terms, clouds must be about 1 km deep, last for about one hour and maintain suitable uplift (ranging from a few centimetres per second to several metres per second depending upon cloud type) for any precipitation to occur. Stratiform clouds generally produce light, steady, quite prolonged rainfall or snowfall (largely dependent upon season), whereas cumuliform clouds may produce showers of rain or hail, often quite heavy in summer.

With this background knowledge of the micro- and macro-physics of clouds and precipitation we are in a position to review the ways in which man may influence them.

3.3 CONSCIOUS MODIFICATION OF PRECIPITATION

Man's conscious attempts to modify precipitation are of two main types: to increase amounts of rain and snow for agricultural and water-supply purposes; and to decrease the amounts of hail which cause so much damage in many agricultural areas. In both cases the modification is concerned with both the macro- and micro-physics of the clouds, particularly the latter.

(a) Increasing rainfall and snowfall

As a result of thirty years of effort stimulated by Schaefer's results (1946), many, but by no means all, meteorologists believe that 'cloud seeding' can indeed increase precipitation. Cloud seeding works in two ways. First, artificial nuclei may stimulate small cloud particles to coalesce; and secondly, cloud seeding with freezing nuclei or solid carbon dioxide (dry ice) may induce freezing and trigger the Bergeron process. In the first case, known as warm seeding, water drops may be introduced into a cloud to start a sweep-out process which otherwise is expected to be too long delayed. Because a tremendous weight of water is required to modify a whole cloud with large drops, finely divided salt or a hygyroscopic liquid mist is usually used instead. For example, experiments by the Oklahoma Bureau of Reclamation used a concentrated water solution of ammonium nitrate and urea, sprayed from an aircraft into a cloud in the form of droplets about 0.02 mm in diameter. Within a minute the nitrate and urea droplets grew by gathering condensation from the vapour to a 0.05 mm size, a factor of 15 greater in mass. The 0.05 mm drops were large enough to start a sweep-out process that may have produced 5 mm diameter drops only 20 minutes later (Kessler, 1973).

The second method, seeding by dry ice or silver iodide, produces the same effects by two different processes. If dry ice is used it has the effect of inducing a massive rapid cooling which freezes the supercooled water. In contrast, silver iodide particles are good nuclei for ice formation because of the close resemblance of their crystal structure to that of ice. The particles do, in fact, act as freezing nuclei (Vonnegut, 1949). Whichever seeding material is used, the result is is the production of ice crystals, which, it is argued, will trigger the Bergeron mechanism.

Hygroscopic particles and freezing nuclei can be admitted to the atmosphere in different ways. In the first field experiments, dry ice was dispersed from a small aeroplane (Schaefer, 1946) and silver iodide was generated at the ground (Vonnegut, 1950). This latter process involved mixing the iodide with acetone and water and spraying the solution into a hydrogen flame. The resultant smoke of potential freezing nuclei was then carried up into a cloud by natural convection currents. More recently silver iodide has been released from aircraft and pyrotechnic devices, with the aim of placing the nuclei more accurately in potentially profitable parts of a cloud. The whole exercise is still rather bedevilled by the finding that sunlight can reduce the power of the silver iodide to act as a nucleant.

Notwithstanding the above difficulties, many meteorologists agree on the possibility of augmenting both orographic and convective precipitation. Indeed, three decades ago Bergeron (1949) concluded that the main possibility for causing considerable artificial rainfall might be found within certain kinds of orographic cloud systems. This conclusion was based on considerations of the steady and often substantial formation of condensate for an extended period of time in a fixed location and the probable accumulation of 'releasable but unreleased' cloud water at levels with temperatures just below the $0°C$ isotherm. The basic criterion for determining whether a seeding potential might exist is the natural precipitation efficiency of the clouds — orographic or otherwise. Seeding is clearly not required when the efficiency is high. On the other hand, seeding may or may not be of value when the natural precipitation efficiency is low. The measure of precipitation efficiency is the percentage of condensate which actually reaches the ground. Whilst precise numerical values are difficult to achieve, a useful basis for evaluating precipitation efficiency lies in the comparison of the removal of cloud condensate by vapour growth of ice crystals with the supply of vapour to the cloud. Chappell (1970) has illustrated this idea by comparing for a broad range of cloud temperatures the average rate of formation of condensate, the average rate of consumption of cloud water by ice crystal growth that would occur from natural concentrations of primary ice nuclei, and actual average rates of precipitation observed at the ground. His model showed that, with cloud-top temperatures of $-20°C$ or colder, the ice crystals had a rate of water consumption equal to or exceeding the rate of formation of cloud condensate. In the main, such clouds should have a high natural precipitation efficiency with little potential for seeding. The observed actual precipitation corresponded quite closely to the rate at which condensate became available to the clouds.

When cloud-top temperatures were higher than $-20°C$, the average rate of cloud water consumption from natural concentrations of ice crystals by natural ice nuclei was much lower than the average rate at which condensate became available to the cloud. In these cases considerable quantities of liquid water condensate can be lost to the precipitation process by evaporation which takes place to the lee of the barrier which causes the cloud. For these cases, when natural precipitation efficiency should be low, observed values of ground precipitation were, in fact, much less than the average amount of condensate available. Clearly, a potential for seeding can exist in these cases of 'warm', orographic clouds.

The existence of a potential does not, of course, mean that it is realized. But two sets of observations suggest that realization does occur: first, changes in microphysical characteristics of seeded orographic clouds; and secondly, increased precipitation at the ground. Microphysical changes were recognized in the first field experiments of the 1940s when clouds rapidly glaciated. Changes in ice crystal concentrations, amounts of riming, crystal size and crystal shape have since been documented. In general, concentrations increase, riming decreases, and crystal size decreases. All these changes are consistent with theoretical expectations. Although overall precipitation changes at the ground have been modest, in 'warm' clouds some fairly substantial increases have been observed. For example, at Climax, Colorado, precipitation increases of 70–80 per cent have been claimed from clouds with top-temperatures between $-11°C$ and $-20°C$. More frequently, increases of 10–20 per cent are reported.

The second most promising type of precipitation, from a modification point of view, is that from cumulus clouds. In the present context, these clouds are meant to be convective clouds existing in relative isolation, as distinct from convective cells embedded in cyclonic systems or convective cells stimulated by extensive orographic lifting of broad currents of air. The natural precipitation efficiency of these isolated clouds is quite low. Mixing with non-saturated air inhibits the growth of precipitation-sized particles and consequently even the largest clouds—those reaching thunderstorm size—exhibit precipitation efficiencies of only 10 per cent (Braham, 1952). The important question, then, is whether isolated cumuli constitute promising targets for artificial nucleation by virtue of their comparatively low natural precipitation efficiency. Within this context, a major difficulty in assessing possible modification is the enormous natural fluctuation in all variables. Unmodified convective rainfall within seedable situations commonly varies by factors of 10 to 1000, while the largest seeding effects claimed by man have never exceeded a factor of about 3.

In examination of seeding potential it has proved useful to subdivide all cumulus clouds into maritime and continental types. Cloud droplet concentrations in these types of cumuli are typically 50 and 400 cm^{-3} respectively. As the cloud liquid-water contents are not greatly different for maritime and continental cumuli, the average droplet radius must be about twice as great in maritime as in continental cumuli. This implies that in maritime cumuli the coalescence process can probably proceed rath rapidly due to a high proportion of initially large droplets: whereas in cumuli containing a continental-type condensation-nucleus population the coalescence process alone would have to operate for a much longer time to result in precipitation. On the basis of this picture, we could conclude that ice nucleants probably offer little potential for stimulating precipitation in maritime cumuli for these clouds can, and evidently do, rapidly develop precipitation by the coalescence process. On the other hand, the same picture suggests that continental cumuli might be artificially modified if one could accelerate particle growth by seeding. This assumes that the continental cumuli have cloud-top temperatures less than 0°C and that natural ice-forming nuclei are so deficient that a substantial part of the cloud water is supercooled.

Attractively simple as the above idea seems to be, it has proved very difficult to test satisfactorily over the last thirty years. Early observation programmes, such as those in both Australia (Warner and Twomey, 1956) and the United States (Braham *et al.*, 1957) were inconclusive. The carefully conducted Project Whitetop carried out in southern Missouri in 1960–64, revealed a 5–10 per cent increase in radar echo frequency in the region lying just downwind of the seeding area, changing to a decrease of about the same size beyond a downwind distance of around 60–80 km and returning to an increase still farther downwind. This induced rain-shadow effect has become known as the problem of 'robbing Peter to pay Paul', and still requires close scrutiny.

In the tropics a few experiments have claimed increases due to seeding, but the general feeling is still that ice nucleants probably are of little effect in initiating precipitation in trade-wind cumuli over the open oceans. In contrast, those cumuli forming over tropical islands may be more promising, but largely because of dynamical rather than microphysical factors. Howell (1960) has discussed briefly the question of the efficacy of using silver iodide to seed such island cumuli in the tropics, suggesting that buoyancy alteration may be the principal purpose of such seeding.

(b) Hail suppression

The main reason for wishing to modify hail is its destructiveness. Damage would be reduced of course by preventing hail altogether, but this is not at present feasible. An alternative is to attempt to reduce the size of hailstones which cause most damage. Thus, the present approach to hail modification is to add freezing nuclei which would then compete with natural nuclei for the same finite supply of supercooled water in the cloud. The end-product should be more, but smaller, hailstones. But this appealing idea has its drawbacks, the main one being as follows. The competition idea implies that the unmodified hail cell is a more or less closed system producing hail at near-capacity rate. In fact, Browning and Foote (1976) have recently demonstrated that large hail cells are both open systems and very inefficient 'hail factories'. They conclude that the addition of nuclei to the main updraught of a storm, which would encourage the production of additional hailstone 'embryos', may increase the number of hailstones of a size equal to that which would have fallen naturally—just the opposite of what is intended. They go on, however, to outline a further way in which seeding may help hail suppression. With the knowledge that a shortage of natural hailstone embryos entering the main storm updraught leads to inefficiency in the production of hail (a natural hail suppression mechanism), Browning and Foote suggest that any action which further decreases the number of embryos should reduce the possibility of large-hailstone growth. One way to achieve this reduction in embryos is to seed the embryo source region on the storm's right flank so as to promote competition during the early growth of the embryos themselves. But first, as Browning and Foote note, we must identify the location of the embryo source more precisely than we are at present able to do. Clearly, the complex and ill-understood interrelations of hailstone and air flow within the hail cloud still provide a significant obstacle to the prospective modifier.

Despite these problems, hail suppression operations have proceeded apace. There are at least eight European, one South American, three African, and eleven Russian projects, quite apart from those in North America. In particular the Russians appear to be convinced that they can mitigate hail and they are doing it routinely, principally be seeding clouds with the aid of rockets. North American visitors to the Soviet operations (Battan, 1977) were impressed by their dedication but remained rather sceptical, particularly in the absence of any breakthroughs in establishing a solid theoretical or physical foundation. In the United States, opinions on the effectiveness of hail suppressions range from 'optimistic' (claims of reductions of 20–48 per cent, but mostly not significant at the 5 per cent level) to 'pessimistic' ('majority of weather modification experts indicate no knowledge of hail suppression capability' (Changnon, 1977)). It is difficult to improve on Changnon's summary:

> Clearly, the current status of hail suppression is in a state of uncertainty. Reviews of the existing results from six recent operational and experimental hail suppression projects are sufficiently suggestive of a hail suppression capability in the range of 20–50 per cent to suggest the need for an extensive investigation by an august body of the hail suppression capability exhibited in these and other programmes. (Changnon, 1977 pp. 26–27).

(c)   Non-meteorological effects of cloud seeding

Despite the many uncertainties involved in cloud seeding, it appears to have increased in scientific respectability in the 1970s. The Russian efforts have been already noted and in the United States a major reappraisal is being made. Although American meteorologists are still cautious in their support for cloud seeding, the number of operations being undertaken is sufficient to stimulate appreciation of possible ecological, social and legal ramifications of seeding.

Over ten years ago a sponsor of a Pennsylvania law designed to curb weather modification said, 'Cloud seeding involves silver iodide, and silver iodide, Mr. Speaker, is highly poisonous'. He then proceeded to paint a gloomy picture which included 'drastically modified weather conditions . . . total pollution and toxic effects upon all fauna and flora'. His oratory succeeded and the bill became law, yet recent analyses (Klein and Molise, 1975) have shown that the silver concentrations in grasslands where seeding has taken place show no significant changes. In parallel with such ecological studies, Davis (1975) has shown that many states in USA have now drafted laws which allow close governmental control over the environmental consequences not only of cloud seeding but also of all weather modification. The social aspects of weather modifications have also been considered (Hass, 1973; Lansford, 1973), the general tone of the discussion being that even if cloud seeding becomes a reliable technology, the public rather than scientists or government will decide whether they wish it to be used.

3.4 UNCONSCIOUS MODIFICATION OF PRECIPITATION

Whereas cloud seeding occupies a well-known, if not widely accepted, position within meteorology, man's inadvertent effects upon precipitation have, until recently been largely unappreciated. But the last two decades have seen a fairly steady increase in evidence of such effects and this work is reviewed below. Up to this point we have investigated only the amount of precipitation: it is now necessary to consider also its chemical composition. It is becoming increasingly clear that man can unconsciously influence not only the quantity but also the quality of the water that falls from the sky. It is worthwhile reiterating here the conditions required for precipitation to occur, because certain factors figure prominently in the assessment of inadvertent effects. Precipitation results from the cooling (usually by uplift) of moist air so that condensation, coalescence, and the ice-crystal mechanisms occur. Clearly, any increases in water vapour content, uplift and condensation, and ice nuclei would tend, at least in a qualitative sense, to favour precipitation formation. If in addition various pollutants existed in sufficient quantities in the right places at the right times it is likely that the quality of the precipitation water would be affected. Because the amounts of water vapour, nuclei, pollutants and, to a lesser extent, the degree of uplift depend on the nature of the earth's surface, man-made changes in this surface may affect these four factors and thus inadvertently affect the nature and amount of precipitation. The remainder of this chapter demonstrates how urban and industrial areas affect the quality and quantity of precipitation. Perhaps more familiar headings are 'acid rain' and 'urban influences on precipitation amounts'. The scales of the phenomena are markedly different: the former spreads over thousands of kilometres; the latter is restricted to within a few kilometres of the initiating urban area.

(a) Acid rain

'Acid rain' is a rather general term which has had widespread use in the press and popular scientific literature, but as yet appears to have no firmly accepted definition. It should be immediately recognized, however, that precipitation of all types, not just rain, comes under its umbrella. The acidity of the precipitation is assessed relative to the global mean of precipitation acidity (pH = 5.6) which, until recently, has been considered to be uncontaminated by man's activities. To cut through possible confusion Marsh (1984) has recently suggested that the term 'acid rain' means precipitation with a pH value of less than 5.0.

The whole issue of 'acid rain' has sprung into public prominence only in the past decade or so. Although the chemistry of precipitation, received attention as long ago as 1872, it was the probable deleterious ecological impact, perhaps most effectively presented by Oden (1968), that demanded scientific attention, with a possible view to ameliorative political action. There is now no doubt that lakes are becoming so acid that they cannot support fish and that large areas of woodland are dying. The immediate scapegoat for these happenings was acid rain but, as research has progressed, it has become clear that there is no simple, clear case against acid rain. It has also become clear that the acid rain is but one facet of the wider problem of long-range transport of pollution (Eliassen, 1980; Fisher, 1975, 1983; Smith and Hunt, 1978). Pollutants emitted in area A are transported and chemically changed before being deposited, either by dry mechanisms or wet (acid rain) mechanisms, in area B. It is now recognized that A and B can be thousands of kilometres apart. The essential problem is thus to establish quantitatively, the nature, strength and location of emissions, the nature of the chemical transformations and the degree of diffusion along the paths of transport, and the mechanisms, amounts and locations of deposition.

On an annual basis, rain and snow over certain regions of the world are currently 5 to 40 times more acid than the lowest value expected (i.e. pH 5.6) for unpolluted atmospheres (Likens and Butler, 1981). Individual storms may be several hundred to several thousand times more acid than expected (Likens, 1976). The whole problem is highlighted in two main areas: North America and Europe. In North America precipitation prior to 1930 had a pH greater than 5.6. By the mid-1950s monthly data revealed acid precipitation throughout the northeastern United States and by the mid-1970s the area extended from northeastern Labrador to Florida, reaching as far west as the Mississippi. Within this area acidities were 10 to 20 times higher than pH = 5.6. Metropolitan Canada no doubt contributes to this picture but Canadians feel quite strongly that much of their problem results from import of pollution from the south (Kurtz and Scheider, 1981).

In Europe observations taken between 1956 and 1976 show statistically significant increases in annual average precipitation acidity, some stations showing increases of 7 per cent per year (Kallend *et al.*, 1983). These increased annual levels of acidity arose from an increased number of intermittent high monthly values rather than a sustained high level. Within the continent the Scandinavians feel, as the Canadians do, that the bulk of the acid is imported, with the UK being the source area. Further south, the Germans, viewing forlorn parts of the Black Forest, are also tempted to blame the UK. Emotions tend to run rather high in this topic but it does appear that two things are agreed upon. First, the acidification of lakes in certain areas is due to the occurrence of acid rain but that, secondly, the tree damage in Germany and elsewhere in central Europe could well be related to high ozone levels, generated in the locally polluted atmospheres in the presence of sunlight.

The acidity of the wet deposition is due to the sulphur dioxide and nitrous oxides and their derivatives within the fall out. Sulphuric acid contributes 70 per cent and nitric acid 30 per cent to the total acidity. In general wet deposition is less important than dry deposition over much of the industrialized central parts of Europe, but it is more important in the remoter areas, especially where the precipitation is heavy as it is over the Scottish and Scandinavian mountains (United Kingdom Review Group on Acid Rain, 1983). The chemistry of the acid rain is only just emerging. Sulphur dioxide dissolves in water. From this sulphites may be formed and in turn they may be oxidized to sulphate. In the case of nitric oxide conversion to a soluble form must occur before incorporation into water droplets can take place. Sunlight provides the necessary force for the photochemistry. Putting these processes into the real atmosphere, we find the following chain of events. Industry emits acid gases (sulphur dioxide and nitrogen oxides). (Power stations emit two-thirds of the sulphur and one-half of the nitrogen. Vehicles emit one-third of the nitrogen.) The hot gases rise in plumes above the chimneys and cool as they are mixed with the air of the planetary boundary layer. Under the influence of sunlight and other chemicals the gases slowly oxidize to sulphuric and nitric acids. The air in which this has occurred is then ingested into clouds and partakes of the normal micro- and cloud-scale physics outlined earlier in the chapter. Also at this stage the oxidation of the gases to acids speeds up. The mechanisms by which particulate and gaseous sulphur and nitrogen compounds may be captured by cloud or rain drops are twofold; first, transfer of material to cloud droplets before they begin their descent as rain drops (in cloud scavenging), and secondly, transfer of material to falling rain drops (below-cloud scavenging). It appears that a dominant process whereby sulphate is incorporated into precipitation is the activation of sulphate particles as cloud condensation nuclei. Whilst all this is going on the winds can transport the air over hundreds of kilometres before the acid precipitation particles fall to earth, probably over upland areas. Good data are difficult to find, but it is estimated that 80–90 per cent of the acid rain measured in Norway and Sweden comes from abroad, with the UK contributing over 10 per cent of that amount. Wet deposition within Britain ranges from $0.1 \, \text{g} \, \text{H}^+ \, \text{m}^{-2} \, \text{year}^{-1}$ over much of Scotland and upland northern England to less than $0.025 \, \text{g} \, \text{H}^+ \, \text{m}^{-2} \, \text{year}^{-1}$ in parts of lowland rural England (United Kingdom Review Group on Acid Rain, 1983). Sulphur is a major source of this acidity. On average Britain sends about 2.45 million tonnes of sulphur per year into the atmosphere, 91 500 tonnes of which reach

Norway and Sweden and 0.24 million tonnes being deposited in Britain by wet processes.

The above outline of events is probably broadly true, but at present we have only a sketchy quantitative picture of the whole process. *Emissions* in UK and Europe have been estimated in recent years. In the UK sulphur dioxide emissions calculated for grid squares of $20 \, km \times 20 \, km$ reach local values as high as 500 000 tonnes per year (United Kingdom Review on Acid Rain, 1983). More representative values for the major industrialized areas are 50 000 tonnes per year. Similar or even higher values are found in southern East Germany, western Czechoslovakia and southern Russia. The *transport* of the pollutants is particularly difficult to measure. Some of the best data have come from airborne measurements of plumes containing tracers. Plumes from power stations have been clearly traced a distance of 650 km downwind, some 24 hours after emission. Oxidation rates vary widely between plumes, but can lead to a loss rate of sulphur dioxide of 27 per cent per hour. The final part of the process is *wet deposition* (the acid rain) and, although this is arguably easier to measure (at the ground) than is the transport, our information remains rather scanty. The primary concern of most experimental programmes is simply to establish the total deposition before tackling the wet-only deposition. Combinations of models and observations suggest that the total sulphur depositions per unit area range from about $1 \, g \, m^{-2} \, year^{-1}$ in southern Scandinavia to $6 \, g \, m^{-2} \, year^{-1}$ over much of England, W. Germany, E. Germany, Poland and Czechoslovakia, reaching values of $12 \, g \, m^{-2} \, year^{-1}$ locally in western Czechoslovakia, and southern Poland. It is probable that these high values are due to dry, as opposed to wet, deposition.

The examination of the 'acid rain problem' is at present being vigorously pursued. The literature is growing rapidly, providing much-needed basic data, revealing spectacular gaps in our knowledge, both observational and theoretical, and suggesting the familiar remedy of 'more research'. There is little doubt that such research will indeed by done, by national and international agencies. In five years' time we may have a well-informed, considered view of the problem and some well thought out and costed possible solutions.

(b) Urban influence on precipitation amounts

Two major ways in which man modifies the landscape are by planting or clearing of trees and by building urban areas. In the nineteenth century, large-scale deforestation was popularly thought to decrease precipitation. Inversion of this idea leads to schemes for increasing regional precipitation by planting trees, and at least one argument has seriously suggested that the states of Oregon and Washington attest to the rain-producing influence of extensive forests. At present, however, little or no evidence exists for significant changes in precipitation amounts due to the present or absence of a forest.

The situation for urban areas is rather different. As early as the 1930s, suggestions emanated from central Europe that urban areas may occasionally induce an increase in precipitation. The first analyses were essentially climatological, relating in a spatial sense mean annual or seasonal values of rainfall amounts or frequency to the form of the urban area (e.g. Changnon, 1961; Huff and Changnon, 1973). In many of these studies, increases in annual amounts of about 10 per cent were indicated and three possible explanatory factors were suggested, namely, increased numbers of condensation and ice-nuclei, increased absolute humidities in cities, and increased uplift over cities due to both thermal and mechanical effects. As a result of detailed case studies and the massive observational Metropolitan Meteorological Experiment (METROMEX) (Changnon *et al.*, 1971) we are now more certain of the absolute and relative effects of the three factors noted above. Before considering each in turn it is worth noting that METROMEX also provided abundant data which essentially sharpened the problem. A detailed analysis of 5-minute rainfall amounts over St. Louis (Huff and Schickedanz, 1974) showed that in 'rain cells' which had passed over the urban-industrial region of the town, the average rainfall volume was 216 per cent greater than in cells outside the urban area. It is important to note that this large increase applied only to distinct rain cells in two summers and it should not be thought of as a representative result for all rainfalls in all seasons over a long period. Nevertheless, the increase leaves little doubt as to the reality of an urban effect — albeit an effect which operates spasmodically.

The problem of urban-induced precipitation becomes more tractable if attention is restricted to convective precipitation. As argued previously (Atkinson, 1968, 1969), convective cloud dynamics often have a clear link with thermal sources at the ground and as such would be expected to respond to changes in surface cover. Within this context we can now profitably review the three factors which affect urban precipitation.

The effects of increases in the concentrations of condensation and ice nuclei are not easy to determine. Apart from being difficult to measure, the importance of such nuclei in determining precipitation amounts is probably far less than that of the air motions within clouds — particularly the updraughts (Battan, 1965). Recent evidence has shown that condensation nuclei numbers increased substantially in the towns of Buffalo, St. Louis, and Seattle, but interpretations differ on the effects that these increases had on the spectrum of sizes of cloud droplets. In Buffalo and St. Louis respectively, Kokmond and Mack (1972) and Fitzgerald and Spyers-Duran (1973) found that the droplet spectrum narrowed markedly in the industrial plume. This means that there were more cloud droplets, but that they were smaller than required for growth of precipitation-sized particles. These findings support the laboratory results of Gunn and Phillips (1957). In contrast, Hobbs *et al.* (1970), working in Seattle, found that the droplet spectrum broadened due to the initiation of many droplets of diameter greater than 30 μm. Such sizes are ideal for the operation of a very efficient coalescence mechanism — thus rapidly producing precipitation-size drops. Braham (1974) was of the opinion that the instrument used in METROMEX was inadequate to measure droplets of 30 μm diameter, but felt that they must have been present as radar observations showed preferential echo growth over the St. Louis area. Braham argued that for this to happen, coalescence must have been a powerful mechanism and this, in turn, requires the large droplets of the kind measured by Hobbs and his colleagues. Later measurements from METROMEX (Braham and Spyers-Duran, 1974) have revealed lower ice nuclei concentrations over St. Louis than in the air upwind of the city. Clearly we are not yet in a position to reach firm conclusions about the effects of urban areas on the concentrations of ice and condensation nuclei and, in turn, their effects on precipitation amounts.

Measurements of humidities in urban areas are as scarce as those of condensation and ice nuclei, but a notable contribution to our knowledge of the former is provided by Chandler's (1967) observations of relative and absolute humidities in Leicester, England. They are particularly attractive in the present context because Chandler found that water vapour pressures on three August 1966 nights were 2 mbar higher within the urban area than in the surrounding rural areas. The measurements by Bornstein *et al.* (1972) in New York confirmed Chandler's results and showed that urban excesses of absolute humidities may extend up to a height of 700 m. Such results, if they have general applicability, imply that urban air has a lower lifted condensation level and a lower level of free convection and thus is more permissive of cloud and precipitation growth than rural air.

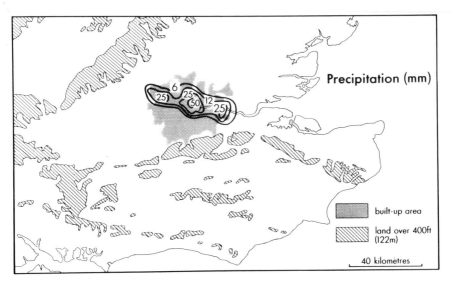

FIGURE 3.1   Total precipitation (mm) over southeast England, 21 August 1959

Important as nuclei and humidities are, the prime factor influencing precipitation amounts is the nature of the cloud updraught. In general, increases in updraught area, depth, and speed lead to increases in precipitation at the ground. Urban areas may affect these updraught characteristics by virtue of their thermal and mechanical influence on airflow. Each is considered below.

The thermal effect of the city upon both stationary and moving thunderstorms has been analysed by Atkinson (1970, 1971). Only the stationary case is used here for exemplification. On 21 August 1959 localized, heavy precipitation fell over London, England (Fig. 3.1), with a maximum amount of 68 mm. Radar evidence clearly showed the growth of deep clouds over the urban area, and this was strongly suggestive of an urban effect. These storm clouds were preceded by surface temperatures of 27–29°C in the city, the highest temperatures

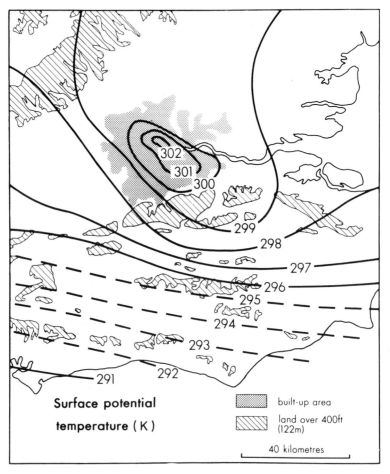

FIGURE 3.2   Surface potential temperature (K) 1200 GMT, 21 August 1959

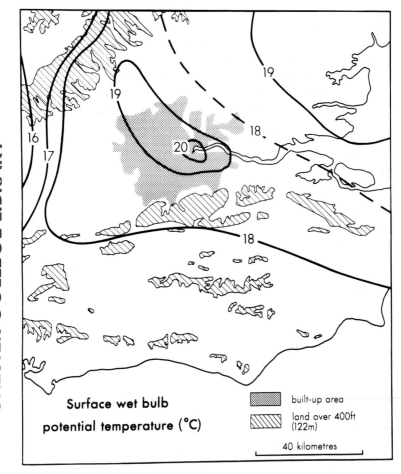

Surface wet bulb

potential temperature (°C)

built-up area

land over 400ft
(122m)

40 kilometres

FIGURE 3.3   Surface wet-bulb potential temperature (°C) 0900 GMT, 21 August 1959

anywhere in southeast England (Fig. 3.2). The lapse-rate was super-adiabatic in the lowest 500 m of the atmosphere and perhaps, in the Finsbury, Shoreditch, Bethnal Green area where surface temperatures reached 29°C, auto-convective instability existed in the lowest 100 m. Such strong instability encourages vigorous vertical motion and near-surface values of the wet-bulb potential temperature of 20°C (Fig. 3.3) meant that air rising adiabatically would reach equilibrium temperature at a height of 11 km. This was in good agreement with the radar observation of echo tops at about 10.6 km in the early afternoon. There appears to be little doubt that in the initially calm atmosphere of that August afternoon, London's heat island played a significant role in releasing the instability which led to such heavy precipitation.

The mechanical effect of urban areas upon airflow manifests itself in two ways: first, as an obstacle to flow, and secondly by causing frictional convergence. Angell *et al.* (1973) found that in strong winds of about 13 m s^{-1} Oklahoma City acted as a barrier to flow and induced upward velocities to air parcels of up to 70 cm s^{-1} at the 400 m level over parts of the city. The upward velocities varied in magnitude with time of day, being strongest between 1200 and 1800, weakest between 1800 and 2100 and between those extremes from 0900 to 1200 local time. Observations of flow around the city also suggested a barrier effect which was not fully compensated by vertical motion. Angell *et al.* (1973) also suggested that frictional convergence occurred, a theme more fully developed by Ackerman (1974).

Ackerman found that in winds of less than 5 m s^{-1} pronounced perturbations occurred in the horizontal flow over St. Louis on two August afternoons. In the lowest 500–600 m of the atmosphere, airflow took on an anticyclonic curvature as it approached the city centre and then cyclonic curvature as it moved toward the downwind city edge. The result of these perturbations was a convergence of air over the city and this necessitated uplift. Ackerman calculated upward velocities of 2–6 cm s^{-1}, apparently small, but, as she

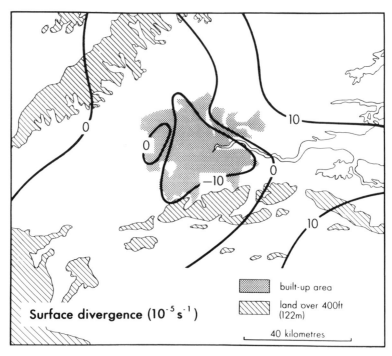

FIGURE 3.4   Divergence of surface wind field (units of 10^{-5} s^{-1}) in southeast England 0900 GMT, 1 September 1960

points out, 'the mean ascent rate could represent the net effect of a number of thermal bubbles or plumes of restricted areal extent but more intense vertical motion, rather than sustained flow. Indeed, vertical velocities deduced from the ascent of pibals indicate that local updraughts of 0.5 to 1 m s^{-1} did occur periodically at many sites . . .' (p. 234).

Applying these principles to a London case study, Atkinson (1975) showed that the precipitation of 1 September 1960 was partially due to frictional convergence over the city. Fig. 3.4 shows an area of strong convergence (10^{-4} s^{-1}) lying over the bulk of the urban area with extreme values of $2 \times 10^{-4} \text{ s}^{-1}$ in the Greenwich area. Such values of convergence, if they existed over a substantial depth of the lowest kilometre as indicated by Ackerman, would result in an uplift of about 10 cm s^{-1}. Consequently, to lift the surface layers through a distance of 1 km would take 10^4 s or about 2½ hours. By 1200 GMT on 1 September the convergence had been going on long enough to lift the bottom 500–1000 m of air to saturation and to a state of absolute instability. It was that air which was drawn into the convective clouds which moved over the city and gave heavy precipitation (Fig. 3.5).

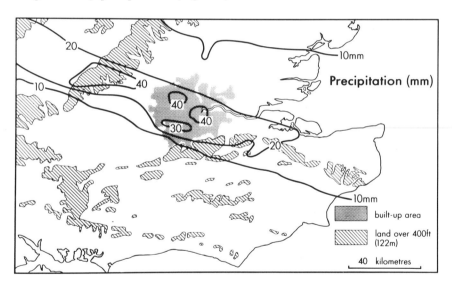

FIGURE 3.5   Total precipitation (mm) over southeast England, 1 September 1960

3.5   CONCLUSION

Man's effects upon precipitation have been suspected for only fifty years, and verification of these suspicions remains incomplete. Yet, as this chapter has attempted to show, such evidence as is increasingly available suggests that both conscious and unconscious effects are real. The magnitude and frequency of occurrence of the effects are small but, nevertheless, may have a significant impact on man's other activities. Thus an artificially induced 10 per cent increase in precipitation in the right place at the right time could be of great value to agriculturalists in areas of low rainfall, such as Australia, Israel and the American Midwest. Conversely a very heavy storm, perhaps triggered by an urban area (Atkinson, 1977), may result in flooding, extensive damage to property and even loss of life. The impact of acid rain is arguably more significant and widespread than both cloud seeding and urban influences, because of its major deleterious ecological effects. The social, economic, and legal implications of all forms of precipitations modification will no doubt ensure a healthy future for scientists interested in the mechanisms of precipitation formation.

REFERENCES

Ackerman, B., 1974, 'Wind fields over the St Louis metropolitan area', *Jnl. Air Poll. Cont. Assoc.*, **24**, 232–236.

Angell, J. K., Hoecker, W. H., Dickson, C. R., and Pack, D. H., 1973, 'Urban influences on a strong day time air flow as determined by tetroon flights', *Jnl. Appl. Met.*, **12**, 924–936.

Atkinson, B. W., 1968, 'A preliminary examination of the possible effect of London's urban area on the distribution of thunder rainfall, 1951–60', *Trans. Inst. Brit. Geog.*, **44**, 97–118.

Atkinson, B. W., 1969, 'A further examination of the urban maximum of thunder rainfall in London, 1951–60', *Trans, Inst. Brit. Geog.*, **48**, 97–119.

Atkinson, B. W., 1970, 'The reality of the urban effect on precipitation—a case study approach', in *Urban Climates, Tech. Note No. 108*, World Meteorological Organization, Geneva, pp. 342–360.

Atkinson, B. W., 1971, 'The effect of an urban area on the precipitation from a moving thunderstorm', *Jnl. Appl. Met.*, **10**, 47–55.

Atkinson, B. W., 1975, 'The mechanical effect of an urban area on convective precipitation', *Occas. Pap. No. 3*, Department of Geography, Queen Mary College, University of London.

Atkinson, B. W., 1977, 'Urban effects in precipitation: an investigation of London's influence on the severe storm in August 1975', *Occas. Pap. No. 8*, Department of Geography, Queen Mary College, University of London.

Battan, L. J., 1965, 'Some factors governing precipitation and lightning from convective clouds', *Jnl. Atmos. Sci.*, **22**, 79–84.

Battan, L. J., 1977, 'Weather modification in the Soviet Union—1976', *Bull. Amer. Met. Soc.*, **58**, 4–19.

Bergeron, T., 1949, 'The problem of artificial control of rainfall on the globe. I. General effects of ice-nuclei in clouds', *Tellus*, **1**, 32–50.

Bornstein, R. D., Lorenzen, A., and Johnson, D., 1972, 'Recent observations of urban effects on winds and temperature in and around New York City', Reprints Conference on Urban Environment and Second Conference on Biometeorology, American Meteorological Society (Boston), pp. 28–33.

Braham, R. R., 1952, 'Water and energy budgets of the thunderstorm and their relation to thunderstorm development', *Jnl. Met.*, **9**, 227–242.

Braham, R. R., 1974, 'Cloud physics of urban weather modification—a preliminary report', *Bull. Amer. Met. Soc.*, **55**, 100–106.

Braham, R. R., Battan, L. J., and Byers, H. R., 1957, 'Artificial nucleation of cumulus clouds', *Meteorological Monog., Amer. Met. Soc.*, **11**, pp. 47–85.

Braham, R. R., and Spyers-Duran, P., 1974, 'Ice nucleus measurements in an urban atmosphere', *Jnl. Appl. Met.*, **13**, 940–954.

Browning, K. A., and Foote, G. B., 1976, 'Airflow and hail growth in supercell storms and some implications for hail suppression', *Quart. Jnl. Roy. Met. Soc.*, **102**, 499–534.

Chandler, T. J., 1967, 'Absolute and relative humidities in towns', *Bull. Amer. Met. Soc.*, **48**, 394–399.

Chandler, T. J., 1970, *The Management of Climatic Resources*, Inaugural Lecture, University College London, H. K. Lewis and Co., London.

Changnon, S. A., 1961, 'A climatological evaluation of precipitation patterns over an urban area', *Air Over Cities, Sec. Techn. Rep. A62-5.*, Laboratory of Engineering and Physical Science of Division of Air Pollution, US Department of Health, Education, and Welfare (Public Health Service, Washington, D.C.), pp. 37–67.

Changnon, S. A., Huff, F. A., and Semonin, R. G., 1971, 'METROMEX: an investigation of inadvertent weather modification', *Bull. Amer. Met. Soc.*, **52**, 958–968.

Chappell, C. F., 1970, 'Modification of cold orographic clouds', *Atmos. Sci. Pap. No. 173*, Colorado State University.

Davis, R. J., 1975, 'Legal response to environmental concerns about weather modification', *Jnl. Appl. Met.*, **14**, 681–685.

Eliassen, A., 1980, A review of long-range transport modelling, *J. Appl. Met.*, **19**, 231–240.

Fisher, B. E. A., 1975, 'The long-range transport of sulphur dioxide', *Atmos. Envir.*, **9**, 1063–1070.

Fisher, B. E. A., 1983, 'A review of the processes and models of long-range transport of air pollutants', *Atmos. Envir.*, **17**, 1865–1880.

Fitzgerald, J. W., and Spyers-Duran, P. A., 1973, 'Changes in cloud nucleus concentration and cloud droplet size distribution associated with pollution from St. Louis', *Jnl. Appl. Met.*, **12**, 511–516.

Gunn, R., and Phillips, B. B., 1957, 'An experimental investigation of the effect of air pollution on the initiation of rain', *Jnl. Met.*, **14**, 272–280.

Hass, J. E., 1973. 'Social aspects of weather modification', *Bull. Amer. Met. Soc.*, **54**, 647–657.

Hobbs, P. V., Radke, L. F., and Shumway, S. E., 1970, 'Cloud condensation nuclei from industrial sources and their apparent influence on precipitation in Washington State', *Jnl. Atmos. Sci.*, **27**, 81–89.

Howell, W. E., 1960, 'Cloud seeding in the American tropics', *Physics of Precipitation, Geophys. Monog. No. 5.*, American Geophysical Union, pp. 412–423.

Huff, F. A., and Changnon, S. A., 1973, 'Precipitation and modification by major urban areas', *Bull. Amer. Met. Soc.*, **54**, 1220–1233.

Huff, F. A., and Schickedanz, P. T., 1974, 'METROMEX': rainfall analyses', *Bull. Amer. Met. Soc.*, **55**, 90–92.

Kallend, A. S., Marsh, A. R. W., Pickles, J. H., and Proctor, M. V., 1983, 'Acidity of rain in Europe', *Atmos. Envir.*, **17**, 127–137.

Kessler, E., 1973, 'On the artificial increase of precipitation', *Weather*, **28**, 188–194.

Klein, D. A., and Molise, E. M., 1975, 'Ecological ramifications of silver iodide nucleating agent accumulation in a semi-arid grassland environment', *Jnl. Appl. Met.*, **14**, 673–680.

Kokmond, W. C., and Mack, E. J., 1972, 'The vertical distribution of cloud and Aitken nuclei downwind of urban pollution sources', *Jnl. Appl. Met.*, **11**, 141–148.

Kurtz, J., and Scheider, W. A., 1981, 'An analysis of acidic precipitation in south central Ontario using air parcel trajectories', *Atmos. Envir.*, **15**, 1111–1116.

Lansford, H., 1973, 'Weather modification: the public will decide', *Bull. Amer. Met. Soc.*, **54**, 658–660.

Likens, G. E., 1976, 'Acid precipitation', *Chemical and Engineering News*, **54**, 29–44.

Likens, G. E., and Butler, T. J., 1981, 'Recent acidification of precipitation in North America', *Atmos. Envir.*, **15**, 1103–1109.

Marsh, A. R. W., 1984, 'Report on acid depositions', The Watt Committee on Energy Ltd, London, The London Science Centre.

Mason, B. J., 1971, *The Physics of Clouds*, 2nd edn, Oxford University Press, Oxford.

Oden, S., 1968, 'The acidification of air and precipitation and its consequences on the natural environment', *Ecological Res. Comm. Bull. No. 1*, Swedish Natural Science Research Council, Stockholm.

Schaefer, V. J., 1946, 'The production of ice crystals in a cloud of supercooled water droplets', *Science*, **194**, 457–459.

Smith, F. B., and Hunt, R. D., 1978, 'Meteorological aspects of the transport of pollution over long distances', *Atmos. Envir.*, **12**, 461–477.

United Kingdom Review Group on Acid Rain, 1983, *Acid deposition in the United Kingdom*, Warren Spring Laboratory, Stevenage, 72 pp.

Vonnegut, B., 1949, 'Nucleation of supercooled water clouds by silver iodide smokes', *Chem. Rev.*, **44**, 277–289.

Vonnegut, B., 1950, 'Experiments with silver iodide smokes in the natural atmosphere', *Bull. Amer. Met. Soc.*, **31**, 151–157.

Warner, J., and Twomey, S., 1956, 'The use of silver iodide for seeding individual clouds', *Tellus*, **8**, 453–459.

PART II

HYDROSPHERE

INTRODUCTION

The waters of the earth which comprise the hydrosphere represent a total volume of approximately 1386 million km^3. According to recent estimates produced by Soviet scientists, about 96.5 per cent of this volume is to be found in the world's oceans and freshwater accounts for only 3.5 per cent. Of the freshwater, storage in glaciers and permanent snow cover accounts for nearly 69 per cent and groundwater storage represents a further 30 per cent. Water in lakes represents about 0.25 per cent of the freshwater volume whilst that in rivers and in the atmosphere accounts for only 0.006 and 0.04 per cent respectively. The hydrological cycle links these various reservoirs within a global closed system and introduces an important unity to the hydrosphere, so that changes in one compartment could have important repercussions on another. The interactions between climatic change and global ice cover, changing sea levels, global patterns of precipitation and runoff and fluctuating lake levels are, for example, well documented and clearly attest to the significance of this interdependence.

Human activity has had an important impact on the hydrosphere, but both the magnitude of this impact and the depth of its study tend to demonstrate an inverse relationship to the volumes of the reservoirs involved. Thus there is abundant evidence of the marked changes in the behaviour of rivers and lakes caused by man's activities and Chapter 3 has already provided a detailed discussion of human impact on atmospheric moisture and precipitation. Much less is known about human impact on the oceans. The vastness of the open oceans has traditionally encouraged views of infinite dilution symbolized by the idiom ' a drop in the ocean' but the potential significance of pollution by human wastes, and more particularly the accumulation of toxic substances, is now being increasingly recognized. In estuarine and coastal areas, however, where there is restricted interchange with the vast water volumes of the open oceans, the potential for human impact is considerably greater. Furthermore, such areas have proved of great attraction to man and current estimates indicate, for example, that approximately one-third of the population of the USA live and work in areas around estuaries. In dealing with the hydrosphere, physical geographers have therefore frequently devoted attention to human impact on terrestrial hydrological processes and to a lesser extent lakes and estuaries, but the oceans have attracted far less attention.

Man's impact on the hydrosphere involves both direct or intentional effects — where, for example, rivers have been dammed to impound reservoirs, lakes drained, coastal areas reclaimed, or ship canals dug to link oceans — and a wide range of inadvertent or indirect effects. The coverage afforded by this section is by necessity selective, but the four chapters have been chosen to focus on specific compartments of the hydrosphere and to provide treatments at different scales and levels of detail. In Chapter 4, D. E. Walling deals with terrestrial hydrological processes and considers both the global and local scales by focusing attention on global water resources and experimental catchment investigations. Chapter 5 by E. C. F. Bird deals with the zone of interaction between the land and the oceans and provides a detailed review of the influence of human activity on the evolution and dynamics of coastal landforms. Faced with the task of dealing with an area representing over 70 per cent of the earth's surface and a vast subject matter, Chapter 6 on ocean processes contributed by C. A. M. King has by design adopted a broad perspective and provides a general review of the main areas in which human activity has been important. Finally, in Chapter 7, P. Sly returns to the more specific topic of lakes and discusses both their importance to man and the wide-ranging impact of human activity on lake systems.

Human Activity and Environmental Processes
Edited by K. J. Gregory and D. E. Walling
©1987 John Wiley & Sons Ltd.

4

Hydrological Processes

D. E. WALLING

Hydrology is no longer limited to the geophysical science of the natural circulation and storage of waters on the earth. While that circulation is still the dominant feature, it is increasingly modified and managed by man, and the hydrologist who does not recognise this will be left back in the nineteenth century!

W. C. Ackermann (1969)

4.1 INTRODUCTION

This view formed the main thrust of an introductory address to an international hydrology seminar held in Urbana, Illinois, in 1969. Faced with the task of highlighting recent trends within the science of hydrology, the speaker emphasized the need for a greater awareness of the impact of man on hydrological processes and added further support to a view which had become increasingly voiced during the 1960s. In response to such pressure the theme of 'the influence of human activity on hydrological regimes' had been included as one of the principal foci of the International Hydrological Decade (IHD) (1965–1974). This period of international cooperation in hydrology, launched by UNESCO, has been succeeded by the International Hydrological Programme (IHP) and several of the projects within this programme have again reflected mounting interest in and concern for the impact of man on the hydrological cycle. Numerous studies, working groups and symposia resulted from this international activity and interest (e.g. IAHS, 1974; UNESCO, 1974, 1980; Keller, 1976) and the vast bodies of evidence and knowledge concerning man's impact on hydrological processes and associated problems that are now available reflect a widespread recognition and justification of the view advanced by Ackermann.

In some instances the effects of man have been largely *intentional*, as in the case of the construction of reservoirs, the development of river regulation schemes and irrigation systems, and the introduction of artificial recharge of groundwater bodies. More widespread, however, are the *inadvertent* effects whereby, for example, forest clearance has markedly changed the flow regime of a river basin or agricultural activity has influenced the water quality of local streams. In both cases a variety of process mechanisms are involved and the hydrological modifications may range from the almost imperceptible to the catastrophic and from the essentially insignificant to the environmentally or economically disastrous. Changes may also vary in both temporal and spatial scale. Some may be only short-term perturbations, whereas others may represent long-term and essentially irreversible modifications. Some, as in the case of large-scale river diversion schemes, may affect hydrological conditions over large areas, while others may be restricted to a local area. Assessments of these changes will, however, frequently pose fundamental problems, for there are few remaining areas of the contemporary world where true wilderness conditions exist to provide a yardstick against which to evaluate change. In the British Isles, for example, any area selected as a natural baseline is likely to have already been subject to human interference over the past 3000 years or more. In many instances, however, there will be clear evidence that the pace of change has increased markedly towards the present in response to increases in population pressure and land use activity and advances in technological capabilities (Fig. 4.1).

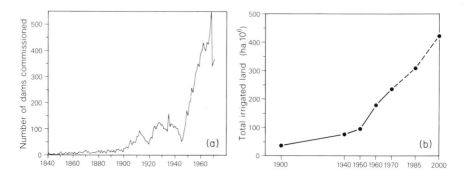

FIGURE 4.1 The changing pace of human impact. Recent trends in world dam construction (a) and total irrigated area (b) are illustrated. (Sources: (a) based on Beaumont, 1978; (b) data in Framji *et al.*, 1981.)

Despite a general recognition of the far-ranging significance of man's impact on the hydrological cycle, few hydrology texts have explicitly covered this aspect of the subject. This chapter attempts to review some of the evidence now available and to provide an insight into the nature and extent of this impact.

4.2 HYDROLOGICAL PROCESSES AND THEIR MODIFICATION

Fig. 4.2 presents a simplified representation of the hydrological cycle operating in an essentially natural drainage basin. Inputs of precipitation are distributed through a number of stores by a series of transfer processes and are output as channel flow, evapotranspiration and deep leakage. Thus, in simple terms, water entering a drainage basin is intercepted by vegetation and subsequently evaporated or reaches the ground surface as throughfall and stemflow. In the absence of a vegetation canopy the precipitation reaches the ground directly. At the ground surface water infiltrates into the soil or is retained in surface storage. The surface water may move downslope as surface runoff or be slowly evaporated. Moisture held within the soil is subject to surface evaporation and plant transpiration, downward percolation to the water table and the groundwater body, and downslope movement as throughflow and interflow. Water stored within the groundwater body is similarly subject to losses by plant transpiration and evaporation when the water table is close to the surface, upward movement into the unsaturated soil by capillary action, deep leakage to adjoining groundwater bodies and slow surface outflow into springs, seeps and river channels as baseflow. The river channel receives variable contributions of surface runoff, throughflow and interflow and baseflow which together contribute to the time-variant outflow hydrograph of the drainage basin.

An attempt has been made to point to the various ways in which this cycle could be modified through man's activities by considering, first, modification of internal processes and, secondly, additional moisture inputs (Fig. 4.2). For convenience, process modification has been taken to include both transfer processes and the functioning of various stores, and inspection of the listed examples of significant human activities clearly demonstrates that in most locations the hydrological cycle is far from a natural system. Specific activities will not in all cases produce similar effects. For example, whereas in many areas cultivation has been shown to reduce surface infiltration and increase surface runoff, there is a significant body of evidence from certain areas of the USSR which indicates that autumn ploughing can cause a decrease in surface runoff associated with spring snowmelt (e.g. Shevchenko, 1962). Similarly, whereas forest clearance will by virtue of decreased evapotranspiration losses lead to increased runoff in most environments (cf. Anderson *et al.*, 1976), there is again evidence from the USSR (e.g. Rakhmanov, 1970) and from the montane forests of Costa Rica (e.g. Zadroga, 1981) to indicate that the reverse may sometimes be true. Additional moisture inputs may further modify the system and a number of potential sources have been listed in Fig. 4.2 and related to individual flow pathways. Although extreme, it would not be difficult to find situations where inputs of irrigation water exceed the natural precipitation input to a drainage basin, or where inputs to channel flow from effluent outflow sometimes exceed natural runoff.

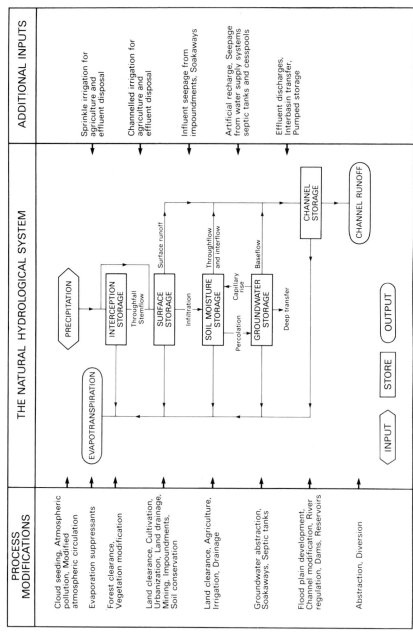

FIGURE 4.2   Man and the natural hydrological system. A simplified diagrammatic representation of the hydrological processes operating in a drainage basin and some examples of human interference associated with process modifications and additional moisture inputs.

Bearing in mind the difficulties associated with any attempt to generalize regarding the nature and extent of process modifications resulting from human impact, Fig. 4.3 provides specific examples for a selection of processes. In the case of evapotranspiration, some of the results of a study by Gleason (1952) aimed at estimating the losses occurring under different land uses within the Raymond Basin, Pasadena, California, are cited (Fig. 4.3(a)). If the areas of lawns and trees are taken to be the most representative of natural conditions, then the extreme case of change to pavement and streets can be seen to cause an 86 per cent reduction in evapotranspiration. Instances of increased losses might, however, be found in other urban areas where sprinkler irrigation is widely used on domestic lawns and recreational areas.

An indication of the impact of agricultural land use on infiltration rates is afforded by Fig. 4.3(b). The cumulative infiltration graphs reported by Holtan and Kirkpatrick (1950) from measurements on Durham, Madison and Cecil soils in South Carolina, USA, similarly lack a natural baseline for comparative analysis, but if the response of old pasture is used for this purpose, reductions in infiltration rates of up to 80 per cent or more are seen to result from various cultivation and grazing practices. Reduced infiltration leads in turn to decreased soil moisture storage and a measure of the potential reduction can be found in the work of Hills (1969) who studied the moisture content of soils under different management practices in the area around Bristol, UK. Measurements made on the heavy soils of the Worcester series (Fig. 4.3(c)) demonstrate decreases in moisture storage by more than 50 per cent in response to the passage of agricultural vehicles and herbicide treatment. These changes were related primarily to reduced infiltration, although the decrease in available storage space caused by compaction could also prove significant in many situations. Infiltration rates exert a critical control over the generation of surface runoff by controlling the relative volumes of water remaining at the surface and passing down through the soil. An example of the impact of grazing activity on infiltration and more particularly surface runoff, is provided by Fig. 4.3(d). Based on the work of Noble (1965) in the intermontane region of Utah, this demonstrates how a reduction in ground cover from good (60–75 per cent cover) to poor (10 per cent cover) as a result of overgrazing, increases the proportion of rainfall appearing as surface runoff from 2 per cent to as much as 73 per cent, an increase of over 35-fold.

Groundwater storage will be influenced by changes in infiltration and recharge and also by modified evapotranspiration rates where the water table is sufficiently close to the surface to render loss by capillary rise or plant uptake significant. The work of Holstener-Jorgensen (1967) at the Danish Forest Experimental Station illustrated in Fig. 4.3(e) provides an example of such changes by demonstrating the effects of clearing a 75-year-old beech stand on groundwater levels. Before clearance, the water table exhibited a marked seasonal variation, dropping during the summer in response to water uptake by the trees, but this pattern almost completely disappeared after clearing.

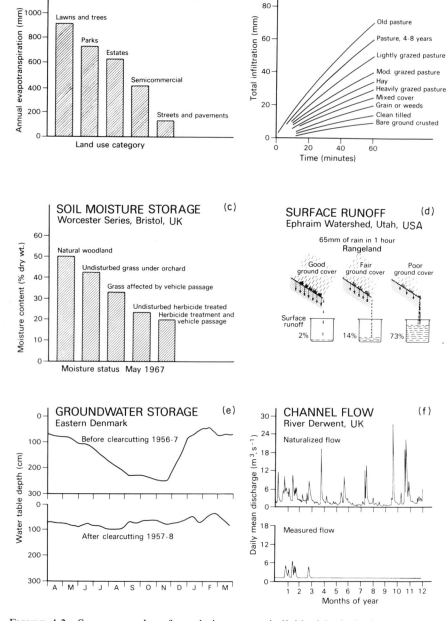

FIGURE 4.3   Some examples of man's impact on individual hydrological processes. (Sources: (a) data in Gleason, 1952; (b) Holtan and Kirkpatrick, 1950; (c) data in Hills, 1969; (d) Noble, 1965; (e) Holstener-Jorgensen, 1967; and (f) Richards and Wood, 1977.)

Finally, Fig. 4.3(f), based on the work of Richards and Wood (1977), presents a graphic example from the UK of the effects of modifying channel storage, in this case by reservoir construction, on channel flow regime. The recorded flow in the River Derwent at Yorkshire Bridge, a site downstream from the Ladybower Reservoir, contrasts markedly with the reconstructed or naturalized record in showing an almost complete absence of peaks and a constant minimum flow throughout 85 per cent of the year.

Fig. 4.3 illustrates some of the process modifications cited in Fig. 4.2 and demonstrates that man can profoundly influence the individual components of the hydrological cycle. In addition, it is clear that such modifications are widespread occurrences, rather than isolated extremes. With the exception of Fig. 4.3(a), however, the examples are all taken from essentially rural locations, and even greater modifications can be found in urban areas. Under tarmac and concrete, infiltration will approach zero and the effective 'roofing over' of the groundwater body will cause rapidly falling water table levels. As noted previously, evapotranspiration losses from buildings and paved areas will be drastically reduced, and the storm runoff dynamics of a system of roofs, gutters, roads, gullies, stormwater drains and culverts will differ radically from those of a natural surface.

Some indication of the change in the relative importance of the major components of the hydrological cycle associated with the transition from rural to urban conditions is provided by Fig. 4.4(a). This depicts simplified water budgets for rural and urban conditions in the area of Moscow, USSR, calculated by Lvovich and Chernogaeva (1977). In that area, the establishment of urban conditions has resulted in a decrease in evapotranspiration and groundwater runoff by 62 per cent and 50 per cent respectively, and an increase in total runoff by 155 per cent.

More detailed investigation of the urban hydrological system reveals numerous features going far beyond simple modification of the natural system shown in Fig. 4.2. Fig. 4.4(b) attempts to depict some of the major components of such a system under fully urbanized conditions and introduces the need to incorporate water import, waste treatment, artificial recharge, storm drainage and effluent disposal components. In some instances a large proportion of the output may be diverted out of the basin to another river.

In developing this review of the nature and extent of man's impact on hydrological processes it is useful now to consider some specific human activities rather than individual processes. To this end, and in an attempt to provide examples of both intentional and inadvertent effects and to embrace both large scale and more local changes, attention will be directed, firstly, to present and future scenarios of water resource exploitation by mankind and, secondly, to the results of a variety of local catchment studies and experiments which have attempted to document the influence of land use activities and changes.

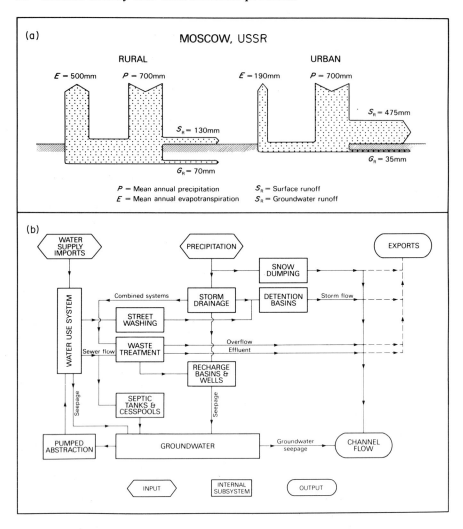

FIGURE 4.4   Hydrological processes in an urban area. (a) compares the water balance of a rural area and a heavily built-up zone of Moscow, USSR. (Source: data in Lvovich and Chernogaeva, 1977.) (b) illustrates some of the additional components associated with the functioning of an urbanized catchment

4.3   WATER RESOURCE EXPLOITATION

Water is essential for life and it is used in vast quantities by man, both to sustain his existence and as a basis for economic activities. Human exploitation of water resources must therefore be viewed as a direct or intentional influence on the hydrological cycle and its magnitude is clearly evident from estimates of global

TABLE 4.1 Estimates of global water use in the 1970s (based on Barney, 1982)

Purpose	Water use $(10^9 \, m^3)$
Irrigation	1830
Energy production	502
Industrial	305
Domestic	201
Total	2838

water use during the 1970s which amount to about $800 \, m^3$ per person per year (Barney, 1982). This figure includes water use for domestic, municipal, industrial and agricultural purposes and for energy production (Table 4.1) which together amount to a total volume of approximately $3 \times 10^{12} \, m^3$ per year. The significance of this volume becomes more apparent when it is noted that it is equivalent to about 8 per cent of the mean annual volume of river runoff from the land surface of the earth to the oceans (ca. $40 \times 10^{12} \, m^3 \, year^{-1}$) which represents the total available water resource. Population growth and advancing technology both lead to increased water use and estimates for the year 2000 point to a situation where the value of 8 per cent may have increased to somewhere in the range of 30–50 per cent (cf. Barney, 1982; Kalinin and Bykov, 1969). These data clearly emphasize that water withdrawal by man now represents an important impact on the hydrological cycle.

Much of the water used by man will of course be returned directly to the rivers and will represent limited disruption of the natural system. Nevertheless, only about 60 per cent of the water withdrawn will be returned in this way and the remainder represents an actual *consumptive use* or an irretrievable water use. This will be largely accounted for by evaporation, and currently represents about 3–4 per cent of the total global runoff. In the USA in 1975 it is estimated that 35 per cent of the annual runoff was withdrawn by man and that the total consumptive use amounted to about 8 per cent of the annual runoff (Barney, 1982).

As can be seen from Table 4.1, irrigation represents by far the most important component of water use by man. The large volumes of water involved may be easier to appreciate when it is recognized that crops consume vast quantities of water and, for example, that $2000 \, m^3$ of water are needed to grow 1 tonne of rice. According to FAO (1978), the total irrigated acreage in the world in 1975 was 223 million ha, representing about 15 per cent of the world's cropland. The total consumptive use by the crops is estimated to be about $1300 \times 10^9 \, m^3$ by Biswas (1983), but this author suggests that the overall consumptive use may be almost as great as $3000 \times 10^9 \, m^3$ when storage and conveyance losses are taken into account. Forecasts suggest that the total irrigated acreage of the globe will increase to 420 million ha by 2000 with an equivalent growth in consumptive use (UNESCO, 1978).

The impact of water withdrawal by man and the associated consumptive use on river flows is strikingly demonstrated by the records of declining flows in several of the major southward flowing rivers of the USSR during the present century (Table 4.2). Water from these rivers has been extensively used for irrigation purposes and Golubev and Vasiliev (1978) estimate that by 1986 their total annual discharge will have declined by 19 per cent, when compared to natural levels. Projections for the year 2000 indicate that this decrease will have increased to 30 per cent. Values for some individual rivers are much higher and in the extreme case of the Amudarya the projections are that the flow will have decreased by 59 per cent in 1986 and by 95 per cent in 2000. It is the reduction in the flow of these rivers that has led to the decrease in surface water inflow and rapidly falling water levels in the Caspian, Aral and Azov seas.

TABLE 4.2   Actual and projected decreases in the flow of selected USSR rivers due to water abstraction (based on Golubev and Vasiliev, 1978)

River	Decrease in annual runoff (%)		
	1970	1981–1986	1991–2000
Volga	6	10	14
Ural	14	23	26
Terek	17	44	78
Kura	2	19	58
Don	18	30	45
Kuban	16	39	60
Dniester	14	32	38
Dnieper	20	26	34
Amudarya	12	59	95
Syrdarya	31	46	92

Increased consumptive use will not always produce an equivalent decrease in the flow of local rivers because, in some situations, agricultural and industrial development may involve a reduction in natural non-productive evaporation (e.g. draining of marshes). Furthermore, the continuity associated with the hydrological cycle means that increased evaporation must result in increased atmospheric moisture leading to increased precipitation inputs and increased runoff, although this effect may only become evident when the water balance of a large area is considered and may have little local influence. Sokolov (1978), for example, provides estimates of the water balance of Europe in 1970 and 2000 and suggests that precipitation and runoff were increased by amounts equivalent to 38 and 13 per cent of the consumptive use respectively in 1970, and that these values will increase to 60 and 21 per cent respectively in 2000. Equivalent estimates of the increases in precipitation and runoff associated with irretrievable water losses for other continents and for the total land surface of the earth are provided in Table 4.3.

TABLE 4.3 The contribution of irretrievable water use to increased precipitation and runoff at the continental and global scales (based on UNESCO, 1978)

Continent	Mean annual river runoff (km³)	Irretrievable water use (km³ year⁻¹)		Increase in precipitation (km³ year⁻¹)		Volume of additional runoff (km³ year⁻¹)	
		1970	2000	1970	2000	1970	2000
Europe	3210	100	240	38	151	13	51
Asia	14410	1130	2000	770	1540	324	650
Africa	4570	100	250	107	214	21	43
N. America	8200	160	280	81	323	35	139
S. America	11760	50	130	0	0	0	0
Australasia	2390	12	30	0	0	0	0
Total land surface	44540	1600	3000	1000	2200	390	880

Dams and reservoirs are an essential component of water resource development in most areas of the world since they provide a means of storing water during times of excess and making it available during times of shortage. Most major dams have been constructed during the present century (Fig. 4.1(a)) and current statistics indicate that there are now over 10 000 major dams on the world's rivers with their reservoirs providing a surface area in excess of 400 000 km² and a total water storage of more than 5000 km³. The significance of this last figure becomes readily apparent when it is recognized that it represents about 12.5 per cent of the annual runoff from the land to the oceans. Looking more specifically at the larger reservoirs with a capacity in excess of 5 km³, an analysis by Soviet hydrologists (Table 4.4) indicates that the proportion of annual runoff stored varies markedly between the continents from a minimum of 1.6 per cent for Australia and Oceania to a maximum of 27.1 per cent for Africa.

Water storage in reservoirs may result in marked changes in downstream river regimes and flow volumes (cf. Fig. 4.3(f)) and the considerable increases in local evaporation that may occur from the reservoirs must be seen as an additional component of irretrievable water use associated with human impact. Current estimates (UNESCO, 1978) suggest that the additional evaporative losses associated with the presence of a reservoir are as much as 8.3 km³ year⁻¹ for Lake Nasser, 4.6 km³ year⁻¹ for the Volta Reservoir and 4.4 km³ year⁻¹ for Lake Kariba and that the resultant overall increase in evaporation from the continents represented about 110 km³ year⁻¹ in 1975. This latter figure is, however, small in comparison to the natural evapotranspiration loss from the land surface of about 70 000 km³ year⁻¹ and the total irretrievable uses detailed in Table 4.3). The construction of reservoirs may not always be associated with increased evaporation, since Shiklomanov and Kozhevnikova (1974) cite the example of the Volga delta in the USSR where the reduction in flooding in the floodplain and delta areas resulting from the cascade of reservoirs upstream

TABLE 4.4   Water storage by major world reservoirs (volume > 5 km^3) in 1972 (based on UNESCO, 1978)

		Major Reservoirs			
Continent	Number	Total volume (km^3)	Useful volume (km^3)	Total annual river runoff (km^3)	Ratio of reservoir volume to river runoff (%)
Europe	25	422	170	3210	13.1
Asia	48	1350	493	14410	9.4
Africa	12	1240	432	4570	27.1
N. America	45	950	210	8200	11.6
S. America	10	286	123	11760	2.4
Australia and Oceania	3	38	10	2390	1.6
Total	143	4286	1438	44540	9.6

has caused a reduction of evapotranspiration loss from this area of 1.4 km^3 year^{-1}.

In some areas of the world, groundwater has been extensively exploited for water supply. The volumes abstracted are considerably lower than those associated with surface water, but are nevertheless still appreciable. For example, Kharchenko and Maddock (1982) report that by 1973 over 1.2 million tubewells had been sunk for irrigation in the USA and that the current groundwater abstraction rate was 50 km^3 year^{-1}. This figure may be compared with a total abstraction for irrigation of approximately 160 km^3 year^{-1}. In terms of total water use, groundwater accounts for about 20 per cent in the USA, 36 per cent in the Federal Republic of Germany, 15 per cent in the UK and 97 per cent in Denmark. Where groundwater abstraction is scientifically managed and related to current recharge rates, it may be viewed as withdrawal of water that might otherwise have contributed to river runoff and therefore as a redistribution of water between the basic components of the hydrological cycle. Where, however, rates of abstraction are considerably in excess of natural recharge or the groundwater body must be viewed as a legacy from a former period of increased moisture availability, major disruption of the hydrological balance will occur and groundwater levels may decline drastically. There are numerous reports in the literature of marked lowering of water table levels. In areas of clay lithology this lowering may be associated with ground subsidence (cf. Chapter 11) and in coastal areas saltwater may encroach into the groundwater body.

As the world's population increases and technology develops, the demand for water will inevitably grow. Man's exploitation of the available water resource will produce an increasing impact on the hydrological cycle at both the local and global scales. Long-distance interregional water transfers are already being

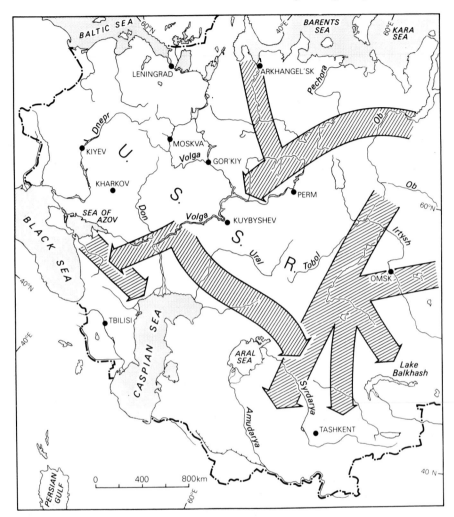

FIGURE 4.5 Long-distance water transfers. An integrated water transfer system proposed for future water resource development in the USSR. (Based on Voropaev, 1978.)

undertaken on a small scale in many areas of the world but such transfers are likely to become increasingly important in the future as regional demand exceeds supply. In the People's Republic of China plans are already being developed to transfer water northwards from the Chang Jiang (Yangtze) River to the more arid areas of north and northwest China (Biswas *et al.*, 1983) and volumes of up to 30 km^3 year^{-1} have been cited. Similarly, major water transfer schemes have been proposed in the USSR to meet a situation where up to 75 per cent of the country's water consumption is based in the southern regions which possess less than 15 per cent of the available runoff and which may be faced with a water deficit of over $100 \, \text{km}^3 \, \text{year}^{-1}$ by the end of the century (Voropaev, 1978). A gigantic canal which would stretch about 25 000 km from the River Ob, in Siberia, across Kazakhstan to the Amudarya is currently being planned and the basis of one proposed transfer scheme is illustrated in Fig. 4.5. It is clear that the implementation of such a scheme would introduce major changes in the hydrology of the 12 million km^2 area involved which would in turn have important repercussions for the global hydrological cycle. Furthermore, environmentalists have also expressed concern that diverting water away from the Arctic might affect the polar icecap or even upset the rotation of the earth.

4.4 CATCHMENT STUDIES AND EXPERIMENTS

The use of catchment studies to assess the impact of human activity on drainage basin behaviour can be traced back to the 1890s and work in Switzerland (Engler, 1919) and to the classic Wagon Wheel Gap investigation initiated in Colorado, USA, in 1910 (Bates and Henry, 1928). The former study involved the instrumentation of two small watersheds to evaluate the contrasts in behaviour resulting from differences in land use and was essentially comparative in approach. The Wagon Wheel Gap study can, however, be classed formally as an experiment. It involved the comparison of two small forested basins for an initial period in order to derive a *calibration* and the subsequent clearance of one of them. The effects of clearance were assessed using the initial calibration to provide estimates of departure from natural response. Similar studies were subsequently undertaken by many other research stations in the USA, for example at Coweeta in North Carolina (Dils, 1957), and in other countries (cf. Pereira, 1973; Rodda, 1976), but it was the Representative and Experimental Basin Programme of the IHD that provided the stimulus for a major upsurge in catchment studies. Whilst representative basins were established primarily to provide data representative of a hydrological region, experimental basins were set up to study the effects of cultural changes on drainage basin dynamics and explicitly involved deliberate modification of the catchment characteristics. The numerous experimental catchment investigations that were carried out within the framework of the IHD and that have been extended during the IHP provide another valuable source of information on man's impact on the hydrological system for many areas of the world. In contrast to the previous discussion of water resource development, most of these results relate to small-scale local investigations, although they can of course be extrapolated to wider areas, and the human impacts involved fall primarily into the category of inadvertent or indirect effects.

The need to provide a baseline or calibration against which to assess changes in catchment behaviour has generated a number of strategies for experimental catchment research. In the first, the classic paired or control watershed study, two essentially similar basins are instrumented and calibrated against each other. The effects of subsequent modifications to one of them can be evaluated using this calibration to reconstruct the natural response of the modified basin from the records of the unaltered control. Fig. 4.6(a) presents an example of this type of watershed calibration. It relates to a paired watershed experiment at the Coweeta Hydrologic Laboratory in North Carolina aimed at assessing the impact of clear-felling on runoff yield. The three-year calibration period was used to derive a relationship between the monthly water yields of the two catchments and the increases in yield consequent upon felling were estimated using this relationship. When compared to the control basin, annual runoff from the clear-felled basin initially increased by 373 mm, but the magnitude of this increase progressively declined as the forest vegetation re-established itself. Twenty-three years later the felling was repeated and the almost identical increase in water yield demonstrated the consistency of the experiment.

Where it is difficult to find suitable twin drainage basins, the single watershed approach has often been adopted. In this a single catchment is calibrated in its natural state against hydrometeorological variables, such as precipitation input, which will not be modified by the subsequent alterations to the basin. Fig. 4.6(b) illustrates the use of this approach in the study of the impact of building activity on the storm runoff response of a small drainage basin on the margins of Exeter, UK (Walling and Gregory, 1970). A multiple regression equation relating storm runoff volume to precipitation character and antecedent moisture status was fitted to the runoff record for the period of undisturbed condition and this was used to estimate the natural response to storm rainfall after building commenced. By comparing these estimated values with the recorded runoff volumes, the change in storm runoff production could be assessed. In this particular case, disturbance of 25 per cent of the surface area of the basin resulted in an average increase in storm runoff volumes of 2.4 times. Similar equations were employed to assess the increase in storm peak discharges which were found to have increased on average by 3.5 times.

Where a number of research catchments are located within a small area it is possible to devise experiments to evaluate the effects of a number of different watershed treatments simultaneously. This strategy is often termed the multiple watershed approach and a good example is provided by work undertaken at the Moutere Soil Conservation Station in South Island, New Zealand, by the New Zealand Ministry of Works (1968). Aimed at studying the effect of different land management practices, this study involved 12 small basins (Fig. 4.6(c)). From 1962 to 1965 the catchments were calibrated and a programme of treatment was subsequently initiated (Table 4.5). Some of the results obtained are presented in Fig. 4.6(c)(i). Here the influence of gorse clearance and subsequent arable cultivation on storm peak discharges from catchment 10 has been evaluated using catchment 5 as a control. Increases in storm peak discharge in excess of one order of magnitude are apparent in the smaller events. In an experiment of

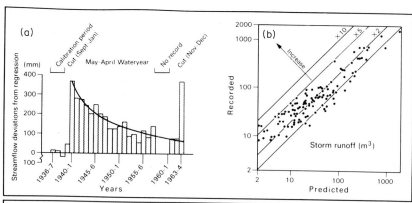

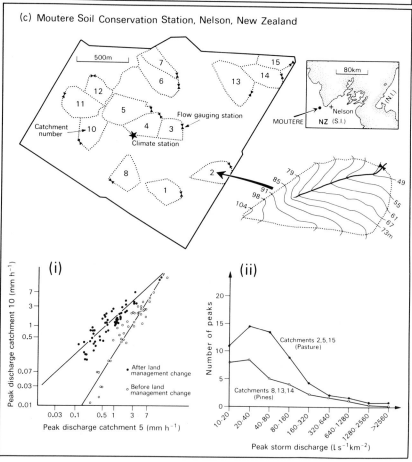

TABLE 4.5 The experimental programme planned for the Moutere catchment in 1966

Treatment	Catchment numbers
1 Mob stocking[1a] plus contouring	2 and 6
2 Set stocking[2b] plus contouring	4 and 15
3 Mob stocking, no contouring	5 and 14
4 Set stocking, no contouring	1 and 3
5 Gorse	8 and 13
6 Cultivated and sown into crops	10 and 12

[a]Intermittent grazing, high animal density.
[b]Continuous grazing, low animal density.
Source: New Zealand Ministry of Works (1968).

this type treatments can be changed once sufficient data have been collected for a particular management practice. In 1970 a further programme of land use change was initiated at Moutere (New Zealand Ministry of Works, 1971), and this included the planting of exotic forest (*Pinus radiata*) on catchments 8, 13 and 14. Peak runoff distributions for the years 1975 and 1976 averaged for three catchments in pasture (2, 5, 15) and the three planted with pines, are presented in Fig. 4.6(c)(ii). Catchments planted with pines yielded fewer peaks in all magnitude classes.

Numerous catchment studies employing strategies similar to those outlined above have been undertaken throughout the world and these have produced a wealth of empirical evidence concerning the effects of human activity on drainage basin response (e.g. IAHS, 1980; Williams and Hamilton, 1982). Although the results relate to small study catchments their significance is readily apparent when it is recognized that the land use change or other human impact under investigation will frequently have influenced vast areas. it would prove a daunting task to review the evidence now available from these studies within the confines of this short chapter, but a number of case studies highlighting particular activities can usefully be introduced.

(a) Land clearance

Over large areas of the globe, the development of agriculture, both by early civilizations and by colonists opening up the new world must have produced

FIGURE 4.6 *(opposite)* Catchment studies. (a) illustrates the results of a paired watershed study of the effects of forest clearance on runoff. (Source: Hibbert, 1967.) (b) shows how data from a single watershed study have been used to assess the increase in storm runoff volumes associated with urban development. (c) provides a plan of the multiple watershed study in progress at Moutere, New Zealand, and presents some results relating to the effects on peak flows of conversion of a small gorse catchment to cultivation and cropping (i) and of forest planting (ii). (Sources: New Zealand Ministry of Works, 1968 and 1978; and Scarf, 1970.)

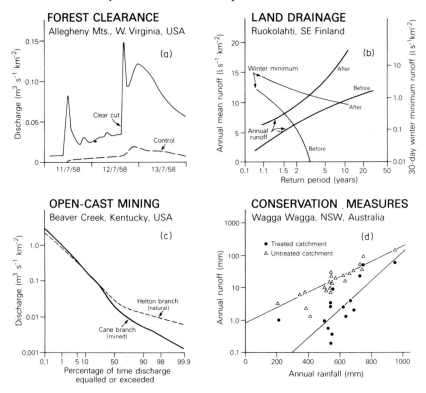

FIGURE 4.7   Man's impact on catchment response. Results of several case studies discussed in the text are illustrated. (Sources: (a) Harrold, 1971; (b) Mustonen and Seuna, 1971; (c) Collier *et al.*, 1970; and (d) data in Adamson, 1974.)

far-reaching changes in drainage basin response. Clearance continues today for extension of agriculture and for timber production and Postel (1984) reports that the annual rate of forest clearance in the tropics is currently 11.3 million ha per year with selective logging influencing a further 4.4 million ha. An indication of the potential magnitude of the hydrological impact is provided in Fig. 4.7(a). This presents some results from a catchment experiment in the Allegheny Mountains of West Virginia, USA, cited by Harrold (1971). The response of a clearcut catchment to a heavy summer storm is compared to that of an undisturbed control basin and an increase in flood magnitude of about six times is evident. This can be accounted for by reduced interception, reduced infiltration, particularly as a result of surface compaction, and by a general reduction in evapotranspiration losses. Higher flood peaks are usually accompanied by increased total annual runoff and increases of more than 50 per cent have been reported for certain locations in the USA by Anderson *et al.* (1976) in their valuable review of the hydrological role of forests.

(b)   Land drainage

In many locations, the development of intensive agriculture and of commercial forests is accompanied by land drainage in order to lower the water table or, in the case of impermeable soils, to remove surface water and Framji *et al.* (1981) estimate that the current area of drained land in the world now amounts to more than 150 million ha. In several locations drainage activities have been linked to increased flooding (e.g. Howe *et al.*, 1967) but the evidence is somewhat contradictory. For example, McCubbin (1938) cites an analysis of the flood hydrographs of the Iowa and Des Moines Rivers in the USA before and after the installation of field drains over a large proportion of the catchment areas, in which no significant changes was detected. Much depends on the complex interaction of groundwater levels, rainfall rate, soil moisture storage and soil transmissibility, and Rycroft and Massey (1975) present evidence which suggests that field drainage can actually reduce flood flows.

More definitive evidence of the impact of drainage on catchment response can be found in the work undertaken in Finland on the effects of peat drainage for forest development. About one-third of the land area of Finland (ca. 10 million ha) is peatland and by 1970 about 3.5 million ha of this had been drained. A catchment study initiated at Ruokolahti in 1935 (Mustonen and Seuna, 1971) has compared two basins, one of which was ditched and drained in the period 1958–60. In this environment, drainage causes lowering of the water table and reduced evapotranspiration and the annual mean runoff for the drained catchment increased by over 40 per cent. Fig. 4.7(b) further demonstrates the effect of drainage in increasing flows by comparing the frequency curves of annual mean runoff and winter minimum runoff for the before and after periods.

(c)   Strip mining

Strip mining provides an example of the severe landscape disturbance that may be associated with the exploitation of mineral resources, and the related surface destruction, deep excavation, and in some cases pumping to lower water levels, can profoundly modify catchment response. A valuable perspective on the hydrological effects of such activity is provided by the work undertaken in the Beaver Creek Basin in Kentucky, USA (Collier *et al.*, 1970). Two small tributary basins, one undisturbed and one with strip mining over 10.4 per cent of its area, were compared over the period 1957–66. The gross runoff from the two catchments was effectively similar, but the detailed temporal distribution of that runoff differed significantly. The mined watershed evidenced increased peak flows and decreased minimum flows and this increased flow variability is clearly reflected in the flow duration curves presented in Fig. 4.7(c). The increased peaks may be attributed to surface disturbance and compaction causing reduced infiltration and the reduction in low flows would in turn be conditioned by reduced infiltration and modified groundwater storage caused by deep excavations. Surface mining activity will, however, not always result in increased flow variability, particularly where extensive pumped drainage is involved. In the latter case, runoff volumes may increase and exhibit a more balanced flow regime with reduced flood peaks and increased minimum flows. This was the situation described by Golf (1967) from a catchment study in the brown coal mining area of Lower Lusatia, GDR, where lowering of the water table is essential for mining operations and where 6.3 m^3 of water are pumped for every tonne of coal mined.

(d) Conservation measures

For the most part, discussions of the hydrological effects of human activity emphasize the essentially detrimental changes, but the potential impact of man's positive action in manipulating basin response to produce a more favourable regime must also be recognized. The planting of forests for flood abatement can reverse the trends discussed above in the context of land clearance, and soil and water conservation measures are widely employed as an essential component of agricultural land management. In many instances man may merely be repairing damage that he himself has caused, but the element of positive control nevertheless remains. Out of many possible case studies an example drawn from catchment studies undertaken at Wagga Wagga, New South Wales, Australia, can be introduced. Here clearance of native woodland for pasture and crops resulted in increased runoff and accelerated erosion and a paired catchment experiment (Adamson, 1974) has been used to evaluate the long-term effects of soil conservation structures and land use management on runoff and soil loss. These measures resulted in an average reduction in runoff of 74 per cent and the marked change in the annual rainfall/runoff relationship is clearly evident from Fig. 4.7(d).

(e) Urbanization

The extreme modification of hydrological processes resulting from urbanization has already been reviewed and the evidence available from catchment studies further highlights this perspective. Increases in flood magnitude have been widely documented and Fig. 4.8(a) illustrates an attempt by Leopold (1968) to synthesize the results of a number of catchment studies in the USA in terms of the effects of impervious area extent and area served by storm sewers on the mean annual flood from a 2.59 km^2 (1.0 mi^2) catchment. Increases in flood peaks of three to four times are associated with highly urbanized areas and may be ascribed partly to increased runoff volumes related to reduced infiltration and partly to the more efficient drainage network. Storm hydrographs from urban

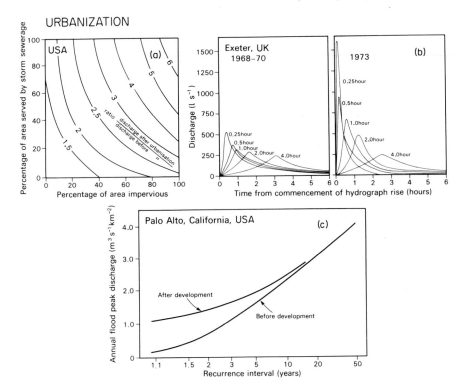

FIGURE 4.8   The effects of urbanization on catchment response. A general relationship between degree of urbanization of a catchment and the magnitude of the mean annual flood from a 2.59 km² basin, proposed by Leopold (1968) is presented in (a). The impact of urban development on hydrograph shape is shown in (b) by comparing unit hydrographs for natural and developed conditions. (c) compares hypothetical flood frequency curves for a small basin under natural and urbanized conditions. (Source: Crippen and Waananen, 1969.)

catchments characteristically exhibit sharper peaks and reduced lag times between rainfall and runoff as a result of this increased efficiency. For example Fig. 4.8(b) portrays some results collected by the author from a small experimental basin on the margins of Exeter, UK, which has been used to evaluate the impact of suburban development on storm runoff response. Unit hydrographs, the hydrographs resulting from 1.0 cm of *effective* rainfall falling over the basin within a specified duration have been derived from the records for the catchment in its natural state and for 1973 when about 25 per cent of the area was being developed. Although primarily relating to the construction phase, the contrasts between the unit hydrographs for these two periods clearly reflect changes in the runoff concentration and transmission characteristics of the basin. The volumes associated with the hydrographs for the two periods are identical (equivalent to 1.0 cm of effective rainfall) and the contrasts in peak flow result specifically from changes in hydrograph shape. The expected decrease in lag times is also evident in 1973.

Also of significance to attempts to assess the impact of urbanization, or indeed any other human activity, on basin response are considerations of the variation of the impact according to the frequency of the events involved. There are many situations where the influence of human activity will be less in relative terms for high magnitude events of low frequency. In the case of urban development this can be accounted for by the contrasts in infiltration properties between an urban and a natural surface being very much reduced under conditions of extreme wetness. A situation of this type is evident in Fig. 4.8(c) which depicts hypothetical flood frequency curves, for a small basin under natural and urban land use in the Palo Alto region of California, derived by Crippen and Waananen (1969). In this case it is tentatively suggested that the curves converge at about the 25-year recurrence interval. Hollis (1974), working the Canons Brook catchment in the UK, similarly concluded that the effects of urbanization on floods with a return period of about 20 years were minimal. Although this frequency threshold will vary from region to region in response to physiographic conditions, it must be borne in mind as a highly significant qualification in any attempt to quantify man's impact.

4.5   QUALITY AS WELL AS QUANTITY

Traditionally, discussion of hydrological processes has focused on *quantity* considerations and assessment of the volumes of water passing through the various components of the hydrological cycle. More recently, however, attention has also been directed towards the physical and chemical *quality* of that water. This interest stems from both the increasing importance of quality parameters to water use (e.g. Renshaw, 1980) and the growing concern for environmental degradation and pollution. The serious pollution of rivers and streams by domestic, municipal and industrial effluents has now been documented in many countries of the world and must be seen as a direct consequence of the increasing proportion of the total runoff from the land surface of the globe which is being used by man. Use of water frequently results in a deterioration of its quality or the addition of polluting substances and, for example, more than 20 per cent of the total length of non-tidal rivers in the UK with a mean flow in excess of $20 \, \text{m}^3 \, \text{s}^{-1}$ was classified as polluted or of doubtful quality in 1975 (Department of the Environment, 1978).

When considering man-induced changes in water quality in more detail, however, a distinction must be made between modified quality and pollution. Many of the changes may be extremely significant from the hydrological or environmental viewpoint and yet not warrant being described as pollution, and in turn changes in water quality may not necessarily involve the addition of polluting substances. Furthermore, although any consideration of man's impact on water quality must necessarily involve the general problem of water pollution by domestic, municipal and industrial effluents, further discussion of such pollution must fall outside the scope of this account, which is primarily concerned with hydrological processes. Discharge of effluent from a culvert or pipe into a river channel, frequently termed *point-source pollution*, provides little opportunity for interaction with catchment-wide processes. Conversely, *non-point pollution*, which has received increasing publicity in recent years and which includes the washing of fertilizer nutrients and pesticides from agricultural land into streams, will be reflected in the quality dimension of many of the hydrological processes operating in a drainage basin and merits inclusion.

The drainage basin has frequently been used as a convenient unit for evaluating the cycling of minerals and nutrients through terrestrial ecosystems (e.g. Likens and Bormann, 1975). Associated measurements of the nutrient and mineral content of water from various locations within the drainage basin system have highlighted the complex interactions between precipitation input, vegetation uptake and release, biomass storage, rock weathering and streamflow loadings, which are involved in regulating the quality dimension of the hydrological cycle (cf. Walling, 1980). Such ecosystems are delicately balanced and human intervention can cause important changes in water quality. Even man-induced changes in precipitation quality related to atmospheric pollution and associated 'acid rain' (see Chapter 3), can be significant. For example, Oden (1976) has shown that in areas such as Sweden with crystalline rocks of low buffering capacity, increased rainfall acidity has produced a clear negative trend in the pH and bicarbonate content of rivers and lakes, and a corresponding increase in sulphate concentration.

(a) Vegetation removal

More widespread are the changes in streamflow quality that may be induced by man's interference with the vegetation cover of a catchment. Perhaps the classic example of the potential impact of vegetation modification is the experiment carried out in the Hubbard Brook Experimental Forest in New Hampshire (Pierce *et al.*, 1970). Somewhat extreme in nature, this involved cutting all the trees and woody vegetation on watershed 2 (Fig. 4.9(a)) during November and December 1965. This material was left *in situ* and soil disturbance minimized. The following June an organic herbicide (bromacil) was sprayed by helicopter over the basin to kill all sprouting vegetation and manual spraying was repeated during the summers of 1967 and 1968 to prevent regeneration. The paired watershed approach was used to evaluate the effects on water quality and in Fig. 4.9(a) post-treatment streamflow quality is compared to that from catchment 6 with an undisturbed forest cover.

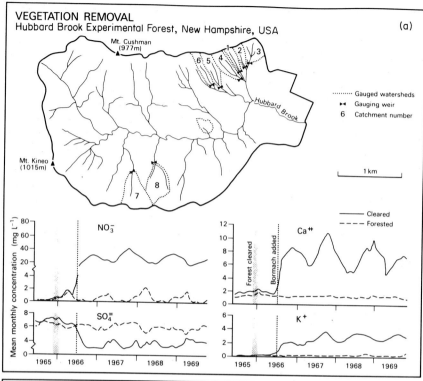

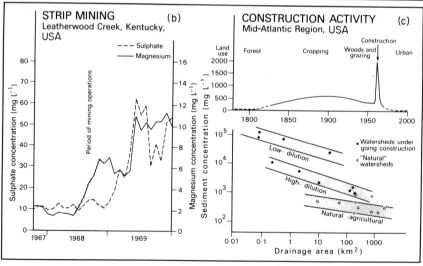

Because clearance occurred in winter, little effect was immediately apparent, but, immediately after snowmelt commenced, marked changes in streamflow chemistry occurred. Nitrate concentrations increased most and exhibited an overall increase of about fifty-fold during the three-year period. Cation concentrations also increased, but not to the same extent as nitrate (Fig. 4.9(a)), but sulphate levels were reduced. The great increase in nitrate levels can be ascribed to the mineralization of organic matter and to the complete lack of nitrate uptake by vegetation. Increases in cation concentrations were closely linked to those of nitrate in that nitrification mobilizes cations through the process of cation exchange. The reduced sulphate levels were attributed to the inhibition of sulphur-oxidizing bacteria by the high nitrate concentrations, to the creation of anaerobic conditions within the soil by increased moisture content and to a general dilution associated wiht increased streamflow volumes.

The treatment involved in this experiment was extreme in nature, but similar, if less marked, increases in nitrate concentrations have been documented in other conventionally clearcut catchments in New Hampshire, USA (Pierce *et al.*, 1972). For example, clearcutting Watershed 101 at Hubbard Brook (Fig. 4.9(a)) caused an increase in nitrate levels of about one-third that described above. Nevertheless, care must be exercised in interpreting the results from New Hampshire as being generally applicable, because, for example, Aubertin and Patric (1972) report only small increases in nutrient levels after clearcutting in West Virginia, USA.

Removal of vegetation may alter the pattern of moisture availability and movement in a soil as a result of reduced transpiration uptake and this may in turn give rise to changes in streamwater chemistry. A useful example of this situation is provided by the work of Peck and Hurle (1973) in southwestern Australia. They describe a situation where clearance of forest for farmland has resulted in increased groundwater recharge and associated leaching of relatively saline groundwater to the streams. Average chloride concentrations in excess of $1000 \, \text{mg} \, \text{l}^{-1}$ were recorded in streams draining catchments with substantial areas of cleared farmland, whereas concentrations in streams draining forested basins were commonly less than $100 \, \text{mg} \, \text{l}^{-1}$.

(b) Soil disturbance

Where human activity involves widespread disturbance of soil and regolith as well as vegetation removal, mineral and nutrient cycles will be further disrupted by acceleration of natural weathering and leaching processes. A useful example of this effect is provided by Fig. 4.9(b) which illustrates some results of a study

FIGURE 4.9 *(opposite)* Man's impact on water quality. Results of the Hubbard Brook vegetation removal experiment (a) and of studies of the effects of strip mining (b) and construction activity (c) are illustrated. (Sources: (a) Pierce *et al.*, 1970; (b) Curtis, 1973; and (c) Wolman, 1967, and Wolman and Schick, 1967.)

of the effects of strip mining on the streamflow quality of a small mountain basin in the Kentucky Appalachians, reported by Curtis (1973). Disturbance of 40 per cent of this 70 ha basin resulted in a nearly five-fold increase in sulphate and magnesium concentrations. In some mining areas the exposure of sulphide-bearing minerals which weather to form ferrous sulphate and sulphuric acid can give rise to extremely acid streamflow and greater increases in solute levels, as a result of both the presence of the acid and its action in modifying weathering processes. Water quality problems associated with acid mine drainage are widespread in the eastern USA (Biesecker and George, 1966).

Surface disturbance will also generally produce increased erosion and therefore higher suspended sediment concentrations in streams. Although watershed sediment yields are often evaluated in terms of erosion rates and soil loss (e.g. Chapters 8 and 13), they also have important implications for water quality. Suspended sediment concentrations and associated turbidity levels exercise important controls over aquatic life and water use. Furthermore, the distinction between sediment and solute loads is somewhat arbitrary in view of chemical absorption and exchange processes (see Golterman *et al.*, 1983). The effects of various agricultural practices in increasing suspended sediment loads are now well documented in the literature (see UNESCO, 1982) but amongst the greatest increases in sediment concentration that have been reported are those associated with construction activity. Work undertaken in the eastern USA (e.g. Guy, 1965; Wolman, 1967) provides a valuable perspective on this aspect of human impact. Wolman and Schick (1967) report suspended sediment concentrations of $3000-150\,000\ \mathrm{mg\,l^{-1}}$ in streams draining construction sites in this area, while in an agricultural watershed the highest comparable concentration was $2000\ \mathrm{mg\,l^{-1}}$. Fig. 4.9(c) compares estimated discharge-weighted mean concentrations for natural watersheds with those for streams draining construction sites in the region of metropolitan Baltimore and Washington, USA. The overall inverse relationship between concentration and catchment area is related to the well-documented decrease in sediment delivery ratio with increasing basin size, but whereas mean concentrations of approximately $500-600\ \mathrm{mg\,l^{-1}}$ are typical of rural catchments, the values for basins where construction sites constitute a large proportion of their area can be as high as $25\,000\ \mathrm{mg\,l^{-1}}$ or more. Fig. 4.9(c) also attempts to interpret the classic diagram produced by Wolman (1967) of the trend of sediment yields over the past 200 years in the Piedmont region of the eastern USA in terms of the discharge-weighted mean suspended sediment concentration of a typical river draining that area. Prior to the development of agriculture, mean concentrations are approximately $50\ \mathrm{mg\,l^{-1}}$ and these increase to $600\ \mathrm{mg\,l^{-1}}$ during the period of intensive farming. With the decline in farming during the period immediately preceding urban expansion, concentrations drop to $300\ \mathrm{mg\,l^{-1}}$ but subsequently rise to $2000\ \mathrm{mg\,l^{-1}}$ during the period of construction activity. Finally with the establishment of stable urban conditions, levels fall to less than $200\ \mathrm{mg\,l^{-1}}$.

Evidence of increased sediment yields and sediment concentrations consequent upon human activity abounds in the literature, but the opposite effects associated with the introduction of soil conservation measures (see Chapter 13) must also be recognized. In some areas of the world these measures have been introduced primarily to reduce sediment loads and associated reservoir sedimentation rather than to improve or maintain soil productivity, although the accelerated erosion which they aim to combat is itself generally the result of human activity. Zhang (1984) reports the example of the Wuding River, a major tributary of the middle reaches of the Yellow River in the People's Republic of China. The highly eroded loess landscape in this region generates extremely high sediment yields, but the introduction of comprehensive soil conservation measures within the Wuding basin reduced the total amount of sediment entering the Yellow River by more than 50 per cent. Considerable effort is now being expended throughout the middle Yellow River basin in order to reduce sediment loads and to alleviate downstream problems of reservoir deposition and channel sedimentation.

(c)   Irrigation

Changes in hydrological processes associated with the development of irrigation can also have important repercussions for water quality. Application of water for crop production in semi-arid areas results in increased evapotranspiration and this can cause accumulation of soluble salts within the soil. In addition, an increase in the level of the water table often permits capillary rise of saline groundwater, again causing salt accumulation. Salinity problems of this type are one of the greatest difficulties facing the development of successful irrigation systems and can be overcome only by careful regulating of groundwater levels and by leaching of accumulated salts from the soil. These measures can have a considerable impact on rivers draining irrigated areas and receiving return water highly charged with soluble salts. Kharchenko and Maddock (1982) indicate that the dissolved solids content of rivers draining areas of intensively developed irrigation may increase by 6–10 times. Fig. 4.10(a) illustrates this impact further by comparing the existing water quality characteristics during the irrigation season of the Lower Yakima River in Washington, western USA, with those that would prevail in the absence of irrigation return flow. Sylvester and Seabloom (1963) estimate that sodium concentrations, which are strongly influenced by ion exchange mechanisms within the soil, show a fifty-fold increase.

(d)   Non-point pollution

To a certain extent, changes associated with irrigation return flow also reflect additions of fertilizer and other chemicals to irrigated areas, but the impact of additional inputs to catchment nutrient and mineral cycles is best seen in more

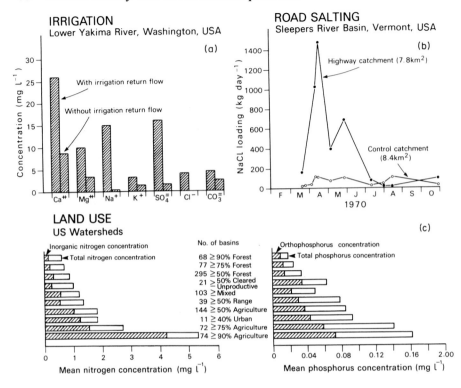

FIGURE 4.10 The effects of irrigation development and non-point pollution on streamflow quality. Results of three case studies discussed in the text are illustrated. (Sources: (a) data in Sylvester and Seabloom, 1963; (b) Kunkle, 1972; and (c) Omernik, 1977.)

obvious examples of non-point pollution from land use activities. In areas where severe winters necessitate the frequent and liberal use of road salt to maintain traffic flow, road drainage can constitute an extremely significant source of non-point pollution to adjacent streams. The total amount of de-icing chemicals used in the USA is probably close to 2 million tonnes per year and there has been considerable concern over the possible effects on receiving streams. Fig. 4.10(b), based on the work of Kunkle (1972) in Vermont, compares the daily NaCl loads of a catchment receiving highway drainage with those of an unaffected control watershed and illustrates the potential magnitude of the changes involved.

Perhaps the most significant form of non-point pollution and one that has attracted increasing attention in recent years, is the change in water quality caused by agricultural activity and more particularly by the loss of nutrients to streams from applications of fertilizers and animal wastes. In view of their high solubility inorganic nitrogen fertilizers are readily leached from the soil and there are many reports that link rising nitrate levels in rivers and groundwater to increased rates of fertilizer application. Based on work carried out in a small watershed in Illinois, Kohl *et al.* (1971) suggested that as much as 50 per cent of the fertilizer applied could be lost to streams and the potential magnitude of the nitrate leaching problem is clearly apparent when one considers that world production of fertilizer nitrogen increased from 22.4 million tonnes in 1966–67 to 45.9 million tonnes in 1976–77 (Golubev, 1980). However, the very high loss rates suggested by Kohl *et al.* (1971) have subsequently been questioned and other studies have documented instances where stream nitrate levels remain relatively low in areas of very high fertilizer application. In spite of such controversy it is generally recognized that non-point agricultural sources provide a significant contribution to increased nutrient levels in streamflow. These may in turn cause excessive algal growth in river channels and where such streams feed lakes and reservoirs increased nitrogen and phosphorus loadings can give rise to problems of eutrophication (see Chapter 7). High nitrate levels may also introduce serious problems in rivers and groundwater used for public water supply since the World Health Organization has recommended that the concentration of $NO_3 - N$ in drinking water should not exceed 11.3 mg l^{-1}. The widespread impact of this problem is emphasized by Vigon (1985) who reports that agricultural non-point pollution significantly affected 68 per cent of the major watersheds in the USA in 1977.

Numerous small catchment studies aimed at evaluating the precise impact of agricultural sources on streamflow quality have been reported in the literature but a useful overview of this general problem is provided by the National Eutrophication Survey undertaken in the USA (Omernik, 1977). This has investigated the relationships between stream nutrient loadings and watershed land use at the national scale. Samples were collected from 928 small drainage basins throughout the country and some of the results are presented in Fig. 4.10(c). Although embracing a considerable variety of physiographic conditions, these results reflect a significant tendency for nutrient concentrations to be lower in streams draining forested watersheds than those draining areas used primarily for agriculture. Mean total nitrogen concentrations, for example, exhibit a nine-fold difference between predominantly forested and predominantly agricultural watersheds, and a large element of this difference must derive from non-point agricultural sources and in turn reflect the impact of man.

4.6 PROBLEMS AND PROSPECTS

Studies of man's effects on hydrological processes can be readily justified academically in terms of the evidence that they provide concerning his importance in the functioning of contemporary environmental systems. Many of these changes will, however, themselves produce adverse effects on human activity and such studies can therefore serve the further purpose of highlighting potential problems that may face man as his impact intensifies. These may include increased flood hazard and other changes in river regime, reduced availability of groundwater and associated saline intrusions in coastal areas, deterioration of water quality, and widespread eutrophication of water bodies and river systems in response to increased nutrient loadings. Much empirical evidence concerning the impact of man on the hydrological cycle has been gathered over the past 20 years. It is to be hoped that this will be used by hydrologists in the future to provide guidelines for avoiding many of the problems that have resulted in the past and to develop models which will permit long-term forecasting of the effects of human activity and the testing of alternative development and management strategies.

REFERENCES

Ackermann, W. C., 1969, 'Scientific hydrology in the United States', *The Progress of Hydrology. Proceedings of the First International Seminar for Hydrology Professors*, Vol. 1, US National Science Foundation, pp. 50–60.

Adamson, C. M., 1974, 'Effects of soil conservation treatment on runoff and sediment loss from a catchment in southwestern New South Wales, Australia', *Effects of Man on the Interface of the Hydrological Cycle with the Physical Environment, Proceedings of the Paris Symposium, IAHS Publ. 113*, pp. 3–14.

Anderson, H. W., Hoover, M. D., and Reinhart, K. G., 1976, 'Forests and Water', *US Forest Service General Technical Rept., PSW-18/1976.*

Aubertin, G. M., and Patric, J. H., 1972, 'Quality water from clearcut land?', *Northern Logger*, **20**, 14–15, 22–23.

Bates, C. G., and Henry, A. J., 1928, 'Forest and stream-flow experiment at Wagon Wheel Gap, Colo. Final report on completion of the second phase of the experiment', *Monthly Weather Review*, Supplement No. 30.

Barney, G. O. (Ed.), 1982, *The Global 2000 Report to the President. Entering the Twenty-First Century*, Penguin, New York.

Beaumont, P., 1978, 'Man's impact on river systems: a world wide view', *Area*, **10**, 38–41.

Biesecker, J. E., and George, J. R., 1966, 'Stream acidity in Appalachia as related to coal mine drainage', *US Geol. Surv. Circular, 526.*

Biswas, A. K., 1983, 'Long-distance water transfer: problems and prospects', in Biswas, A. K., Dakang, Z., Nichum, J. E., and Changming, L. (Eds), *Long Distance Water Transfer: A Chinese Case Study and International Experiences*, Tycooly International, Dublin, pp. 1–13.

Collier, C. R., Pickering, R. J., and Musser, J. J. (Eds), 1970, 'Influences of strip mining on the hydrological environment of parts of Beaver Creek Basin, Kentucky, 1955–66', *US Geol. Surv. Prof. Paper, 427-C.*

Crippen, J. R., and Waananen, A. O., 1969, 'Hydrological effects of suburban development near Palo Alto, California', *US Geol. Surv. Open File Rept.*

Curtis W. R., 1973, 'Effects of strip mining on the hydrology of small mountain watersheds in Appalachia', in Hutnik, R. J., and Davis, G. (Eds), *Ecology and Reclamation of Devastated Land*, Gordon and Breach, New York, pp. 145–157.

Department of the Environment, 1978, *River Pollution Survey of England and Wales, updated 1975*, HMSO, London.

Dils, R. E., 1957, 'A Guide to the Coweeta Hydrologic Laboratory' *US Dept. of Agric., Forest Service Rept.*

Engler, A., 1919, 'Einfluss des Waldes auf den stand der Gewasser', *Mitt. Schweiz anst fur das Forsliche Versuchswesen*, **12**.

FAO, 1978, 'Water for agriculture', in Biswas, A. K. (Ed.), *Water Development and Management, Proceedings of the United Nations Water Conference*, Part 3, Pergamon Press, Oxford, pp. 907–941.

Framji, K. K., Garg, B. C. and Luthra, S. D. L., 1981, *Irrigation and Drainage in the World. A Global Review*, International Commission on Irrigation and Drainage, New Delhi.

Gleason, G. B., 1952, 'Consumptive use of water, municipal and industrial areas', *Trans. Amer. Soc. Agric. Engineers*, **117**, 1004–1009.

Golf, I. W., 1967, 'Contribution concerning flow rates of rivers transporting drain waters of open-cast mines', *IAHS Publ.*, **76**, 306–316.

Golterman, H. L., Sly, P. G., and Thomas, R. L., 1983, *Study of the relationship between water quality and sediment transport*, UNESCO Technical Papers in Hydrology No. 26, UNESCO, Paris.

Golubev, G., 1980, 'Nitrate leaching hazards. A look at the potential global situation', *International Institute for Applied Systems Analysis Working Paper WP-80-89.*

Golubev, G., and Vasiliev, O., 1978, 'Interregional water tranfer as an interdisciplinary problem', *Water Supply and Management*, **2**, 67–77.

Guy, H. P., 1965, 'Residential construction and sedimentation at Kensington, Md.' in *Proceedings of the Federal Inter-Agency Sedimentation Conference, USDA, Misc., Publ.*, **970**, pp. 30–37.

Harrold, L. L., 1971, 'Effects of vegetation on storm hydrographs', in *Biological Effects in the Hydrological Cycle. Proceedings of the Third International Seminar for Hydrology Professors*, US National Science Foundation, pp. 332–346.

Hibbert, A. R., 1967, 'Forest treatment effects on water yield', in Sopper, W. E. and Lull, H. W. (Eds), *Proc. Int. Symp. on Forest Hydrology*, Pergamon, Oxford, pp. 527–543.

Hills, R. C., 1969, 'Effects of agricultural treatment on the quantity of water in the soil', in *The Role of Water in Agriculture, Aberystwyth Symposia in Agricultural Meteorology Memorandum*, **12**, pp. F1–F7.

Hollis, G. E., 1974, 'Urbanization and floods', in Gregory, K. J., and Walling, D. E. (Eds), *Fluvial Processes in Instrumented Watersheds, Institute of British Geographers Special Publication, No. 6*, pp. 123–139.

Holstener-Jorgensen, H., 1967, 'Influence of forest management and drainage on ground-water fluctuations', in Sopper, W. E., and Lull, H. W. (Eds), *Proc. Int. Symp. on Forest Hydrology*, Pergamon, Oxford, pp. 325–334.

Holtan, H. N., and Kirkpatrick, M. H., 1950, 'Rainfall infiltration and hydraulics of flow in runoff computation', *Trans. Amer. Geophys. Union*, **31**, 771–779.

Howe, G. M., Slaymaker, H. O., and Harding, D. M., 1967, 'Some aspects of the flood hydrology of the upper catchments of the Severn and Wye', *Trans. Inst. Brit. Geog.*, **41**, 33–58.

IAHS, 1974, *Effects of Man on the Interface of the Hydrological Cycle with the Physical Environment, Proceedings of the Paris Symposium*, International Association of Hydrological Sciences Publication No. 113.

IAHS, 1980, *The Influence of Man on the Hydrological Regime, Proceedings of the Helsinki Symposium*. International Association of Hydrological Sciences Publication No. 130.

Kalinin, G. P., and Bykov, V. D., 1969, 'The world's water resources, present and future', *Impact of Science in Society*, UNESCO, Paris, pp. 143.

Keller, R. (Ed.), 1976, *Man-Made Transformation of Water Balance*, Geographisches Instituts, Freiburg, FRG.

Kharchenko, S. I., and Maddock, T., 1982, *Investigation of the Water Regime of River Basins affected by Irrigation*, UNESCO, Paris.

Kohl, D. H., Shearer, G. B., and Commoner, B., 1971, 'Fertilizer nitrogen: contribution to nitrate in surface water in a Corn Belt watershed', *Science*, **174**, 1331–1334.

Kunkle, S. H., 1972, 'Effects of road salt on a Vermont stream', *Jnl. Amer. Waterworks Assoc.*, **64**, 290–294.

Leopold, L. B., 1968, 'Hydrology for urban land planning—a guidebook on the hydrologic effects of urban land use', *US Geol. Surv. Circular*, *554*.

Likens, G. E., and Bormann, F. H., 1975, 'An experimental approach in New England landscapes', *Ecological Studies*, **10**, 7–29.

Likens, G. E., Bormann, F. H., Pierce, R. S., Eaton, J. S., and Johnson, N. M., 1977, *Biogeochemistry of a Forested Ecosystem*, Springer-Verlag, New York.

Lvovich, M. I., and Chernogaeva, G. M., 1977, 'The water balance of Moscow', *Effects of Urbanization and Industrialization on the Hydrological Regime and on Water Quality, IAHS Publ.*, *123*, pp. 48–51.

McCubbin, G. A., 1938, 'Agricultural drainage in south-western Ontario, its effects on stream discharge' *Engineering Journal*, **21**, 66–70.

Mustonen, S. E., and Seuna, P., 1971, 'Metsaojituksen vaikutuksesta suon hydrologiaan', *Vesientutkimuslaitoksen Julkaisuja*, **2**.

New Zealand Ministry of Works, 1968, *Annual Hydrological Research Report for Moutere*, *1*, Wellington.

New Zealand Ministry of Works, 1971, 'Moutere, IHD experimental basin No. 8', *Hydrological Research Annual Report*, *20*.

New Zealand Ministry of Works, 1978, 'Research and survey, annual review 1977', *Water and Soil Technical Publication*, *9*.

Noble, E. L., 1965, 'Sediment reduction through watershed rehabilitation', *Proc. Federal Inter-Agency Sedimentation Conference. USDA Misc. Publ.*, *970*, pp. 114–123.

Oden, S., 1976, 'The acidity problem—an outline of concepts', *Proc. First Int. Symp. on Acid Precipitation, US Forest Service Rept.*, NEFES/77-1, pp. 1–36.

Omernik, J. M., 1977, *Nonpoint Source-Stream Nutrient Level Relationships: A Nationwide Study*, Corvallis Environmental Research Laboratory.

Peck, A. J., and Hurle, D. H., 1973, 'Chloride balance of some farmed and forested catchments in southwestern Australia', *Water Resource Res.*, **9**, 648–657.

Pereira, H. C., 1973, *Land Use and Water Resources*, Cambridge University Press, Cambridge.

Pierce, R. S., Hornbeck, J. W., Likens, G. E., and Bormann, F. H., 1970, 'Effects of vegetation elimination on stream water quantity and quality', *Results of Research on Representative and Experimental Basins, Proceedings of the Wellington Symposium, IAHS Publ.*, *96*, pp. 311–328.

Pierce, R. S., Martin, C. W., Reeves, C. C., Likens, G. E., and Bormann, F. H., 1972, 'Nutrient loss from clearcuttings in New Hampshire', *Watersheds in Transition, American Water Resources Association Proceedings Series.*, *14*, pp. 285–295.

Postel, S. 1984, 'Protecting forests', *State of the World 1984*, Worldwatch Institute, Norton, New York, pp. 74–94.

Rakhmanov, V. V., 1970, 'Dependence of streamflow upon the percentage of forest cover of catchments', *Proceedings of the Joint FAO/USSR International Symposium on Forest Influences and Watershed Management*, FAO, Rome, pp. 55–64.

Renshaw, D. C., 1980, 'Water quality objectives and standards', in Gower, A. M. (Ed.), *Water Quality in Catchment Ecosystems*, Wiley, Chichester, pp. 229–241.

Richards, K. S., and Wood, R., 1977, 'Urbanization, water redistribution and their effect on channel processes', in Gregory, K. J. (Ed.), *River Channel Changes*, pp. 369–388.

Rodda, J. C., 1976, 'Basin studies', in Rodda, J. C. (Ed.), *Facets of Hydrology*, Wiley, Chichester, pp. 257–297.

Rycroft, D. W., and Massey, W., 1975, 'The effect of field drainage on river flow', *Ministry of Agriculture, Fisheries and Food, Field Drainage Experimental Unit Tech. Bull.*, 75/9.

Scarf, F., 1970, 'Hydrologic effects of cultural changes at Moutere experimental basin', *Jnl. of Hydrology (N.Z.)*, **9**, 142–162.

Shevchenko, M. A., 1962, 'Effects of various methods of tillage on the reduction of snowmelt runoff from sloping land', *Soviet Hydrology*, 27–33.

Shiklomanov, I. A., and Kozhevnikova, V. P., 1974, 'Poteri stoka v Volgo-Aktubinski poime i delte Volgi i ikh izmenenie pod vliyaniem khozyaistvennoi deyatelnosti', *Trudy GGI*, **221**, 3–47.

Sokolov, A. A., 1978, 'The water balance of Europe', *Water Balance of Europe*, UNESCO, Paris, pp. 3–20.

Sylvester, R. O., and Seabloom, R. W., 1963, 'Quality and significance of irrigation return flow', *Proc. Amer. Soc. Civil Eng. Jnl. Irrig. and Drainage Div.*, **89**, pp. 1–27.

UNESCO, 1974, *Hydrological Effects of Urbanization*, UNESCO, Paris.

UNESCO, 1978, *World Water Balance and Water Resources of the Earth*, UNESCO, Paris.

UNESCO, 1980, *Casebook of Methods of Computation of Quantitative Changes in the Hydrological Regime of River Basins due to Human Activities*, UNESCO, Paris.

UNESCO, 1982, *Sedimentation Problems in River Basins*, UNESCO, Paris.

Vigon, B. W., 1985, 'The status of nonpoint source pollution: its nature, extent and control', *Water Resources Bull.*, **21**, 179–184.

Voropaev, G. V., 1978, 'The scientific principles of large-scale areal redistribution of water resources in the USSR', *Water Supply and Management*, **2**, 91–101.

Waananen, A. O., 1969, 'Urban effects on water yield', in Moore, W. L., and Morgan, C. W. (Eds), *Effects of Watershed Changes on Streamflow*, University of Texas, Austin, pp. 169–182.

Walling, D. E., 1980, 'Water in the catchment ecosystem', in Gower, A. M. (Ed.), *Water Quality in Catchment Ecosystems*, Wiley, Chichester, pp. 1–47.

Walling, D. E., and Gregory, K. J., 1970, 'The measurement of the effects of building construction on drainage basin dynamics', *Jnl. Hyd.*, **11**, 129–144.

Williams, J., and Hamilton, L. S., 1982, *Watershed Forest Influences in the Tropics and Subtropics*, East–West Environment and Policy Institute, Honolulu.

Wolman, M. G., 1967, 'A cycle of sedimentation and erosion in urban river channels', *Geografiska Annaler*, **49**, Ser. A, 385–395.

Wolman, M. G., and Schick, A. P., 1967, 'Effects of construction on fluvial sediment, urban and suburban areas of Maryland', *Water Resources Res.*, **3**, 451–464.

Zadroga, F., 1981, 'The hydrological importance of a montane cloud forest area of Costa Rica', in Lal, R., and Russell, E. W. (Eds), *Tropical Agricultural Hydrology*, Wiley, Chichester, pp. 59–73.

Zhang, S., 1984, 'Effects of comprehensive soil conservation measures on the reduction of sediment yields in the Wuding River Valley', *J. of Sediment Research*, **1984**, 1–12.

Human Activity and Environmental Processes
Edited by K. J. Gregory and D. E. Walling
©1987 John Wiley & Sons Ltd.

5

Coastal Processes

E. C. F. BIRD

5.1 INTRODUCTION

Coastal features are the outcome of a variety of processes working on the available geological materials in the zone where the land meets the sea. Cliffs and rocky shores have been shaped largely by erosional processes, and beaches, spits, and marshlands by depositional processes. The outlines of erosional and depositional features are related to the patterns of waves generated by winds blowing over the sea surface, and currents, particularly those associated with the rise and fall of the tide.

Geological materials available at the coast include both solid rock formations that outcrop in cliffs and along the shore, and unconsolidated sediments, ranging in size from large boulders and cobbles down through pebbles and sand to silt and clay. Some of the sediments have come from the breaking-down of coastal rock formations outcropping in cliffs and shore platforms; others have been brought by rivers from the hinterland; and others have been washed in from the sea floor. Sand and pebbles are moved to and fro along the shore by waves and currents, and deposited as beaches and spits, whilst the finer sediments accumulate in sheltered inlets and estuaries as mudflats, which may be colonized by vegetation and built up as marshes. There is much variation in the way in which coastal processes operate, in the types of rock outcrop and unconsolidated sediment present, and in the kinds of landforms developing around the world's coastline. Accounts of the various relationships between landforms and processes can be found in textbooks of coastal geomorphology (e.g. Zenkovich, 1967; King, 1972; Komar, 1976; Davies, 1980; Bird, 1984).

Many coastal features have been shaped entirely by natural processes, but others have been modified, directly or indirectly, by man's activities. Some

Dr Bird is Chairman of the International Geographical Union's Commission on the Coastal Environment, and this paper uses examples taken from the Commission's project on world-wide coastline changes (Bird, 1985).

modifications take the form of either accelerating or slowing down of the natural course of coastal evolution, and these can be difficult to detect. Others, often more drastic, lead to features that would not have developed in the absence of man's activities. This chapter examines some of the ways in which coastal processes have been modified by the impact of man.

5.2   DIRECT AND INDIRECT EFFECTS

There are not many examples of man's attempts to modify directly the processes at work in coastal waters. It is possible to disrupt wave motion, and weaken the energy of waves moving in towards a coast, by injecting 'air bubble curtains' that rise from perforated piping laid across the sea floor and fed with compressed air. This technique was used on a small scale to reduce wave action while engineering works were in progress at the entrance to Dover harbour several years ago, but it is too complicated and expensive to use as a means of reducing or halting coastal erosion. In Florida experiments have been carried out with fields of 'plastic seaweed', vertically-projecting structures attached to the sea floor, in the hope of reducing the energy of waves approaching the coastline, but this technique has not yet been perfected.

More common are attempts to deflect or resist the effects of waves or currents by introducing structures such as sea walls, breakwaters, or groynes (e.g. Fig. 5.1); by importing sediments to create or replenish beaches; or by planting vegetation. Some schemes have tried to halt coastal erosion; others have sought also to promote deposition. In each case, success depends on an understanding of the processes at work; the sources, quantities, and patterns of flow of coastal sediment; and the consequences of attempting to modify the coastal system in a particular way. Unfortunately, there have been failures. Around the world's coastline are many examples of derelict and abandoned structures, and elaborated works that remain unsuccessful. Some of these failures stem from assumptions that have been widely held by engineers and laymen but are now considered questionable. These include the view that coastal erosion is always an evil to be resisted; that the natural condition of a coastline is one of stability, if not active accretion; and that, given the right design of structural works, accretion will automatically be induced on previously eroding coastlines. But coastal erosion in one place is commonly the source of coastal accretion in another, and the demand for anti-erosion works usually arises where there has been unwise development behind the coast. Evidence from a recent world-wide survey indicates that during the past century erosion by natural processes has been dominant on the world's beach-fringed coastlines (Bird, 1985). And if sediment is not being supplied to the coastal system, structural works cannot produce beach accretion.

FIGURE 5.1   Beach outlines modified by the insertion of groynes on the southern shores
of Botany Bay, New South Wales (E. C. F. Bird)

Beach erosion has been initiated or intensified on coasts where sand or gravel has been extracted from the beach or from the nearshore area. Such beach quarrying has been widespread during the past century, although the severity of ensuing erosion has in many cases brought it to a halt. On the south coast of England gravel has been dredged from several shingle beaches, and the extraction of certain kinds of pebbles is still occurring, despite protests, at such places as West Bay and Seatown Beach on the shores of Lyme Bay, in Dorset. Cliff erosion is already extensive here, and it would be wiser to augment these beaches by dumping gravel on them, thereby reinforcing them as natural barriers to storm wave attack. This would be aesthetically and scenically more acceptable than the progressive extension of sea walls, already in evidence at West Bay.

Indirect effects of man's impact include those resulting from structural works elsewhere, usually on an adjacent part of the coast. Such works can lead to changes in nearshore topography, wave and current regimes, and patterns of coastal sediment flow. In addition, there have been coastal changes that can be traced to man's activities in the hinterland area, or offshore; changes which are the indirect outcome of activities that were not expected to modify coastal processes or alter coastline features. Some examples of each of these situations are given below.

5.3   EFFECTS OF SEA WALLS

The most obvious response to coastal erosion is to build a protective structure, usually a wall of masonry or reinforced concrete, designed to prevent waves from attacking the land margin. This is usually only justified where development has raised land values in the immediate hinterland, so that roads and buildings are in danger of being undermined if cliff recession continues. Sectors of cliffed coast at seaside resorts have generally been stabilized by the building of sea walls, sometimes incorporating some kind of promenade or public seafront area for recreational purposes.

Bournemouth, on the south coast of England, is a typical example. The seaside resort developed during the past century, spreading out from an initial 'watering place' at the mouth of the Bourne valley, and expanding laterally along the crests of receding cliffs of gravel-capped Tertiary sands and clays which lined the shore of Bournemouth Bay. These cliffs, subject to undercutting by storm waves and recurrent slumping, were fronted by a broad beach of sand and shingle derived from their erosion. A short section of sea wall, built in the form of a promenade in the nineteenth century, has gradually been extended east and west from the mouth of the Bourne valley, along the base of the cliffs, which were then graded into artificial coastal slopes and stabilized by planting vegetation (Fig. 5.2). There is now little left of the cliffs that once lined Bournemouth Bay, and the developed coastal property is no longer threatened by cliff recession.

FIGURE 5.2　Coastal engineering works near Alum Chine, Bournemouth (E. C. F. Bird)

The promenade at Bournemouth is appropriate for a seaside resort, and provides easy access to the beach, but as the sea wall extended the beach became depleted. The sand and shingle have been carried away, mainly eastwards, to Hengistbury Head and beyond, by the longshore drifting generated by southwesterly waves moving in from the English Channel, and without continuing natural replenishment from the eroding cliffs, the volume of beach material has dwindled. Insertion of a series of groynes extending out at right-angles to the sea wall, breaking the beach into distinct compartments, reduced the rate of loss of sand and shingle, but the wastage continued. Here, as elsewhere, it was found that erosion became severe when waves were able to break against the sea wall during storms. Such waves were reflected, and generated a scouring of beach material offshore beyond the limits of the groynes. While some of this beach material was later carried back to the shore by gentler wave action in calmer weather, a proportion moved away eastwards as the result of the unhindered longshore drifting out beyond the ends of the groynes. By the 1960s the Bournemouth beaches had been depleted to such an extent that the local authority decided to introduce artificial beach nourishment, using sand and gravel dredged from the sea floor, and piped in to the shore (Fig. 5.3).

Depletion of beaches as a sequel to sea wall construction along the base of eroding cliffs can be illustrated at other seaside resorts in Britain and elsewhere. In Australia, for example, the cliffs of Tertiary sandstone and clay which used

FIGURE 5.3   Beach nourishment in progress at Bournemouth (E. C. F. Bird)

to line the northeastern shores of Port Phillip Bay, where the coastal suburbs of Melbourne have developed, have also been largely stabilized by sea wall construction and graded to artificial slopes; and here, too, the beaches have been depleted. The effects of wave reflection by sea walls, also evident here, are a good illustration of the way in which an introduced structure can modify the processes at work on a coast.

Reflection scour begins only when storm waves break directly against a sea wall. As long as a broad protective beach is maintained in front of the sea wall the process is not effective, but if the sediment supply is cut off, and the beach becomes depleted so that storm waves can reach the sea wall, then the process of reflection scour sets in, and beach erosion accelerates (Fig. 5.4).

Coastal engineers have experimented with a variety of designs and materials in sea wall construction in the hope of producing structures which do not set up reflection scour: for example, permeable groynes, and heaps of boulders or caged stones (gabions) placed along the eroding shoreline, intended to absorb storm wave energy rather than reflect it. Much use has been made, in Japan and elsewhere, of precast concrete structures, such as tetrapods, dumped on the shore to protect the coastline and absorb wave energy (e.g. Fig. 5.5). But a simple structural solution to the problem of halting erosion of the land margin without losing the adjacent beach has proved elusive.

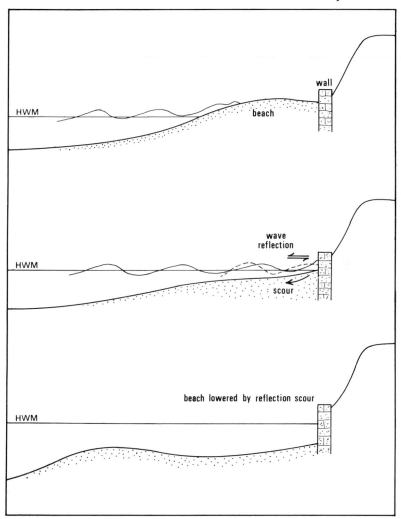

FIGURE 5.4 Sea walls and erosion. A broad, high beach prevents storm waves breaking against a sea wall and will persist, or erode only slowly, but once waves are reflected by the wall, scour is accelerated and the beach is quickly removed

On some coasts the construction of sea walls has followed the domino principle: the halting of cliff erosion on one sector has been followed by beach depletion and lowering of the shore profile, allowing larger waves to move in and intensify erosion on the adjacent, unprotected sector; and when the sea wall has been extended to combat this, cliff erosion is accentuated on the next adjacent sector. The sequence of sea wall construction at Point Lonsdale, at the entrance to Port Phillip Bay, is an example (Fig. 5.6). In 1900, a sea wall

FIGURE 5.5   Concrete tetrapods dumped on the shore at Atami, Japan, to protect the esplanade from the impact of large waves generated by typhoons (E. C. F. Bird)

was built to stabilize the shore in front of the township of Point Lonsdale, at the southern end of a gently curving sandy bay. The beach became depleted, and as each adjacent sector has shown intensified erosion the wall has been extended northwards, producing a sequence of slightly offset structures which commemorate the engineering ideas of the past seventy years, culminating in the recent addition of a large boulder wall (Fig. 5.7). If this approach continues, the whole of Point Lonsdale Bight will eventually consist of a walled shore, fronted by a depleted beach. On the other hand, if there had been no development immediately behind the Point Lonsdale coast, sea wall construction would have been unnecessary, and the relatively slow processes of natural erosion would have maintained a broad beach along this bay shore line, a recreational and scenic asset conserved at the cost of secular losses of the sandy hinterland. Such losses, would in any case have been offset by gains in sandy depositional land at Queenscliff, to the north. It is up to coastal zone managers to ensure that development is not permitted behind eroding sandy coastlines; for the existence of such development leads to the demand for erosion control, the addition of artificial structures, and the consequent depletion of the beach.

Where sea walls (dykes) have been built to enclose tidal marshlands and mudflats as a prelude to land reclamation, as in the Netherlands and on the shores of the German and Danish Wadden Sea, they are fronted by broad, gently shelving, intertidal sandy and muddy areas which reduce the effects of storm

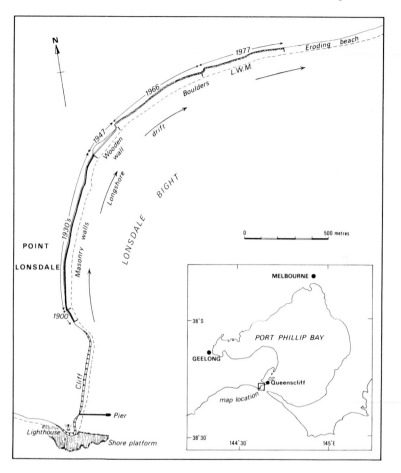

FIGURE 5.6   Sequence of sea wall construction at Point Lonsdale, Victoria, Australia
(see Fig. 5.7)

waves, so that reflection scour does not usually take place. Instead, there is often continued accretion of mud, colonized by marsh plants, on the seaward side of such walls, and in places beaches of shelly sand accumulate at about high tide level. By excluding tidal waters, such sea walls modify the range and extent of tidal rise and fall, and may change the pattern and strength of current flow associated with tidal movements. At the mouth of the Rhine, tidal channels have locally been deepened as the result of changes in configuration due to the enclosure of formerly tidal areas by walls and dykes, and in some places the deflection of channel flow has led to scour on coastlines where it had not previously occurred.

FIGURE 5.7   Boulder walls near Point Lonsdale, Victoria (J. McArthur)

Land reclamation schemes have also advanced the coastline of The Wash in eastern England in such successive stages, continuing intertidal mud accretion preparing the way for each phase of enclosure, drainage and reclamation. Other coastal regions modified by large-scale land reclamation include the fringes of Singapore Island, the infilled embayments around Hong Kong, and the extension of deltaic land at the head of Tokyo Bay.

5.4   EFFECTS OF BREAKWATERS

Breakwaters built out from the coast to provide a harbour sheltered from the effects of strong wave action may intercept the longshore drift of beach material that occurs where waves arrive at an angle to the shoreline. On the south coast of England, for example, there is eastward movement of sand and shingle as the dominant southwesterly waves move in to beaches of southerly aspect, and on the east coast of England the drift is southward in response to the dominant northeasterly waves arriving from the North Sea. Such trains of migrating beach material tend to deflect the mouths of rivers: the Adur and the Ouse in Sussex have had a long history of eastward deflection, while the East Anglian rivers have been diverted southward by the growth of such spits as Orford Ness at the mouth of the Ore. Such deflections impede navigation, and where the river mouths are used as harbours it has been necessary to stabilize a navigable entrance by means of breakwaters.

The mouth of the Sussex Ouse had been repeatedly deflected eastwards by drifting shingle, but in the eighteenth century the port of Newhaven became established sufficiently to necessitate the stabilization of a navigable entrance. A breakwater built on the western side in 1731, and subsequently extended, excludes southwesterly waves from the river mouth and prevents the drifting of sand and shingle into the harbour entrance. As a result, beach material has accumulated to the west of the breakwater. Such accretion on the updrift side of a structure is generally followed by the onset of a sediment deficit, marked by beach depletion downdrift. East of the stabilized river mouth at Newhaven the beach at Seaford has been reduced, and cliff erosion on Seaford Head accelerated. The coastline at Seaford has been maintained only by the building of sea walls of masonry and concrete, to form an esplanade which is subject to recurrent storm damage as the beach dwindles (Fig. 5.8).

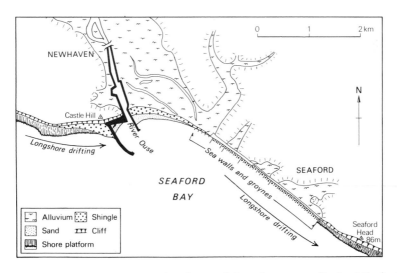

FIGURE 5.8 Accretion of drifting beach material on the western flank of the harbour breakwater at Newhaven, Sussex, and intensified erosion along the drift-starved sector to the east of Seaford

This pattern of updrift accretion and downdrift erosion as a sequel to breakwater construction is seen at many such harbours around the world. On the Florida coast, there has been accumulation of southward-drifting sand alongside the breakwaters built to stabilize South Lake Worth Inlet and give navigable access to a lagoon behind the coastal sand barrier. On the Nigerian coast, the breakwaters built to stabilize the entrance to the harbour in Lagos Lagoon have accumulated eastward-drifting sand on Lighthouse Beach to the west and induced rapid erosion on Victoria Beach to the east (Fig. 5.9). At Durban, in South Africa, and Madras, in India, northward drift of sand has

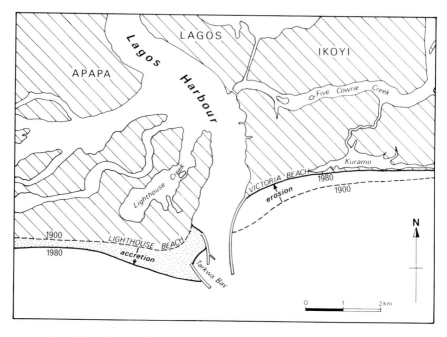

FIGURE 5.9   The effects of breakwaters at Lagos Harbour, Nigeria. Breakwaters were built in 1900, and subsequently extended, to maintain and protect the entrance to Lagos Harbour, Nigeria. Longshore drifting of sand from the west has subsequently led to accretion on Lighthouse Beach, while Victoria Beach, to the east, deprived of this sediment supply, has been cut back by marine erosion. Map based on surveys by Etop Usoro

been arrested to prograde the shoreline south of the harbour breakwaters, with beach erosion ensuing to the north; and at Santa Barbara, in California, a breakwater has intercepted sand drifting from the north and depleted the beaches to the south. On the east coast of Australia the breakwaters built to stabilize the mouth of Tweed River have intercepted sand drifting northward and resulted in beach depletion at the seaside resort of Coolangatta, to the north. In some cases an attempt has been made to solve the problem by introducing sand by-passing schemes, whereby some of the sand accumulating on the updrift side is pumped through pipes under the harbour entrance and used to replenish the eroding beach snowdrift: such scemes are active at Durban; at South Lake Worth Inlet; at Surfside, near Los Angeles; and at Salina Cruz in Mexico.

A somewhat different pattern has developed at Lakes Entrance, in south-eastern Australia, where breakwaters were built alongside an artificial entrance cut through the coastal sand barrier to provide a permanent navigable channel into the Gippsland Lakes, replacing an earlier, impermanent natural outlet which was frequently deflected, and occasionally closed, by sand accretion (Bird, 1978). After the artificial entrance opened in 1889, sand accretion began on either side of the twin breakwaters, and led to the growth of cuspate forelands linked by an underwater sand bar which developed off the new cut. In this case, there have been alternations of eastward drifting resulting from southwesterly wave action, almost balanced by westward drifting produced by southeasterly wave action, and natural bypassing of sand probably takes place in either direction along the underwater sand bar. The breakwaters have thus modified coastal processes by establishing a localized nearshore outflow from the Gippsland Lakes which has interacted with wave action and available sediment in such a way as to develop a looped sand bar off the entrance. This bar, in turn, has established a new pattern of wave refraction, which has shaped the cuspate forelands east and west of the entrance (Fig. 5.10). Other examples of sand accretion on both sides of breakwaters built at lagoon entrances or river mouths include those at Rogue River, Oregon; Newport, California; the Swina River in Poland, and Umuiden on the Netherlands coast.

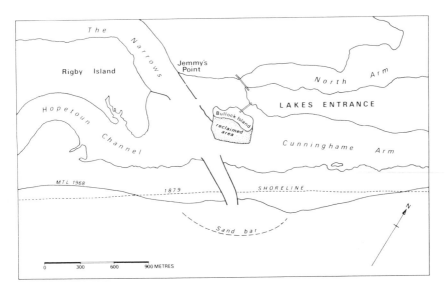

FIGURE 5.10   The shoreline at Lakes Entrance, Victoria, Australia in 1968. This shows the pattern of accretion on either side of the artificial entrance. The 1879 shoreline, mapped ten years before the artificial entrance was opened, is a basis for measuring subsequent changes

On the eastern shores of Port Phillip Bay (Fig. 5.6), seasonally alternating patterns of beach drifting have posed problems in harbour maintenance. The embayment at Hampton, for example, is subject to northward drifting of beach material in summer and southward drifting in winter, in response to the predominance of westerly wave action in the summer months, and southwesterly in the winter season (Fig. 5.11). Offshore breakwaters constructed in 1909, and

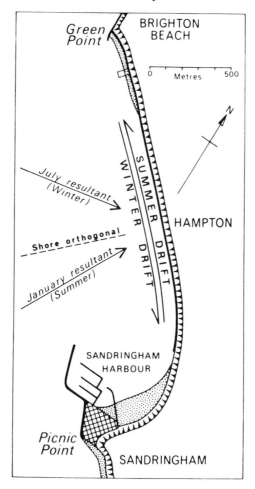

FIGURE 5.11   Coastal compartment at Hampton, Port Phillip Bay, Victoria. Onshore wind resultants, and the seasonal pattern of beach drifting produced by consequent wave action are illustrated. Following the building of Sandringham Harbour (Fig. 5.12) the beach that formerly lined the coastal compartment has been dispersed by seasonal drifting, the bulk of the beach material having been deposited within the harbour.

extended in 1935–39, caused some sand accretion on the adjacent shore, but when a large stone breakwater was built (1949–54) to shelter Sandringham Harbour, the sand moving southward each winter was trapped in its lee, within the sector no longer exposed to the southwesterly waves that previously generated northward drifting. As a result, Hampton Beach has disappeared and Sandringham Harbour has been reduced and shallowed by sand accretion (Fig. 5.12). Demands for the artificial restoration of Hampton Beach can only be met

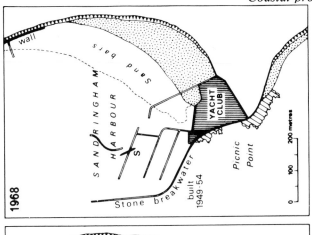

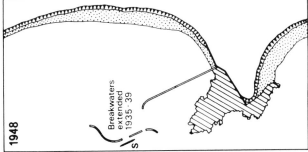

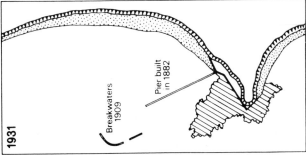

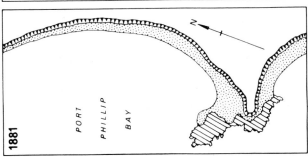

FIGURE 5.12 Stages in the evolution of Sandringham Harbour and associated shoreline changes (see Fig. 5.11)

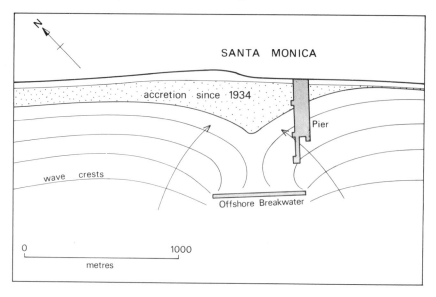

FIGURE 5.13   Coastline accretion at Santa Monica, California, following the insertion of an offshore breakwater in 1934

if an additional breakwater is built on the northern side of Sandringham Harbour to prevent the restored beach also drifting south into the harbour each winter.

An offshore breakwater has also been used at Santa Monica, in California, to protect a harbour adjacent to the pier. Built in 1934, this has modified the patterns of waves approaching the coast in such a way as to converge them to the lee and shape the deposition of a sandy cuspate foreland (Fig. 5.13). In a similar way, cuspate forelands have developed in the lee of ships grounded offshore, for example in Sukhumi Bay on the Soviet Black Sea coast (Zenkovich, 1967: Fig. 193), and at Port Hueneme in California. In each case, localized trapping of sand has been followed by beach erosion downdrift, and at Santa Monica it has been necessary to replenish artificially the depleted beaches to the south on the Los Angeles coast. One answer to this problem would be the multiplication of such offshore barriers in such a way as to protect and augment beaches along the whole length of a sandy coastline. This has been done on parts of the Italian coast, notably at Rimini, where several kilometres of resort beaches are now protected by a chain of offshore breakwaters consisting of dumped boulders. An alternative could be the use of floating breakwaters, moored a short distance offshore and parallel to the coastline until they had induced local beach accretion of the kind seen at Santa Monica. Such floating breakwaters could be moved to and fro along a sandy coastline to produce a succession of induced cuspate beaches, thereby artificially replenishing a beach without the expense of pumping or trucking sand.

Breakwaters built in the nineteenth century to enclose a bay area as a harbour, as at Portland on the south coast of England, can modify the wave regime and intensify wave refraction in such a way as to change the pattern of wave energy, and also the angle of wave incidence on the adjacent coast. This may explain the erosion of beaches and the redistribution of shingle on the shores of Weymouth Bay, to the north-east, in the period following the construction of the Portland break-waters. Similar changes have followed the construction of an enclosing breakwater to form a harbour at another Portland, in southeastern Australia, the sequel here being the onset of severe beach erosion at Dutton Way, to the north, where the energy of refracted waves reaching the shore has apparently increased (Fig. 5.14).

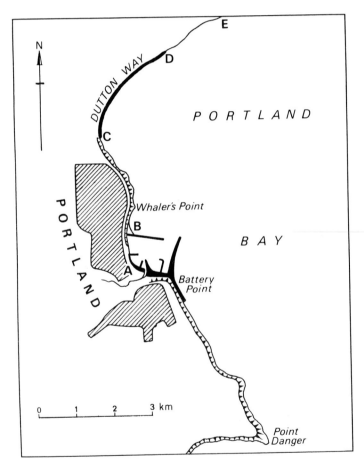

FIGURE 5.14 Coastal features at Portland, Victoria, Australia. The construction of the large harbour breakwater at Battery Point in 1957–59 was followed by rapid erosion of the shoreline at Dutton Way (CD). This is now lined by a boulder wall, but shoreline erosion has intensified in the next sector (DE)

5.5   EFFECTS OF DREDGING AND DUMPING

In many places the approaches to ports have been dredged to provide or improve a navigable channel. Dredging can modify the pattern and velocity of currents, which in these environments are generated mainly by tidal action and river discharge. As a rule, deepening of a channel is likely to be followed by a weakening of the currents, and where the approach is to an estuarine or lagoonal system the increased cross-sectional area of the dredged channel may also modify the range and extent of tidal rise and fall within that system. On the other hand, a channel dredged on an alignment which is shorter than the course of pre-existing natural channels may intensify current flow, and increase tidal ventilation within an estuary or lagoonal system. The onset of erosion on the sandy coastline of Matakana Island, near Mount Manganui in the North Island of New Zealand, followed the dredging of the approaches to Tauranga Harbour, and may have been caused by larger waves arriving through the water deepened by this dredging.

Material dredged from such channels is commonly dumped either offshore, or in adjacent shallows, or carried onshore for use in land reclamation or beach nourishment works. Dredged material dumped offshore creates a new sea floor topography, which can modify wave patterns and the direction and strength of current flow, whereas material dumped onshore establishes a new coastal outline. In either case, there is usually dispersal and reworking of dredged material. An example is the Cairns area, in north-eastern Australia, where a channel has been dredged through extensive mudflats to give access to a port at the mouth of Trinity Inlet. The dredged mud, dumped in adjacent shallows, has been dispersed across Cairns Bay by wave action, and some of it has been deposited on bordering sandy shores, as at Machan's Beach and at Yarrabah in Mission Bay.

Dredging of the sea floor to obtain sand and gravel or other mineral resources results in localized deepening of the water, which enables larger waves to move in towards the adjacent coastline. This can result in a change from progradation or stability to erosion, or the acceleration of erosion that was already in progress. Erosion has become severe on the western shores of Botany Bay, near Sydney, Australia, following the dredging of a deeper channel through the bay entrance, which permits larger waves to move through into the bay, and break upon its western shore. The situation has been complicated by the extension of the airport runway out into the sea, a protrusion that may have intensified wave energy on the adjacent sandy shoreline in much the same way as the Portland breakwaters.

Changes can also result from the cutting of new outlets from rivers on deltaic coastlines. During the sixteenth century the Po delta in northeastern Italy was growing northward towards the Venetian Lagoon, and in order to prevent siltation which would have reclaimed this lagoon, and linked the island of Venice to the mainland, a canal was cut to divert the outflow east from Contarina. The modern Po delta has grown substantially, outward from this diverted river mouth. Other examples of delta growth following river mouth diversion include the Brazos delta in Texas, and the Cidurian delta in Java. In both cases earlier deltas built at and around the previous river mouth have been cut away by marine erosion.

5.6   EFFECTS OF SUBSIDENCE

Submergence of coastal regions may take place as a result of natural changes such as a rise in sea level or tectonic lowering of the land margins (as in the Netherlands), but on some coasts subsidence has been due, directly or indirectly, to man's activities, notably underground mining or the extraction of groundwater. The best documented example is in the Venice region, where tide gauge records show that mean sea level has risen up to 27 cm since 1871. In Venice, more than half of this change has been ascribed to subsidence resulting from the extraction of groundwater by pumping from wells to provide for urban and industrial water demand. The submergence is also partly due to changes in tidal regime in the Venetian Lagoon caused by the reduction of the lagoon area by the impoundment of fishponds and the deepening of entrance channels by dredging, as well as to tectonic subsidence and a regional rise in sea level in the northern Adriatic region in recent decades. It has resulted in the recurrent tidal flooding ('acqua alta') of the city of Venice when atmospheric depressions accompany high spring tides, and the frequency and depth of this flooding has increased during the past century. In addition to the damage to Venice, the rising sea level has increased beach erosion along the Adriatic coastline, necessitating the building of sea walls and boulder armouring, and has also led to changes in the salt marshes ('barene') of the Venetian Lagoon, including erosion of their seaward margins and waterlogging of low-lying areas. In recent years the extraction of groundwater has been halted, and recharging of subterranean aquifers has diminished the rate of subsidence. Nevertheless, the sea continues to rise relative to the land in this region, and floodgates that can be raised to exclude storm surges from the tidal entrances have been proposed as a means of saving Venice from further submergence.

5.7   EFFECTS OF SEDIMENTOLOGICAL CHANGE

Apart from modifications to coastal processes caused by the addition of artificial structures there have been changes resulting from man-induced alterations in the nature and quantity of sediment available at the coast. Some of these have been direct, as in the cases where waste from coastal quarries spills directly into the sea, and is reworked by waves and currents to build beaches that would not otherwise have formed. An example of this is the chalk quarry at Hoed, on the east coast of Jutland in Denmark, where during the past two centuries unwanted flint nodules separated from the chalk have been dumped on the coast. The outcome has been local progradation of a beach ridge plain (Fig. 5.15). Coastal quarrying on the Lizard Peninsula in Cornwall has resulted in gravelly waste spilling into the sea, and substantial beach accretion has ensued in the cove at Porthoustock. On the Cumbrian coast, dumping of waste material from a colliery and a steelworks has augmented local beaches with gravelly deposits.

In recent years, depletion of beaches, particularly at seaside resorts, has prompted coastal engineers to attempt artificial replenishment of the lost sand and shingle with material brought from inland or offshore sources. Mention has been made of the restoration of eroded resort beaches downdrift from harbour breakwaters at Durban and Santa Monica. On the Soviet Black Sea coast, beaches were formerly quarried to supply sand and gravel for building

FIGURE 5.15   Prograded beach ridges south of gravel waste tips at Hoed Quarry, Jutland, Denmark (F. Hansen)

purposes, but by the early 1960s it was obvious that this extraction of beach sediment was resulting in accelerated erosion. The coast at Sochi, for example, began to suffer severe storm wave erosion as its beach became depleted. The procedure was then reversed, with loads of sand and gravel from inland quarries being brought down to the coast and dumped on the shore to restore the beach, improving a recreational resource at the same time as countering storm wave erosion. Similar beaches have been emplaced on the coastline near Odessa.

A few resorts have the unusual problem of excessive sand accretion, with complaints from seaside residents and hoteliers that the sea is retreating, and that dunes are building up to obscure their view. Examples of this are found at Seaside in Oregon and at Malindi in Kenya, in both cases with beach sand being augmented by deposits washed into the sea from a nearby river. More generally, beach erosion is the problem. In the United States there have been a number of beach nourishment projects, based on the dredging of sand from the sea floor and its delivery by pumping or dumping on to the eroding shore. The beach at Atlantic City, New Jersey, was restored in this way between 1935 and 1943; those at Palm Beach, in Florida, and West Haven, Connecticut, in 1948; and others subsequently at Virginia Beach, south of Cape Henry, Harrison County, Mississippi, and several sites on the coast of California. Waikiki Beach, at Honolulu, has been restored by the importation of sand dredged from beaches on other Hawaiian islands, notably from northern Oahu and western Molokai. Mention has already been made of the artificial replenishment of Bournemouth beach, in southern England, with sand brought in from the sea floor, and similar projects are active elsewhere, for example at Mentone, on the shores of Port Phillip Bay, Australia (Fig. 5.16).

Less direct have been the effects of deforestation and cultivation of hinterlands within the catchments of rivers draining to the coast. As a rule, such rivers show increased discharge and become more flood-prone; they deliver larger quantities of sediment to the coast, often with an increased proportion of coarser material. The outcome is accelerated growth of deltas, and progradation of beaches supplied with fluvial sands or gravels. It is believed that the reduction of vegetation cover by clearing, burning and grazing in the Mediterranean region, which became extensive about 2000 years ago, resulted in rapid siltation of bays and inlets at and near river mouths, especially on the Adriatic coast between Ravenna and Trieste, and on the shores bordering the Gulf of Corinth and the Gulf of Euboea, in Greece. There is, however, some doubt about the often-quoted rapid advance of the Mesopotamian deltaic coastline, at the mouths of the Tigris and Euphrates rivers, where progradation by rapid deposition of sediment has been offset by continuing subsidence. More recently the onset of severe soil erosion in river catchments around Chesapeake Bay has been correlated with shallowing and marsh encroachment within that bay, an effect which is repeated on a larger scale in the humid tropics, especially at Jakarta Bay in Indonesia. On the south coast of Java, Indonesia, the Segara Anakan

FIGURE 5.16   Artificial beach emplaced in front of the sea wall at Mentone, Port Phillip
Bay by the method shown in Fig. 5.3 (N. Rosengren)

is a broad estuarine embayment being rapidly shallowed by sedimentation,
notably from the Citanduy River, so that fringing mangroves are advancing
rapidly on to accreting mudflats (Fig. 5.17).

In New Caledonia there has been rapid extension of mudflats and mangrove
swamps (for example at the mouth of the Népoui River) in recent decades as
a result of augmented supplies of sediment, initially clay and silt, then sand
and gravel, from rivers whose upper catchments have been made unstable by
open-cast hill-top nickel mining (Bird *et al.*, 1984). Similar changes have been
noted on the west coast of Bougainville as the result of downwashing of tailings
from a copper mine, and on several Malaysian rivers where sediment loads have
been augmented by dredging for alluvial tin. In Cornwall, mining waste from

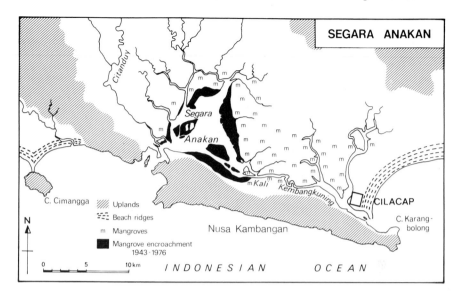

FIGURE 5.17   Silting of Segara Anaken, Java. Segara Anakan, an estuarine embayment in southern Java, is silting rapidly because the sediment yield from the Citanduy River has been increased by soil erosion in its headwater region. As siltation proceeds, bordering mangroves are spreading forward on to accreting mudflats: between 1943 and 1976 an area of about 30 km^2 was occupied by encroaching mangroves. If the present rate of siltation is maintained, the embayment will be almost filled by mangrove swamp by the year 2000, but if soil conservation in the Citanduy catchment diminishes sediment yield from the river, this infilling will take longer

the Hensbarrow china clay quarries has led to local progradation of beaches at Par and Pentewan.

A reverse sequence is seen where dams have been constructed to impound reservoirs, which act as sediment traps (see Chapter 9). Diminished sediment yields from rivers thus modified has resulted in a reduction of sand supply to the beaches of Southern California. These beaches show southward longshore drifting, the sand being lost into the heads of submarine canyons close inshore, as at La Jolla, and as the fluvial input is reduced the beaches are becoming depleted. Several of the world's major deltas now show rapid coastline erosion, the sediment supply that formerly maintained and prograded them having diminished as the result of dam construction. The most notable is the Nile, where the delta coastline is receding rapidly, up to 40 m a year, erosion having accelerated in consequence of the reduction in sediment supply following the completion of the Aswan High Dam in 1970. Other major deltas with eroding coastlines as a sequel to diminished sediment yield following dam construction includes the Rhône, the Dneiper, the Dneister, the Volga, the Volta, and the Zambesi. In Italy and Greece, extraction of sand and gravel from the lower

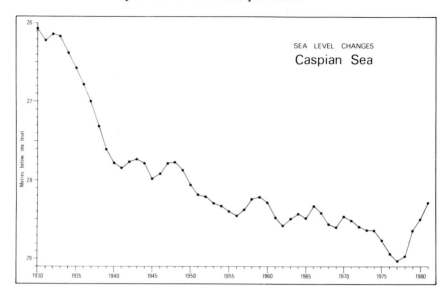

FIGURE 5.18   Changing levels of the Caspian Sea. The level of the Caspian Sea was lowered between 1930 and 1977, partly because of reduction of inflow from rivers by dam construction and reservoir impoundments, and partly because of a trend toward more arid conditions in this region. In the last few years there has been some recovery, possibly related to the damning of the Kara-Bogaz-Gol embayment, formerly an area of high evaporation losses on the east coast of the Caspian Sea. Based on information supplied by O. I. Leontiev

reaches of rivers has similarly diminished fluvial sediment yields to their mouths, and resulted in coastline erosion.

Reduction of fluvial discharge by dam construction is thought to have been a primary cause of the lowering of level of the Caspian Sea, which fell by 2.87 m between 1930 and 1977 (Fig. 5.18). In consequence, there was extensive emergence of bordering shores, accompanied by widespread deposition, as the sphere of coastal processes was withdrawn from the earlier coastline (Fig. 5.19). Since 1977 there has been a renewed rise in the level of the Caspian Sea, possibly related to the reduction of evaporation losses following the damming of the Kara-Bogaz-Gol embayment, into which much water previously flowed to compensate high evaporation losses. The outcome is the submergence of part of the land area that had emerged during the phase of falling Caspian Sea level.

5.8   EFFECTS OF VEGETATION

Coastal processes can also be modified as the result of changes in vegetation. On the one hand, man's activities can reduce the extent and vigour of the vegetation cover, and result in the onset of erosion of terrain that was previously

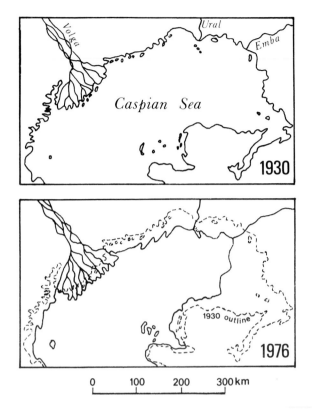

FIGURE 5.19   The changing coastline of the Caspian Sea. The coastline of the northern part of the Caspian Sea advanced rapidly between 1930 and 1976 as a consequence of lowering of sea level (Fig. 5.18) and deposition of sediment, mainly from the Volga, Ural and Emba Rivers. Based on information supplied by O. K. Leontiev

stabilized. On the other, the introduction of species that were not previously present, or the improvement of vegetation cover by artificial means (such as the addition of fertilizers) can influence the pattern of sedimentation and promote stability.

The first sequence is well illustrated where coastal dunes, previously stabilized by a cover of grasses, scrub or woodland, become unstable as the result of a reduction in vegetation cover, for example by deliberate clearance, burning, grazing by introduced animals (sheep, cattle, goats, rabbits), trampling, or vehicle damage. In New Zealand a phase of coastal dune mobilization has been correlated with the arrival of the Maori invaders, probably in the fourteenth century. In southeastern Australia the coastal dune fringe is typically interrupted by blowouts where the vegetation cover has been removed or destroyed, and sand is being moved inland by onshore winds. Some of these may have developed naturally behind eroded beaches; others may be the outcome of Aboriginal impact, notably the lighting of fires which weakened the retaining vegetation cover and allowed the sand to be drifted by the wind. Many have alignments that originated where deflation began along trackways to the shore where trampling (especially by holidaymakers visiting the beach) had destroyed the vegetation cover. In some sectors the vegetation has become so weakened that the dunes have developed into mobile sand sheets: the active dunes at Cronulla, south of Sydney, exemplify this. In recent decades dunes have been activated on the east coast, and on parts of the southwestern coast, of Australia, by open-cast extraction of heavy mineral sands, notably rutile and ilmenite.

The sequence is reversed in places where coastal dunes that were previously active (either naturally, or because of prior destruction of vegetation) have been stabilized by the planting of retentive vegetation, notably the European marram grass, *Ammophila arenaria*. The introduction of marram grass to the rear of broad sandy beaches has been followed by the accretion of sand blown from the beach to form high foredunes. Elsewhere, previously mobile dune topography has been arrested by planting grasses, shrubs and forest trees, for example in the Landes region of south-western France, at Culbin on the Scottish coast, and in the Danish 'Klit', the coastal fringe of stabilized dune topography where a century ago the sand was spilling inland across farming country. Generally such stabilization of drifting coastal sands has been regarded as beneficial, but on the Oregon coast the development of high foredunes as a consequence of marram grass introduction has recently been criticized on the grounds that it has greatly diminished sand supply to the hinterland, so that the area of drifting sand dunes has been much reduced. It has been necessary to remove marram grass and remobilize coastal sands to ensure the preservation of active dune areas as open spaces within the Oregon Dunes National Recreation Area. In some places the planting of vegetation has stabilized dunes that were drifting *towards* the coast and nourishing beaches. Thus the stabilization of dunes on Cape Recife, near Port Elizabeth in South Africa, has been followed by depletion of beaches that were formerly maintained by the arrival of wind-blown sand.

Vegetation has a similar effect in coastal marshlands, tending to promote the accretion of sediment within colonized zones in such a way as to build relatively stable salt marshes. In 1870, hybridization between a European and an American species of *Spartina* in the marshlands of Southampton Water yielded a new and vigorous species, now known as *Spartina anglica*. This species was able to spread forward on to mudflats below the level of other salt marsh species, and its vigorous growth provided a filtering matrix which rapidly accreted sediment to build a new prograded salt marsh terrace. It was soon introduced to Poole Harbour (Fig. 5.20) and other British and European estuaries and inlets, and eventually to similar areas as far away as the Tamar in Tasmania, Anderson's Inlet in southeastern Australia, and several New Zealand

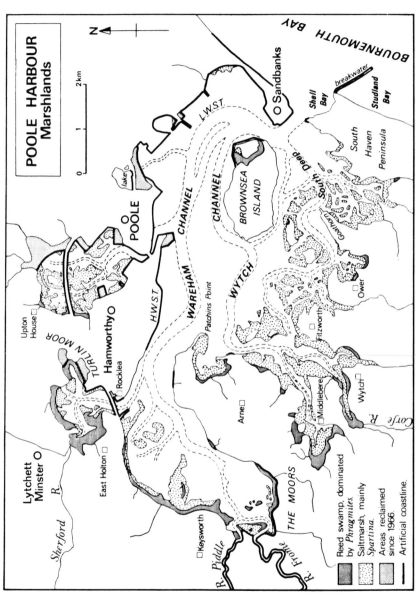

FIGURE 5.20 Siltation of Poole Harbour, Dorset. The tidal mudflats bordering Poole Harbour, Dorset, have been modified by the development of depositional terraces where sediment has been trapped by *Spartina anglica*, which spread rapidly after its introduction to this estuarine embayment in 1899. In places, reedswamp is colonizing the rear of the *Spartina* terraces. Poole Harbour has also been reduced in area by land reclamation and port development, and long sectors of its northern shoreline have been stablished by the building of sea walls. Map based on surveys by Vincent May

estuaries, to act as an agent in marshland development, preparing the way for reclamation. It provides a good example of the impact of man on coastal processes, for the dense growth of *Spartina* greatly diminishes wave and current action as the tide rises and falls in marsh-fringed estuaries, with the result that sedimentation is facilitated within the *Spartina* zone. In some estuaries, however, the *Spartina* terraces built during the past century are now showing decay and erosion, a consequence of marginal die-back of the *Spartina*, for reasons which have not been precisely determined. At Lymington, for example, the muddy sediment previously trapped by *Spartina* growth is now being laid bare and eroded away by wave and current scour along the channel margins. It has thus become evident that the phase of *Spartina* terrace building was accompanied by the narrowing and deepening of intervening tidal channels, and that these are now widening and shallowing as the formerly trapped sediment is released.

Mangroves fringe estuaries, tidal inlets and embayments in tropical regions, especially on muddy substrates. They occupy the upper part of the inter-tidal zone, spreading forward as accretion proceeds: where man's activities have accelerated accretion as on the shores of Cairns Bay, northeastern Australia (Fig. 5.21) and in the Segara Anakan, southern Java, mentioned previously. Mangroves with dense subaerial root structures, such as the pneumatophore

FIGURE 5.21 Mangroves advancing on the shores of Cairns Bay, Northeastern Australia. The advance has been more rapid around the mouth of a drainage channel cut through the mangrove fringe because of accelerated deposition of sediment to this sector (E. C. F. Bird)

networks of *Avicennia*, also trap and retain sediment in much the same way as salt marsh vegetation, building up depositional terraces. Clearance or destruction of such mangroves is followed by the release of sediment and the degradation of these depositional terraces. Conservation of mangrove and salt marsh vegetation is thus important in maintaining the stability of estuarine environments, and the navigability of associated channel systems.

5.9 CONCLUSIONS

The various examples quoted show how coastal processes can be modified by the addition of artificial structures, by changes in the nature and quantity of available sediment, and by modifications in the vegetation cover, all due directly or indirectly to man's activities. In some sectors the modifications have resulted in deposition and coastal progradation that would not otherwise have occurred; in others, they have initiated or accelerated erosion. Sequences have been noted where man's impact has stimulated coastal deposition and further interference has resulted in erosion of newly deposited terrain; or where an initial erosional response has subsequently been reversed. Man's impact on coastal processes has been strongest and most obvious on densely populated and intensively used coastlines, but minor and indirect consequences of human interference may be detected elsewhere, particularly on coastal dune systems, which may show man-induced erosion (by way of stock grazing on dune vegetation, for instance) in areas removed from intensive usage. The presence of erosional features in coastal dunes on sparsely populated sectors of the Australian coastline, such as Discovery Bay, Cape Howe and Fraser Island, shows how widespread are some of the impacts of man's activities on coastlines.

In recent years many coastal countries have set up agencies to plan and manage the utilization of their coastlines, deciding where urban, industrial, port and recreational developments should be sited, and where natural or cultural features should be preserved. In England and Wales, for example, sectors of high quality coastal scenery have been designated Heritage Coasts, subject to strict planning controls and management procedures, while other sectors have become Nature Reserves of various kinds, managed for the conservation of coastal ecosystems.

The management of coastal resources requires an understanding of changes taking place in coastal environments and the processes that lead to these changes. It also requires an assessment of the impacts of man's activities, past and present, within these environments, and an awareness of the possible consequences of further modification of coastlines, especially where engineering works are contemplated. Coastal management can be seen as a deliberate attempt to stabilize coastal features, or to guide their continuing evolution. Geomorphological studies which elucidate the relationships between man and coastal processes therefore provide a scientific basis for coastal management.

REFERENCES

Bird, E. C. F., 1978, *The Geomorphology of the Gippsland Lakes Region*, Ministry for Conservation, Victoria, Publication No. 186.

Bird, E. C. F., 1984, *Coasts*, Blackwell, Oxford.

Bird, E. C. F., 1985, *Coastline Changes*, Tycooly, Dublin.

Bird, E. C. F., and Schwartz, M. L. (Eds), 1985, *The World's Coastline*, Van Nostrand Reinhold, Stroudsburg, Pennsylvania.

Bird, E. C. F., Dubois, J. P., and Iltis, J. A., 1984, *The Impacts of Open-cast mining on the Rivers and Coasts of New Caledonia*, United Nations University, Tokyo.

Davies, J. L., 1980, *Geographical Variation in Coastal Development*, Longman, London.

King, C. A. M., 1972, *Beaches and Coasts*, Arnold, London.

Komar, P., 1976, *Beach Processes and Sedimentation*, Prentice-Hall, New Jersey.

Zenkovich, V. P., 1967, in Steers, J. A. (Ed.), *Processes of Coastal Development*, Oliver & Boyd, Edinburgh.

Human Activity and Environmental Processes
Edited by K. J. Gregory and D. E. Walling
©1987 John Wiley & Sons Ltd.

6

Ocean Processes

C. A. M. KING

6.1 INTRODUCTION

The oceans cover over 70 per cent of the earth's surface and exert a strong influence on many aspects of the geography of the earth, both physical and human. Some of the basic characteristics of the oceans will be mentioned first, leading to a consideration of the relationships between man and the oceans, which cover a wide range of topics, as the influence of the oceans on mankind is all-pervading.

The distribution of land and sea. It is now known that on a geological time scale the pattern of land and sea is ephemeral. Continents drift across expanding or contracting oceans. The present pattern of land and sea, however, has many important consequences for mankind. There are four main oceans, and from the oceanographic point of view a fifth may be identified. The Arctic ocean is almost land-locked apart from its important link with the Atlantic ocean. It is largely ice-covered at present. The largest ocean is the Pacific, which covers nearly half the globe, while the Indian ocean is essentially a low latitude ocean. At their southern extremes the three major oceans, Atlantic, Pacific and Indian, link to encircle the high southern latitudes. This continuous belt of water can usefully be referred to as the Southern ocean.

The ocean receptacles are closely related to the formative processes of sea floor spreading and the consequent continental drift. The major oceanic morphology consists of abyssal plains, submarine ridges and curving island arcs and elongated deep sea trenches. Smaller, but significant, features include sea mounts, guyots and submarine canyons and associated abyssal fans. From the human point of view the continental shelf and slope are particularly important. They link the continental and oceanic elements of the earth and are rich in mineral and biological resources.

Geophysical work has shown that there is a fundamental difference between the continental and oceanic crust, the former being much thicker and less dense, while the latter is thinner and denser. Granitic material is lacking in the deep oceans, but makes up the bulk of the continental crust. One of the most important discoveries resulted from the recognition of magnetic reversals, which allowed the ocean crust to be dated. Its age indicated that the ocean basins are young features, most of which have formed since the Cretaceous; the oldest ocean crust is only 200 million years old, whereas continental rocks about 3700 million years old still exist. The theory of sea floor spreading very cogently linked many aspects of ocean morphology into a consistent whole.

The main ocean ridge, some 60 000 km long, bisects the Arctic ocean between the dead Lomonosov ridge and the north Asian coast and continues down the central Atlantic and round into the Indian ocean half way between South Africa and Antarctica. The volcanic activity of Iceland and other mid-Atlantic islands demonstrates clearly its character and function. It is the line along which oceanic crust is being created by eruption of mainly basaltic material. As lava is ejected and cooled it is magnetized with the polarity of the time, but as more material upwells at the ridge crest the older crust moves away. The process is revealed by the paired alternating strips of polarity in the oceanic crust on either side of the ridge crest. One branch of the mid-Atlantic oceanic ridge runs up into the developing ocean of the Red Sea, while another bisects the distance between Australia and Antarctica. The ridge can be traced across the south Pacific to run eventuality into the Gulf of California. Thus the Atlantic is a spreading ocean, expanding at a rate of 2 to 3 cm/year, while the South Pacific rise shows expansion rates of as much as 5 to 6 cm/year in places.

While some oceans are expanding others must be contracting as the earth's surface is of constant area. The process of subduction occurs where the ocean crust is diverted downwards on coming into contact with the thick continental crust. The morphological results of this process are the island arcs and deep sea trenches, most of which occur around the Pacific ocean, only pushing through into the Atlantic in the Caribbean and Scotia arcs in the centre and south respectively. The subduction zones are marked by the curved deep sea trenches. At these positions the Mohorovic discontinuity, which divides the crust from the mantle, plunges downward from an average of 6 km below the ocean floor to 20 to 30 km below the continents. It is also a zone of deep focus earthquakes. The island arcs arise on the continent side of the trench where the continental crust is thin or lacking. Where it is thicker, mountain ranges can occur; the Andes of South America are being thrust up while offshore the Peru–Chile trench is the longest one. The trenches are the deepest part of the ocean, and have been considered to be the safest repository for toxic wastes.

At present the Pacific Ocean is contracting. The oldest ocean crust is in the northwest of this ocean, while the Atlantic and Southern oceans are expanding and a new ocean is forming along the Red Sea rift. This fundamental process determines the nature of the oceanic–continental boundary, which has important repercussions from the human point of view. It affects the nature of the continental slope, shelf and coastline, as will be further considered when resources are discussed (pp. 128–136).

Sea water. The oceans contain about 97 per cent of all the water on earth, and much of the rest is locked up in the solid state as ice in Antarctica and Greenland and other ice masses. The characteristics of sea water that are most important are its salinity and temperature, the two together determining its density.

Temperature in the ocean varies between $-2°C$ and $+30°C$. At the lower limit ice forms. The ocean water is densest at $-2°C$. Ice has a lower density and, therefore, floats, with important repercussions on ocean circulation. The distribution of temperature at the ocean surface reflects the pattern of heat input from the sun. One important feature is that at any given latitude the northern hemisphere ocean is warmer than the southern, and the oceanic thermal equator is permanently north of the geographic equator. Another important fact concerning heat in the ocean is the great capacity of the sea to absorb and retain heat, thus providings its moderating influence on climate worldwide. The oceans provide a great store of heat. A fall of $2.2°C$ in ocean temperature it has been claimed would cause extensive glaciation. The bulk of ocean water is cold, with more than half colder than $2.3°C$ and only 8 per cent is warmer than $10°C$, 75 per cent is between $0°C$ and $4°C$. A thin layer of warm water is separated from a very deep, thick layer of cold water by the thermocline.

Salinity. Sea water contains a very wide range of elements, but the dominant one is common salt, sodium chloride. One cubic kilometre of sea water contains about 40 million tons of salt. The salinity of sea water is expressed in parts per thousand, ‰, grams/kilogram, and has an average value of about 35‰, of which 30‰ is sodium chloride, with sulphate, magnesium and calcium contributing nearly a further 4.5‰. The proportion of the different elements remains very uniform throughout the ocean, but the total salinity varies within rather narrow limits, 75 per cent of the water has a salinity between 34.5 and 35‰. The Atlantic has a greater variation than the Pacific, in which about half the water has a salinity between 34.6 and 34.7‰.

The *density* of sea water depends on its temperature and salinity, increasing as the temperature decreases and the salinity increases, so that cold salty water is densest and occupies the deeper ocean basins, and is overlain by fresher and/or warmer water. Distinct types of water, described by their temperature/salinity relationships, are recognized as *water masses*. The surface water masses reflect external conditions at the surface including precipitation/evaporation ratio and temperature. They lie above the thermocline and are generally only about 200 m thick, reaching a maximum of about 900 m in the Sargasso Sea. The intermediate and deep water masses are fewer in number, but much larger in volume. Their characteristics are related to the surface conditions and the deep water circulation.

Ocean currents within the surface water masses are responsible for significant distribution of heat over the earth's surface. The surface pattern is mainly determined by the wind field and the Coriolis effect. The trade winds generate east to west flowing currents in low latitudes, forming the north and south equatorial currents, separated by the equatorial countercurrent, which flows east just north of the equator. The water reaching the western side of the oceans is diverted north or south in the northern and southern hemispheres respectively. It then turns along the coasts to form in the northern hemisphere the well known Gulf Stream of the Atlantic, and the Kuro Shio of the northwest Pacific, while the Brazil current and east Australian currents are the southern hemisphere equivalents. The North Atlantic drift continues the Gulf Stream into the westerly wind belt in the Atlantic, and then turns south as the Canaries current to complete the clockwise gyre typical of the northern hemisphere oceans. The equivalent in the Pacific is the California current. The gyres in the southern hemisphere rotate anticlockwise. All around the Southern ocean the west winds drive a continuous current around Antarctica, and from this the Peru or Humboldt current flows north along the west South American coast, while the Benguela current is the south Atlantic equivalent. The Indian ocean has generally similar currents in the south, but in the northern Indian ocean the monsoon causes seasonal shifts of currents.

One important respect in which the ocean circulation differs from the air circulation, to which it is closely related, is the western intensification of the current. This is due to the Coriolis effect. It varies with latitude and acts to divert moving water to the right in the northern hemisphere and to the left in the southern. The latitudinal variation is responsible for the western intensification. The result of this is rapid flow in the Gulf Stream and other western boundary currents. The Gulf Stream system derives its driving force from the fact that the Gulf of Mexico has a higher water level than the Atlantic owing to water carried in by the west-flowing north equatorial current. The first part of the system is the Florida current which flows fast eastwards through the Straits of Florida. It merges into the Gulf stream to flow north along the coast of the United States. Then it becomes the meandering North Atlantic current, which flows east towards Europe. The system carries a lot of warm salt water northwards. It forms a boundary between the warm Sargasso water on its right and the cold shelf water on its left. The warming of the northwest Atlantic countries due to its influence is well known. The system also derives water from the southern hemisphere as the south equatorial current flows partly north of the equator, owing to the position of the thermal equator north of the true equator.

The eastern boundary currents are more diffuse than the western ones, but they provide some of the most fertile waters on earth, owing to the upwelling that takes place within them. Upwelling is due to the wind along the coast and the Coriolis effect, and applies particularly to the Benguela and Peru currents.

The Coriolis effect is also responsible for a divergence along the equator in the south equatorial current, and at about 10° north where the equatorial counter-current flows adjacent to the north equatorial current. These divergences bring more fertile water to the surface. Another result of the change of the direction of the Coriolis effect at the equator is the equatorial undercurrent which flows east under the south equatorial current along the equator as a narrow ribbon of water. It is 4° latitude wide but only about 200 m deep; it is most marked in the Pacific, where it flows at up to 150 cm s^{-1}.

The *deep circulation* beneath the shallow surface currents moves much more slowly, but in very large volumes. There is a fundamental difference between the Atlantic and Pacific oceans, in that both the major sources of deep water occur in the Atlantic, which, as a result, has more diverse water masses. They are arranged in a sandwich, and the water is better oxygenated because it is younger. By the time the water reaches the Pacific it is very uniform, and has less oxygen. One of the two deep water sources is in the south Atlantic in the Weddell Sea where freezing of surface water increases the salt content of very cold water. This densest water of all, the Antarctic Bottom water, moves north under the next densest, the North Atlantic Deep water. This second deep water mass originates on either side of Greenland as warm, salty water from the Gulf Stream system is cooled in winter and mixes with the cold water of the east Greenland and Labrador currents. Its formation is intermittent, occurring during specially cold winters. The salinity is enhanced by the addition of dense, very saline and warm water from the Mediterranean, which flows out over the Gibraltar sill, and flows south with it. Eventually this water mixture rises to the surface to the south of the Antarctic convergence, which lies in about 52°S latitude. The convergence zone forms the effective northern boundary of the Southern ocean, with its strong west to east current. The water is cold, but fertile owing to the rising of the water from below to form the Antarctic Circumpolar current and water mass. The top layer of the Atlantic sandwich is the Antarctic Intermediate water. This relatively fresh and cool water mass sinks at the subtropical convergence and moves north below the surface currents at relatively shallow depths. The deep water circulation is slow and C-14 dating of the deep Pacific water has given ages of 1350 to 1910 years, which is a mean velocity of 15 km year^{-1}. Atlantic water at 2000 m depth is estimated to be 50 to 1000 years old. Even in the deep oceans, however, western boundary currents appear to exist in the Atlantic and the south Pacific, while the west to east flow around the Southern ocean continues to the bottom, thus carrying water from the south Atlantic to the Indian and Pacific oceans.

Waves. Although ocean currents move much water and play a vital role in helping to distribute heat more evenly over the earth, the waves, which are another result of the wind blowing over the water, are more conspicuous. They play an important role in shaping the coastline, while in the open sea they affect shipping and other offshore activities. Waves are generated by the wind blowing over the water and imparting energy to it, in the form of undulations which grow longer and higher as the wind velocity, duration and fetch increase. In the generating area the waves are known as 'sea', and consist of a whole spectrum of waves of differing length, height and direction of movement, especially when the wind is strong. The spectrum can be analysed to reveal the components of which it is made. Each wave in deep water travels at a rate dependent on its period and length, and longer waves travelling faster than the shorter. Thus when the waves move out of the generating area they become swell, as the longer waves outrun the shorter. The longest wind-generated waves may have periods up to about 20 seconds and are over 600 m long. ($L_m = 1.56\ T^2_{sec}$). They can travel immense distances. Waves, for example, have been recorded on the coast of California which have originated between Kerguelen and Madagascar nearly half way round the earth. Short waves, however, are attenuated much more rapidly and they are often steeper than the long swells, as their height to length ratio is greater. As waves approach shallow water, that is when it is less than half their length deep, the open circular orbits of deep water gradually become modified to open ellipses. They, therefore, still retain their slow mass transport in the direction of travel. They become shorter and steeper and their velocity is reduced as the water becomes shallower, until finally they break. This occurs when the increasing orbital velocity of the water particles becomes greater than the decreasing velocity of the wave form. The reduction in velocity in shallowing water also leads to wave refraction, which causes concentration of energy on headlands, and plays an important part in coastal development. Wave recording can now be carried out by ship-borne wave recorders, as well as in shallow water. Nevertheless, wave forecasting, which depends on accurate meteorological data, as well as details of the present state of the sea over a wide area and the bottom configuration in shallow water, is still far from being an exact science. Reasonably accurate wave forecasting diagrams are, however, available for many areas.

Tides. Waves are variable and difficult to predict, but the tide is one of the most predictable of oceanic phenomena. The tide results from the mutual gravitational attraction of the sun, moon and earth. The moon's effect is larger than that of the sun because, although it is much smaller, it is also nearer. The tide-producing force is proportional to the cube of the distance between the bodies. The tide is generated by the tractive force, which is the resultant between the attractive gravitational force and the centrifugal force and operates parallel to the earth's surface. It creates two crests and two troughs around the earth, causing the semidiurnal tide to be dominant over much of the ocean. The semidiurnal pattern is dominant when the lunar declination is zero, but as it increases the diurnal element becomes more important on an ideal completely water-covered earth. The actual shape of the oceans, however, influences their response to the tide-producing forces.

The response of the oceans is based on the nature of the tidal wave, which is basically a stationary wave influenced by the rotation of the earth to produce an amphidromic system. A stationary wave results from the reflection of a progressive wave from a barrier in its path. In a rectangular basin it is high water at one end and low at the other, while currents reach their maximum when the surface is flat at intermediate states. The diversion of these currents by the Coriolis effect creates the amphidromic system. This consists of the tidal wave rotating round the amphidromic point of zero range in a clockwise direction in the southern hemisphere, and anticlockwise in the northern hemisphere. Each amphidromic system has a natural period of oscillation that is determined by the length and depth of the basin. When this natural period is similar to the tide-producing force period the basin responds to produce high tidal ranges. Some basins respond to the semidiurnal forces and others to the diurnal forces of sun or moon. The smaller Atlantic has predominantly semidiurnal tides, whereas the larger Pacific ocean responds more readily to the diurnal forces, and many places have mixed tides and some predominantly diurnal tides. Tahiti has a solar tide as it is near the amphidromic point of the lunar system but far from that of the solar system. The Mediterranean is too deep for its length, and the Baltic too shallow, to respond to the tide-producing forces, but the North Sea, English Channel and Bay of Fundy are the right size to respond, and therefore, have high tidal ranges. Over most of the oceans, which are divided into a series of amphidromic systems, tidal ranges are relatively low. Only in certain very restricted areas are macrotidal ranges experienced, but in these areas the tide can exert an important influence in several ways.

Life in the ocean is extremely varied, every environment having species adapted to its special characteristics. Life evolved in the oceans and ranges from minute one-cell plants to the largest animal ever to exist, the great blue whales. As on land, life must be created from the chemical elements in conjunction with sunlight by photosynthesis. Thus the basis of life at sea is the plants. They provide sustenance for the higher forms of life, which live in complex ecosystems. Photosynthesis can only take place near the surface of the sea to which sunlight can penetrate. This may be about 100 m in the clear seas of the tropics, but is much less near the coast, where suspended matter prevents light penetration, where it may only be 15 m. Life can only be generated where light and nutrients occur together, so fertility depends on the distribution of nutrients near the surface. The most fertile areas are where nutrient-rich water upwells from below, or is stirred up by wave or other action. The North Sea is more fertile in spring after winter storms, and so is ready for the spring flowering of the diatoms, minute plants, when the light intensity and warmth increase. By autumn the nutrients are used up as the summer warmth causes stratification of the water. In the open ocean the most fertile areas are determined by the surface currents and circulation of the deep water. Highly fertile areas occur round the North Atlantic and North Pacific where stormy conditions maintain nutrients near the surface. These cold waters do not become stratified, as do the subtropical waters of the Sargasso Sea, for example, which is a marine desert, as are many subtropical waters. The equatorial divergence provides a strip of fertile water extending westwards from central America along the equator. The most fertile areas, however, are along the west coasts of South America and southern Africa in the upwelling of the eastern boundary currents, the Peru and Benguela currents respectively. Another very fertile zone is to the south of the Antarctic convergence, where the nutrient rich, cold, deep north Atlantic water, with an admixture of Antarctic Bottom water, Antarctic Intermediate water and Mediterranean water rises to the surface to form the Antarctic Circumpolar water. This rich water provides the sustenance of the great whales that used to feed in this area in great numbers until hunted to near extinction by the uncontrolled and irrational greed of man.

Marine life is adapted to the extreme range of conditions in the oceans, from the violent wave action and tidal emersion on rocky shores to the immense pressures of the cold, dark ocean depths. The organisms include the free-living pelagic creatures and the bottom-living or benthic forms. Some are mobile and others are sessile.

Marine plants include both the sessile seaweeds and the mobile phytoplankton; the former include some of the largest organisms, such as the giant kelps off California, and the smallest, which are the diatoms and algae. There are also bacterioplankton that play an important part in the deeper sea. Bacteria are also common in the substrate and provide sustenance for filter feeders outside the photosynthesis food chain. By the process of photosynthesis carbohydrate and oxygen are formed from carbon dioxide and water by the sun's energy and chlorophyll. From this basis larger molecules and new living matter can be formed, but it is not an efficient process. Only 1 per cent of the available solar energy is fixed as new living matter and 99 per cent goes to heat the sea. It has been estimated that 40 000 million tons of organic matter are created each year. This is roughly equivalent to the production on land, which is achieved on $2/5$ of the area of the ocean. There are also great losses at each step in the food chain. There are 16 000 million tons of phytoplankton, 1600 of zooplankton, 160 of the consumers of zooplankton and of the next predators only 16 million tons.

The phytoplankton are the broad base of the marine food chain. The term 'plankton' means that which is made to wander or drift. Because these tiny plants must remain in the upper layers of water to feed and reproduce, they must be small to prevent stinking. Their surface area is large in proportion to their volume, which is advantageous as they absorb nourishment through their surface. They also often have complex appendages to help them to float. They can reproduce very rapidly under favourable conditions and do so by splitting into two. Light, temperature, salinity and, above all, nutrient supply determines their reproduction. When reproduction is especially rapid the water can become discoloured. The term 'bloom' applies to these events. The Red Sea, for example, can at times be coloured red by great increases of *Trichodesmium erythraeum*.

Red tides also occur in the Gulf of Mexico; elsewhere white water can occur. On these occasions all the oxygen is used up and toxic conditions can kill off fish and other life, but in moderation white water can be a sign of an abundance of herring. The water in this case is coloured by many coccoliths.

The other marine plants are the seaweeds that grow around rocky shores. They are more productive than phytoplankton, but their distribution is restricted. The giant kelp beds off California are especially prolific, and they grow at a rate 10 times that of phytoplankton in fertile waters. Seaweeds are important round the western coasts of Scotland and Ireland which are rocky and exposed. Another ecosystem in which marine algae play an important part is the coral reefs of the warm waters of the Pacific and Indian oceans especially; although the corals themselves are animals they need algae in the reef community.

Marine animals are very varied, filling every available niche, and living together in complex ecosystems. The zooplankton are the drifting members of the community. They consist of permanent plankton as well as the larval stages of many fish and invertebrates and feed on phytoplankton. Their success depends on the availability of phytoplankton, which in turn depends on marine fertility and other factors, including light and warmth. In some areas zooplankton have a seasonal rhythm, as in the North Sea. The zooplankton provide sustenance for the next trophic level. Some organisms have a short food chain and feed directly on zooplankton, notably the herring family and the rorqual whales which live off euphausiids, or krill, a shrimp-like creature. These are, therefore, good value from the human point of view. The trophic level above the plankton can be divided into pelagic creatures, or nekton, that live in the open water, and benthic creatures that live in or on the bottom. The benthic creatures include the epipelagic fauna which live in shallow water on the continental shelves and bathypelagic fauna of the deep oceans, which must be carnivores. Of the former, the coral reefs and mangrove swamps provide two prolific ecosystems of the warmer tropical seas. In the colder waters of higher latitudes the numbers of genera and species are much reduced, but the number of individuals may be very high indeed, as these colder waters are generally more fertile. The benthic fauna include a wide range of invertebrates, such as shellfish, starfish and worms. There are also bottom-living fish, which include flat fish, such as plaice, turbot and sole, and round fish, of which cod are the most important commercially in northern waters. Other fish belong to the pelagic communities. These include schooling varieties, such as herring, sardines and anchovies, all of which have a short food chain. They are very important commercially, especially in the fertile waters of the North Sea and eastern boundary currents off south Africa and Peru. There are also free-swimming fish, such as tuna and sharks who do not form large schools, and which, like cod, have a longer food chain in that they prey on other fish or invertebrates. Mention must also be made of the marine mammals, which belong to the order *Cetacea*. They are warm-blooded and feed their young on milk. They include the whales and dolphins, and must be fairly large to maintain their body temperature. The large whalebone whales are filter feeders, living on krill, while the toothed sperm whales prefer large squids. Whales, unlike fish which are very prolific, reproduce very slowly and are, therefore, much more in danger of extinction when over-hunted.

6.2 MAN AND THE OCEANS

(a) Exploration

The oceans have for a very long time lured the more adventurous sailor to seek what lay beyond the next headland or the horizon. Over 2000 years ago Phoenician seamen were exploring the coasts of Africa and circumnavigated the continent. Thor Heyerdahl has shown that even primitive craft were sufficiently seaworthy for long oceanic crossings in both the Pacific and Atlantic. The best-known period of ocean exploration was in the late fifteenth and sixteenth centuries when European sailors explored all the oceans and sailed round the globe. By 1600 the general outlines of the continents were fairly well known. The southern seas were still largely unknown, until the voyages of Captain Cook between 1768 and 1780. He made accurate charts of his discoveries and overcame one of the major problems of the early navigators, scurvy; this allowed long voyages to be undertaken safely.

(b) Research

Although soundings were made as early as the sixteenth century, ocean science did not begin seriously until the nineteenth century. Problems of recording ocean depth, temperature and other properties were very severe, with only hemp ropes on which to suspend instruments. Observations of surface phenomena were easier to make. The relations of tides to the moon were known to Pliny. The ocean currents were of particular interest to the early navigators, especially in the days of sailing ships. By 1786 Ben Franklin had charted the Gulf Stream. Such charts, however, could not be drawn accurately until longitude could be measured, and this required the development of an accurate chronometer. This was achieved in the early eighteenth century. Latitude could be measured much more easily, by use of the quadrant or sextant, and was known to the early Greeks. The use of the magnetic compass also greatly aided ocean navigation. One of the most influential works on the oceans was Matthew Fontaine Maury's *Physical Geography of the Sea*, published in 1850. His work was based on a large collection of records of ocean currents and other data. Maury's book allowed sailing times to be greatly reduced by making best use of currents and winds, cutting the passage from England to Australia and back from 250 to 160 days. Naval architects also helped by improving ship performance.

Soundings were still scarce and Maury produced a map of the north Atlantic based on 800 soundings in 1857. The use of steam winches and piano wire rather than hemp lines speeded up deep sea scientific observations. These developments also had a practical purpose in that deep sea cables were beginning to be installed, and these needed accurate records of sea bottom configuration. By the mid-nineteenth century the main features of the north Atlantic were known, including the mid-ocean ridge and the deep basins. With deep water observations so difficult and slow to make, it is not surprising that it was thought that there was no life in the cold, dark deep oceans. It was not until 1868 and 1869 that W. Carpenter and Wyville Thomson showed that a variety of organisms were living in depths up to 5000 m. Wyville Thomson was dredging on either side of the ridge that bears his name and which separates the cold deep waters of the Norwegian basin from the warmer waters of the north Atlantic to the south. The difference in the fauna in the two water masses was clearly apparent.

Modern oceanography dates from soon after this work, and is, therefore, little over a century old. The Royal Society of London, aided by the Royal Navy, fitted out an oceanographic cruise in the *Challenger* under the command of Wyville Thomson, which lasted from 1872 to 1876 and extended to all the oceans of the world. This first major ocean research effort brought back a vast amount of material on water character and bottom deposits. Trawls and dredges revealed a great deal of information on marine organisms. The *Challenger* had a displacement of 2306 tons, with sail and an auxiliary engine. She covered 130 000 km and made 362 stations for sounding, dredging and recording sea water temperatures and salinities. 700 new genera and 3500 new species were identified. The greatest ocean depth was plumbed in the Challenger Deep off the Mariana Islands. Following the successful *Challenger* cruise many other nations sent out oceanographic expeditions, such as the *German Meteor* of 1925–27, which worked in the Atlantic. The Southern ocean was exhaustively studied by the *Discovery* expeditions from Britain in the first half of this century. Although the United States was later in the field of ocean research, their ships and techniques are now amongst the most advanced. In recent years international cooperation has led to great advances in knowledge, for example, with the Indian ocean exploration of the 1958 to 1965 expeditions under the auspices of ICSU, in which ships of many countries participated to study a relatively neglected ocean. The potential resources of the Indian ocean could profitably be developed to assist the advancement of the peoples of the Indian subcontinent, Malaysia and various East Indian islands, and the developing countries of east Africa. The name Challenger has been given to a much more recent research ship, the *Glomar Challenger*, which has been specially designed for deep sea drilling, and started work in 1968. Since then the *Glomar Challenger* has worked in all the oceans and amassed a great deal of material concerning the nature of the sea floor sediments and ocean crust. The Deep Sea Drilling Program, now

International Program of Ocean Drilling, began in 1964. The work of the *Glomar Challenger* is supported by the National Science Foundation of the USA, although scientists of 21 countries have taken part in the many cruises. Over 400 holes have been drilled in depths of up to 6000 m, through more than 1000 m of bottom sediments and 600 m into the hard underlying rock. Many of the techniques now used for ocean research were developed in the 1939 to 1945 war, and have been put to good use since, including navigational aids.

An international decade of ocean exploration is concerned with (1) environmental quality and factors affecting it, (2) environmental prediction, particularly in the geochemical field, (3) living resources and the control of their productivity and harvesting, and (4) seabed assessment for mineral resources. The field covered is wide and complex and requires the latest equipment. This includes deep sea submersibles, some of which can accommodate three passengers. The *Archimede* is designed to go to depths of 11 000 m. The French have been particularly active in the field of deep sea exploration. For shallow water the development of the aqualung has greatly facilitated research into marine life in this important zone. The deeper submersibles are useful in pursuing the objects of the International Geodynamics Project, initiated by ICSU with 50 nations participating. The object is to obtain more information concerning the plate tectonics model, and to locate resources associated with it. As well as these large-scale international projects there are many less ambitious but significant contributions made in a wide range of oceanographic research. The use of satellite data to study wave generation and propagation in relation to meteorological phenomena is one field of study of great potential value. Work on tidal prediction and surges, and tidal morphology in shallow water, also provides useful data. Wave and current recording instruments have been developed for use in ships, while the FLIP ship was designed to carry out a wide range of recording with different instruments. The ship sails to the desired position at which point it turns through 90 degrees to become a stationary observation platform, which is 106 m long, extending well below the normal wave base. Research into the living resources of the sea is carried out by marine biologists, using a great range of techniques to cover the wide variety of marine life.

(c)   Exploitation

(i)   Food resources

The oceans cover a very large proportion of the earth's surface, yet they produce only a very small proportion of the food eaten. In 1970 about 2892 million tons of food were produced on land, but the harvest from the sea was only about 60 million tons, and of this only about 36 million tons was used for human food, the rest being used to make fish meal for animal consumption. Thus the oceans only provide about 1 per cent of all food consumed, but their share of animal protein is higher, being 5 per cent directly and 5 per cent via the fishmeal fed to livestock. As on land, marine productivity is very variable in the oceans, but a few systems are as productive as good land, for example the Georges Bank in the northwest Atlantic. A wide range of living resources are harvested from the sea, including fish, molluscs, crustaceans and marine plants. In 1970 the total fish catch was nearly 91 per cent of the total of 62.2 million tons. Marine molluscs accounted for nearly 8 per cent, and marine plants 1.4 per cent, while the freshwater catch was about 7 million tons. Although fishery dynamics is a highly sophisticated and mathematical study, the actual fishing process is still at the hunting stage, the method depending on the type of fish. The fish consist of four main groups, first the pelagic planktivores, which include herring, sardines, pilchards, anchovies and anchovetas. The second group, the pelagic carnivores, include tuna and mackerel, and the third group, the demersal carnivores, include cod, Alaska pollack, flounder, haddock, sole, halibut and hake. The fourth group are diadromous fish, which migrate between salt and fresh water, and include salmon, capelin and eels. The first group is composed of the schooling fish which have a short food chain, as they live directly on zooplankton. Herring, for example, feed mainly on calanus, and their life style depends on the behaviour of their prey. The calanus rise upwards at night when they are preyed on by the herring shoals. They can, therefore, be caught in drift nets, shot at night. Echo sounders now help to locate the shoals. The most important fish in this group are the anchovies and anchoveta that make up the very rich fishery of the upwelling waters off Peru. This fishery rose very rapidly to a climax in 1970, when nearly 12.3 million tons were caught. It was at that time the largest single fishery in the world. In 1969 Peru produced 45 per cent of the world supply of fish meal. Before the fishery became so large the fish were preyed on by huge flocks of pelicans, cormorants and gannets. Up to 40 million birds consumed about 1000 or more tons each day of anchovies. By 1972 the anchovy catch by fishermen was down to 4.5 million tons. It is estimated that the total bulk of fish is 15 to 20 million tons at the height of the annual cycle. The sudden collapse of the Peruvian fishery in the early 1970s was partly due to overfishing, but it was also seriously affected by the El Nino, which prevents the upwelling of cold fertile water, and has many other effects to be mentioned later. Overfishing was at least partly responsible for the fall in catch of anchovies, and it is also a problem in several of the next most important groups of fish to be exploited. These are the demersal carnivores. They are bottom-living fish, and therefore, have to be caught by trawling, dragging a net across the bottom. This method of fishing could not be carried out very effectively until power was available for fishing boats. Trawling started in the southern North Sea, and has for long been important all around the North Atlantic, on the wide continental shelves characteristic of this expanding ocean. Bretons, Normans and Basques have a long tradition of fishing. The

Newfoundland Banks were exploited from the seventeenth century by fishermen from northeast North America. The main fish in this group are cod, haddock and hake. Cod are also important round Iceland, and there have been many problems in the attempts that have been made to regulate these fisheries. One way in which fishing catches are regulated is by controlling the mesh size. A larger mesh has the advantage of catching only the more mature fish, while most of the young can escape to breed. These fish produce very large numbers of eggs, but these are very vulnerable to predation and unsuitable conditions when they hatch and join the plankton. Fish have very variable year broods, which make fishing control difficult.

One demersal fish stock which has been successfully managed is the halibut of the northeast Pacific. The waters are fished jointly by the United States and Canada, who have agreed a suitable catch since 1923 to provide the maximum sustainable yield. The fishery started in 1890, and owing to a small market did not suffer over-exploitation for 35 years, but increasing fishing led to the necessity for control. Intensive investigation of the biology of the fish stock initiated satisfactory control. Catches increased to 1960, with larger fish gradually becoming more numerous in the catch. This indicates that the yield is sustainable, as nursery grounds were protected, allowing the fish to mature, and fishing methods were also controlled. Hook and line produced larger fish than trawling, which was not allowed. This example shows that sensible control of fishing can reduce fishing effort, as well as maintaining a sustainable yield of larger fish. Only two governments were involved in this instance, while in many areas the human situation is much more difficult, as indicated by the great difficulty the EEC has experienced in reaching a satisfactory control of fishing in the North Sea and around European coasts. The cod, plaice and herring have all suffered from overfishing in this fertile, but heavily fished, area, which is so close to density populated countries with all the latest fishing techniques available for the exploitation.

One of the worst examples of overcatching is that of the great whales. The control of their catching is more difficult in some respects than fish. One important difference is that fish produce a very large number of eggs. A mature cod can produce up to a million eggs at each annual spawning. On the other hand, blue whales are monogamous and only breed once in several years, and then only produce one calf. The whalebone whales, or rorquals, also migrate great distances, going to warm waters to breed. They feed during the southern summer on the rich krill of the Southern ocean. The first whales to be exploited were the sluggish north Atlantic Right whales (*Balaena glacialis*). Whaling was mentioned as early as A.D. 890 in a report to King Alfred, and Norsemen and Basques also hunted whales in the tenth and eleventh centuries. The development of the explosive harpoon in 1865 made the larger, more active rorquals and sperm whales vulnerable to hunting. After the northern rorquals had been hunted to near-extinction, the attention of the whalers turned to the southern hemisphere.

The first southern whaling station was set up on South Georgia in 1904. By 1911 there were eight whaling stations and many whalers were operating. The whalebone whales include the largest of all animals the blue whale, which can reach 30 m length and weigh 120 tons; next in size is the fin whale, which is about 26 m. The smaller whales of this group are the sei and humpback, both about 15 m long. The sperm whale (*Physeter catodon*) is a toothed whale. It is about 18 m long and prefers large squids to eat. The sperm oil is a liquid wax which allows the whale to sound rapidly and deeply after its prey.

Various attempts have been made since the 1930s to regulate whaling because already by this time scientific advice indicated that too many whales were being caught. The first whaling convention was signed in 1931, and another in 1946, when the blue whale unit (BWU) was introduced (1 blue = 2 fin = 2.5 humpback and 6 sei), and restrictions on size, season and area were imposed. By 1946–47 the agreed catch was 16 000 BWU. By 1960, however, blue whales had almost entirely disappeared, and they have been completely protected since 1965, largely because there were too few to be economically exploitable. The same fate has befallen the smaller whales in turn, with catches rising to a maximum, and then falling off rapidly, as the whales became almost extinct, and not worth hunting. Sperm whales are still being hunted, and there are still some sei whales about. The main whaling nations are now the USSR and Japan. It is very shortsighted to continue whaling, and it is to be hoped that whaling will be completely stopped, to give any surviving whales a chance to reproduce, and build up a new stock. This would take at least 50 years, and probably longer, but eventually a substantial sustainable yield could be obtained if the stocks are not too small to rebuild.

Seaweeds are amongst the most easily exploited living matter from the oceans. They are attached to the bottom close to the shore in territorial waters, and can, therefore, be relatively easily harvested. Seaweeds contain iron, iodine and vitamins. They have long been used as fertilizers on the rocky coasts of western Ireland and Scotland. One of the most useful elements is alginic acid, and about 15 000 tons year^{-1} are harvested, about half of it coming from the United States. The rich kelp beds of the Pacific coast could produce 3 million tons year^{-1}. The density of the standing crop is about 10 to 16 tons ha^{-1}. Nova Scotia is also rich in kelp beds, with a standing crop of over 1 million tons on 520 km coastline. In Scotland there are over 1 million tons on 865 km of coast. Norway also produces a great deal of seaweed, but by far the largest amount comes from Japan. One of the problems of harvesting seaweeds is the storminess of the exposed, rocky coasts on which they grow. This makes mechanical harvesting difficult, and much is collected as it is washed up on the shore. More could be collected if suitable methods of gathering it were available. There is considerable potential for further development of the seaweed industry.

Sea farming would produce considerable benefits if it were further developed. It is estimated that the present fish catch could probably be doubled without depleting stocks, if controls were enforced for optimum catches, allowing conservation of stocks. The advantages of fish farming are (1) exclusion of predators, (2) population can be manipulated to optimum level, (3) undesirable animals can be excluded, and (4) fertility can be increased with manuring, for example, sewage could be deposited in barren areas of sea, or nutrient-rich lower water could be pumped to the surface. Experiments have been carried out in Scottish lochs to add fertilizer to the water. Species that are amenable to farming are the sessile organisms, such as oysters and mussels, and these are successfully farmed by Japan especially, and elsewhere. Mussels can produce $300\,000\,\mathrm{kg\,ha^{-1}\,year^{-1}}$. In southeast Asia, scrap fish are used to make a high protein edible paste, and another method of increasing the nutrient derived from the sea is to use lower trophic levels. The vast amount of krill no longer consumed by the whales in the Southern ocean is a rich potential if it could be economically collected and made palatable for human consumption. But how much better it would have been to allow the whales to do the concentration for us. However, in many ways the food harvest of the sea could be considerably increased to provide protein so necessary in many developing countries.

(ii)   Resources of minerals and water

Practically every mineral and element exists in sea water, but only some of them can be considered as resources as many are not extractable economically. For example the *German Meteor* expedition of 1925–27 tried to extract gold from the ocean. It is, however, present in such small quantities that this is not possible. One cubic mile of sea water contains 17 kg gold. It can be recovered from placer deposits offshore, off South Africa and New Zealand, for instance. A resource is defined as a concentration of naturally occurring solid, liquid or gaseous material in such a form that economic extraction is feasible. As far as the oceans are concerned resources can be on the ocean floor, beneath the sea floor or in the water. The water itself is also a resource of considerable importance in some areas.

Resources on the ocean floor include low value, high bulk commodities, such as sand and gravel. These construction materials are becoming difficult and expensive to extract and transport in some areas; in these places marine sources are worth exploiting. Over 50 million cubic yards of sand and gravel are taken from the continental shelves each year. The Netherlands, for example, obtains 90 per cent of its needs from the North Sea, in which ice sheets left large deposits of sands and gravels. Sand from the sea bed can often provide the best means of sea defence for beach nourishment and replenishment. It must, however, be taken from beyond the depth of wave disturbance. Old shell reefs provide a valuable source of lime for many areas, such as those on the Gulf coast of the United States and off Iceland. Bulk transport of these bulky materials by sea is often the cheapest means of transport. In some areas heavy minerals have been concentrated by currents to provide valuable placer deposits, especially in shallow water where old beaches and river channels have been submerged by the Flandrian transgression. Gold, diamonds, iron and titanium have been mined from these deposits and tin is economically significant in channels off Thailand and Indonesia. Metals are also important in some deep sea sediments. The deep sea basins of the Red Sea, which is a young developing rift system, are rich in metals. The top 10 m contain over a million tons of copper, as well as zinc, silver, gold, lead and iron, with traces of cobalt, nickel, arsenic and tin. The sediments can be pumped to the surface. They provide a potentially rich metal harvest if the metallurgy can be developed to exploit them economically. The hydrothermal activity of this active splitting zone may also be repeated in other similar situations along the major oceanic ridges. The deep sea drilling project has revealed iron-oxide dominated, metal-rich sediment layers above the basalt ocean crust. There are also more discrete metal resources on the ocean floor in the form of nodules, which are widespread and especially common in the Pacific. Amongst the most important resources of minerals in the sea are the ferromanganese nodules that have been found to be very widespread, particularly on the floor of the Pacific. They were first dredged up by *Challenger*. There are problems of gathering the nodules in the deep ocean basins, and also difficulties concerning their ownership, as they are outside territorial waters. Various harvesting devices have been suggested, including undersea tractors with suction dredges. Nodules frequently occur in depths of about 4000 m, although they are in shallower water on the Blake Plateau off southeast United States. The total tonnage of nodules is estimated at about 10^{11} to 10^{12} tons. It is likely that they accrete at a rate greater than 6 million tons year^{-1}. The reserves available include about 2200 million tons of manganese, 98 million tons of nickel, 80 million of copper and 20 million of cobalt, assuming the average nodule is 3.7 cm diameter and the average abundance is 19 kg m^{-2} with an average recovery rate of 20 per cent. One problem in the use of the nodules is the heavy requirement of water, energy and chemicals in the extraction of the valuable minerals. Another useful mineral obtainable from the ocean floor in nodular form is phosphorite, which mainly occurs close to the continental margin. This is another potential resource, although some has already been recovered. The nodules are of organic origin and provide valuable material for fertilizers, chemicals and detergents.

Resources from beneath the sea floor include sulphur, which occurs in association with salt domes. These occur where evaporation of sea water has caused salt deposits that have been subsequently buried under sediments and then migrated upwards into a dome because of its low density. The most important undersea resources, however, are hydrocarbons, oil and gas, which are now being actively sought and exploited in many areas. In 1975 about 20 per cent of world production came from offshore sources. Exploration drills have been made in depths exceeding 650 m. Twenty eight countries now obtain oil and gas from offshore, and Britain, producing from the North Sea, is now fifth largest producer in the world. Since 1946 10 000 wells have been drilled off the United States, and more than 6.5 million acres of the outer continental shelf had been leased by 1969, although only 10 per cent had been surveyed in detail. Important recent finds have been made off north Alaska, Mexico, Trinidad, Brazil, Dahomey, Australia, Taiwan and the North Sea. There may also be deeper supplies on the continental rise in depths of 1525–5500 m. Drilling has already been carried out to 3570 m in the Gulf of Mexico. Floating platforms will be needed to exploit these deeper resources, which will be necessary eventually as the average annual rate of oil and gas accumulation is rated at 1×10^{10} kg cal, or 1×10^{-11} of the influx of energy from the sun. The rate of extraction is 3×10^{16} kg cal or 3 million times the accumulation rate. Resources of oil to be economically extractable require several conditions to be fulfilled, namely (1) there must be a rich source of organic material which must be preserved until it is buried by sediments; (2) temperature and pressure conditions must be right to produce the various types of oil and gas; a minimum of 1000 m sediments is needed; (3) the gaseous or liquid hydrocarbons must migrate from the fine-grained sediments to coarser ones from which they can be extracted; (4) a trap for the hydrocarbons must exist with an impermeable cover; (5) these conditions must apply in the correct order. Salt domes, faults and folds can all provide suitable reservoir conditions.

Resources from sea water itself are also important. Associated with the Red Sea and other ocean rifts there are hydrothermal brines, which are potential sources of many minerals. The Red Sea brines are up to 62°C and only remain at the bottom on account of their very high salinities, reaching 270‰, with a density of 1.18 to 1.23 g cm^{-3}. They contain much more salt than normal sea water, as well as zinc, copper, lead and nickel, together with calcium, magnesium, iron and manganese. The problem the brines pose is how to retrieve them from the sea bed, and then how to extract the required minerals. Magnesium is already extracted from sea water at Freeport, Texas, by a complex process that requires salt, sulpher, oyster shell, calcium carbonate and energy. The products include magnesium metal and salts, chlorine and bromine. 65 per cent of world needs of magnesium are extracted from the sea.

The most valuable material to be obtained from the sea, and the earliest to be extracted, is common salt (NaCl). It can be obtained by simple evaporation of sea water in warm, shallow ponds. San Francisco Bay has a complex series of evaporation ponds forming a sequence in which different salts can be obtained. The first to precipitate is calcium carbonate, followed by gypsum, anhydrite, magnesium salt, and finally sodium chloride, common salt, which is 99.6 per cent pure.

The oceans also contain an inexhaustible supply of an essential element of all life, pure fresh water. This can be obtained by various processes of desalination, by condensation or freezing for example. Ships have for a long time practised desalination. One of the problems of desalination is the energy needed to heat or freeze the sea water. At Kuwait where there is an abundant supply of surplus gas and a shortage of fresh water, desalination is economically practicable. Many of the Middle East dry areas use desalination to supplement their fresh water supply, including Israel. An ingenious scheme has been suggested to supply water to islands in the West Indies trade wind belt. For the Virgin Islands it has been suggested that the moist trade winds are used to pump cold, fertile water from a depth of about 1000 m, with a temperature of about 5°C to a condenser, where fresh water is extracted from the moist warm air. The cooled dry air can be used for air conditioning. The fresh water is led off to a covered reservoir. Some of the water would be allowed to generate power through a turbine. The remainder of the now warmer sea water, after passing through the turbine, would be fed into coastal lagoons, where its nutrients, derived from below the photic zone, would fertilize the water for fish farming. Only the last stage has so far been installed, but the system is ingenious, and would provide several benefits.

Another ingenious method of supplying fresh water to dry areas without adequate land sources, such as the Middle East, Australia and California, has been suggested. This is to tow large icebergs, which are formed of fresh water, from the Antarctic to the localities concerned. The large tabular icebergs are formed mainly of snow that has fallen on the ice shelves around the Antarctic continent, and are therefore, pure and fresh. When they calve off and drift north they could be towed to their destination. Several high-powered international conferences have been convened to discuss this possibility, and to assess its economic feasibility. The main problem is the expense involved, and the difficulty of containing the fresh water as the ice gradually melts in the warmer waters through which it is towed. The water would only benefit coastal people without costly pumping operations. Many of the processes discussed in this section require a great deal of energy, so that it is worth considering the oceans as a source of energy.

(iii) Power from the oceans

Two main methods of power production have been suggested that make use of the energy in waves and in the tide. One of the problems of wave energy is that the largest waves with the highest energy, as far as Britain is concerned, are on the exposed west coast which has the lowest demand and the least population. Long transmission distances would be required with consequent losses *en route*. A feasible method of extracting energy from the waves has been tested experimentally with reasonable success, but the basic problem still remains.

Tidal power has been used in a small way at suitable coastal sites for a long time. A tide mill was built at Woodbridge in Suffolk in 1170, and there were tide mills in Devon and Cornwall. There are plans to utilize tidal power to generate electricity on the north coast of the USSR. One of the first successful tidal power schemes was built in the Rance estuary of northern France recently. The tidal range at the site is 13.5 m at spring tide and 11.4 m at mean tide. The volume of water flowing through the barrage is 18 000 m³ at flood and ebb with a basin capacity of 184 million m³. The barrage is 15 m above mean sea level, with the thrust in alternating directions, and output is 500 million kWh. The possible sites for tidal power stations are limited to areas with a macrotidal range. These include the White Sea of northwest USSR and a few areas around Britain, of which the Severn estuary is the most suitable. A scheme has already been considered for the Severn and it could be viable. Other macrotidal areas include northwest and northeast India, a few isolated parts of southeast Asia, northwest Australia, the extreme southern tip of South America, part of the coast of Brazil, the Bay of Fundy, south Baffin Island and a few places on the west coast of North America. Most of these areas are far from centres of population, and therefore, unlikely to be developed. Wave and tidal power could, however, provide permanent renewable sources of energy, which may eventually become attractive economically.

(d)   Strategy

The oceans have played an important part in strategy for a long time. Britain has been saved from invasion by the narrow surrounding seas since the Norman and Scandinavian invasions a long time ago. The Spanish Armada in 1588 demonstrated clearly the value of the sea to Britain. In more recent times invasion from the continent was prevented, although air power was as vital as sea power by this time in the Second World War. The complex planning of the invasion on the Normandy coast from England in this war illustrates the value of the sea for defence. Naval strategy has changed a great deal over the centuries as ship-building and armaments have progressed. The first large naval ships were those of the classical Mediterranean, with their many oarsmen. The well-built, seaworthy ships of the Viking adventurers, which carried them safely all around the North Atlantic were of advanced construction. They controlled the sea by virtue of their fine sailing ships around A.D. 1000. The navies of Britain, Spain, Portugal and the Netherlands and other European seafaring countries vied with each other to control the sea during the period of ocean exploration and colonization. Rich treasure ships from the newly discovered lands of Latin America on their way to Spain and Portugal were intercepted by the British and French ships, some of whom were operating on the fringe between naval action and piracy. Naval strategy changed radically as steamships and armaments developed. The great naval commanders, such as Drake and Nelson, in the days of sail, had cannon, but were very dependent on the wind and weather. By the time of the First World War in 1914–1918 large battleships were more or less immune to the wind and weather, but they were very vulnerable to a new marine development, the submarine. It was the U-boats that caused a danger of starvation to Britain in both world wars, when the vital shipping was constantly threatened. A new dimension has now been added to naval strategy in the form of aircraft and their carriers, while helicopters can also play a vital role, as was seen in the recent Falklands conflict, which again demonstrated the importance of naval strategy. The deployment of submarines with nuclear power and atomic warheads or ballistic missiles has again changed naval strategy. Fleets of these ships are controlled by the United States and the USSR, and they play a very significant part in the balance of power on which the peace of the world is hoped to rest. Their deterrent power lies in their manoeuvrability and the difficulty of keeping in touch with them even with modern sonar and radar sensors, a problem which does not apply to land-based missiles.

(e) Recreation

The oceans provide for recreation of various forms; most of this takes place around the coasts in the form of seaside holidays, with the beach being the main attraction. Increasing numbers of people, however, are becoming interested in aqualung diving for pleasure, treasure or archaeology or to study the fauna and flora. Other people enjoy surfing and yachting. The latter pastime covers a whole range from local trips in small boats to trans-Atlantic races, either single-handed or with a crew. At the far extreme are the few lone sailors who intrepidly sail alone all around the globe, and who brave the roughest seas on earth around Cape Horn and in the Southern ocean. Less hardy people can still enjoy the ocean, and are doing so in increasing numbers and comfort by going on cruise liners for trips to the Mediterranean, West Indies or Norwegian fjords. The latest cruise liner launched in 1984 is expected to cruise mainly off the west coast of America from California. It is an extremely luxurious, large ship with every modern convenience. Relatively few people now travel by sea merely to get to their destination, but nevertheless shipping is still vital in other respects.

(f) Transport

Movement of goods by sea is becoming more important as the developing countries expand their trade. 600 million tons of cargo were carried in 1950, and the United States Department of Transportation estimates 20 times this volume by the year 2000, although over the past few years recession has halted the growth in total deadweight tons (dwt). Air transport only accounted for less than 0.5 per cent of total ocean tonnage in 1970, because air transport is 75 times more expensive than sea transport. Crude oil constitutes much of the volume carried. Bulk carriers of ever-increasing size have been specially built for this trade. In 1949 the average tanker was 12 800 dwt; the average size of tanker was 27 000 dwt in 1965 and 39 000 dwt in 1970, with the largest tankers now being about 300 000 dwt. Dry bulk carriers are also growing in size and 16 per cent of them in the late 1970s were more than 80 000 dwt. The large increase in ship size causes problems of harbour maintenance and the possibility of serious damage and pollution if they run aground. Dredging of harbour approaches used to be to 6 m 100 years ago; now 9 or 12 m are required or more. The problem then arises of what to do with the dredge spoils. Some can be used for land reclamation, but there are problems if the waste is dumped at sea in relation to fishing, and other ecological repercussions may ensue. The size of ships must also be related to canal size if they are to trade through the Suez or Panama canals, or use the St Lawrence Seaway, which can take ships up to 20 000 dwt. A ship of 60 000 dwt can use the Panama canal. Cargo ships are designed either to carry bulk liquids or dry cargo, while others are designed for both purposes. This reduces journeys made in ballast. Another development in the late 1960s has been the building of special container carriers, which save much handling of cargo. This has reduced dock labour forces. Some ships now carry both general cargo and containers. The great cost of these large new ships has resulted in the formation of consortia. Ships registered with OECD countries are declining, being 44 per cent in 1983, while open registry under flags of convenience are increasing to 27 per cent, as are those of the market economy developing countries, with 17 per cent, while the eastern bloc contributes about 8 per cent and is static. The remainder of the world has only about 4 per cent. Safety regulations and wages and taxes are often lower under flags of convenience, which partly accounts for the increase in this group, thus giving them an advantage. The United Kingdom deadweight tonnage declined by a half between 1975 and 1982, while the fleets of Saudi Arabia, Hong Kong, Malaysia, Malta and Venezuela grew larger. Some of the world tonnage is at present laid up, amounting to 11 per cent at the end of 1982, or 74 million dwt, mainly due to recession and reduction in oil shipments; it is likely that supertankers will gradually disappear. In order of size of fleets in deadweight tonnage the largest is Liberia, mainly consisting of flags of convenience ships, totalling 123 million dwt; next comes Greece with 65, Japan 64, Norway 32, United States 29, USSR 27, United Kingdom 26, Italy 15 and China 13 million dwt. The busiest routes are from western Europe to many destinations, especially round Africa to the Middle East, and through the Suez Canal to southeast Asia and Japan. Australia also exports a lot to Japan and many other routes also carry freight around the world. The large volume of traffic, especially oil tankers, is one source of a major problem of the ocean environment, that of pollution.

6.3 CONCLUSIONS—PROBLEMS AND OPPORTUNITIES OF THE OCEANS

Pollution is one of the major problems of the oceans, because the ocean is the ultimate sink to which most waste from land finds its way. It can take many forms, one of the most obvious being oil pollution. This was exemplified by the *Torrey Canyon* disaster, when this large tanker ran aground in 1967 on the Sevenstone Reef off Cornwall. Another example was the blow-out off Santa Barbara in California 1969. At least some useful lessons have been learnt from these disasters. In the case of the *Torrey Canyon* the detergents used to disperse the oil caused more damage than the oil itself, which given time is biodegradable. Sinking the oil by using chalk may help the surface fauna and flora, but is adverse to benthic life. About 1 million tons of oil are spilt into the ocean each year. Various attempts have been made to reduce oil pollution by international agreement. New techniques of cleaning ship's tanks should help to reduce the problems, but accidents will no doubt continue to pose a threat from this source of pollution. A total of 13 million tons of evaporated petrol, waste oil and other solvents are probably added to the ocean each year from all sources. One of the main questions concerning oil pollution is whether the oceans will be able to keep pace with the added oil through biogradation by means of bacteria. Alternatively it is possible that eventually the bacteria will be killed, and the oceans will become dead as have some more restricted water bodies, such as lakes and stretches of rivers.

Other pollutants, such as sewage and fertilizers lead to excessive growth of phytoplankton. This causes bacteria to deplete the oxygen in the water as they break down the dead phytoplankton, and in turn other organisms are killed. Some blooms of phytoplankton produce toxic substances that kill other fauna. Overfertilization can produce red tides. Another serious source of pollution is chlorinated hydrocarbons and pesticides, such as DDT. The main danger of these substances is that they become concentrated at each level of the food chain. In estuary water there may be 0.00005 ppm DDT, the zooplankton has 0.04 ppm, shrimps 0.16 ppm, minnows 0.50 ppm, needlefish 2.00 ppm, terns 2.8 to 5.17 ppm, cormorant 26.4 ppm and an immature gull 75.5 ppm. The latter value can be lethal, as it results in soft-shelled eggs that cannot hatch. Ospreys have suffered severely in this way. The use of DDT is now restricted so this problem may be reduced in time. Heavy minerals also tend to become concentrated up the food chain. They include mercury, which is very toxic, lead and copper also cause problems. The Baltic has suffered from mercury poisoning, and it led to many deaths in Japan after the consumption of polluted shellfish. Radioactive substances and plastics can also cause pollution owing to their longevity and indestructibility. Pollution occurs in many ways of which oil pollution is one of the most obvious and common. Other problems arise through deliberate, accidental or natural interference with the delicately balanced ecosystems in the sea. One of these is the tropical coral reef.

In the 1970s the Pacific reefs, including the Great Barrier Reef of Australia, were threatened by an explosive growth of the Crown of Thorns starfish, *Acanthaster planci*. These 16-armed starfish prey on the living coral. The cause of the population explosion is not known, but it might have been due to mortality among the planktonic predators of the Crown of Thorns larvae as a result of organochlorine pollutants. The starfish have destroyed large areas of coral including around Guam and the Great Barrier Reef. The depredations have been widespread in the Indo-Pacific area. In 1963 the Great Barrier Reef near Cairns was affected. Another possible cause of the starfish outburst has been suggested to be the depletion of the large triton shell, *Charonia tritonis*, as this species preys on *Acanthaster planci*. The large shell is in great demand by shell collectors and tourists. Coral blasting for lime is another possible cause, as it kills the filter feeder predators of the shellfish larvae. The population explosion may also be a natural phenomenon which will eventually be controlled through the ecosystem interactions. One result of the coral problem was the realization of the lack of knowledge of the details of the operation of the complex coral reef ecosystem, and work to remedy this is being actively undertaken. Conservation of the unique qualities of the Great Barrier Reef and other coral area is now becoming recognized as essential, if permanent damage is not to be done by exploitation, including drilling for oil, blasting coral for lime and shell sand dredging for the same purpose. Tourists who come to enjoy the beauty of the living reefs pose problems of their own, by disturbance of the environments with motor boats, fishing and shell collecting, and other results of too many people in a delicate ecosystem that is easily disturbed. Not all ecological problems are man-made, however, and one good example, which does have very far-reaching human repercussions is the El Nino.

The El Nino occurs in the Humboldt current off Peru. It prevents the upwelling of cold fertile water and raises water temperature. The El Nino occurs somewhere between 2 and 10 years apart. One occurred in 1976–77, but 1982 produced a very strong one which had worldwide repercussions. It caused drought in Australia, Indonesia, central America, south India and southern Africa, while there were fierce storms around Hawaii and Tahiti. Serious flooding occurred on the west coast and Gulf coast of the United States, in Ecuador and north Peru. Marine and bird life off Peru was seriously affected by the warm water. The cause of the El Nino is the intensification of the pressure systems, which piles up warm water to the west as the trade winds strengthen. Then the system collapses and a Kelvin wave moves east, taking 60 days to cross the Pacific. When the trade winds fail they are replaced by west winds, the surface currents reverse and warm water moves towards South America. The reason for these changes is not known, but may well be related to the monsoon system. It does, however, have a worldwide effect. It was one of the major factors in the decline of the Peruvian fishery, as a severe El Nino in 1972 came after several years of overfishing. By 1970 14 million tons of anchovies were caught by 1500 fishing boats off Peru. In 1972, when warm water offshore herded the fish close to the coast, as many as 180 000 tons a day were caught, but neither the fish nor the guano birds survived this slaughter, and the anchovy fishery has not recovered. A ten-year programme of research starting in 1985, called Tropical Ocean Global Atmosphere (TOGA) will involve three satellites, many research vessels and instruments to monitor both ocean and atmosphere. The cause of the El Nino may be found in the western Pacific at the seasonal change of the monsoons. It is a very complex system, which involves the whole ocean–atmosphere interaction. Solving this problem could lead to a greater understanding of global weather and its baffling abnormalities, in which the oceans play a vital part.

The laws of the sea are nearly as complex as the physical interrelationships, and as difficult to formulate successfully. The law of the sea is based on two conflicting principles, the freedom of the high seas, and the concept of territorial waters for coastal states. Much of the law of the sea has developed through years of custom, which has gradually become codified. The ancient people of the Mediterranean had their own laws governing marine matters, such as commerce by sea. These old maritime codes recognized the right to use the sea for commerce, and this became part of Roman law, but was lost in the succeeding 1000 years. The principle had to be rediscovered and imposed in Elizabethan times when great voyages of exploration and trading were being made. The first modern work on the freedom of seas was written by Hugo Grotius in 1609, *Mare Liberum*. The other major strand of maritime law was advocated by John Seddon in 1635, in his *Mare Clausum*, advocating territorial waters for coastal states. Already fishery problems were arising between Britain and the Netherlands. The same problems still remain. Great Britain as the leading maritime state of the nineteenth century advocated freedom of the seas. This involved four freedoms, (1) of navigation, (2) of fishing, (3) of cable laying, and (4) of aerial circulation. Several attempts have been made internationally to resolve the conflict of the two main opposing laws, freedom of the high seas and territorial waters. At the 1958 and 1960 Geneva conferences delegates from 86 countries attended. A new problem was raised in 1967, that of rights over the seabed and its resources. Another conference was called for 1973 to deal with this problem, and to establish an international regime and machinery to implement it. Three seabed subcommittees were set up, the first to deal with the international seabed regime, the second to consider territorial waters, innocent passage, functional zones, continental shelf and high seas, fisheries, straits, etc., and the third dealing with marine pollution and scientific research. The zones for which agreements are sought include territorial sea, functional zones, continental shelf and high seas. The territorial seas claimed by different countries vary greatly, but none are less than 3 nautical miles. Territorial waters are subject to free passage by merchant vessels of all nations. The 3 n. miles arose originally from the range of cannon shot, and was accepted by most maritime nations at the beginning of this century; but after the 1939–45 war many nations increased their claim to 12 n. miles. A meeting in 1958 failed to agree a specific limit, and another conference in 1960 also ended in failure to agree. At the end of 1972 only 25 states had a 3 n. mile limit, 6 opted for

6 n. miles, but 59 claimed 12 n. miles or over, with 7 claiming 200 n. miles. The idea of a functional approach is new and is concerned with the need to protect and control resources. Economic concern became paramount in the 1960s. The Latin American countries in particular were worried about fishery problems. Elsewhere non-living resources were the focus of attention. The developing countries played a significant part in this approach, including the Organization of African Unity. An economic zone of over 200 n. miles was recommended. There is still much to be discussed before agreement on these issues can be achieved. The definition of the continental shelf has also given rise to much argument. The one agreed in 1958 was open-ended, and this led to subsequent arguments. The reference to depth of exploitation of natural resources is not clearly defined, although a depth of 200 m or 12 miles, whichever was furthest, gave more precision. Again many differing claims have been proposed; 200 miles have been proposed, and other claims vary from 12 to 100 miles. The economic zone concept may eventually provide a possible answer to this difficult problem.

The high seas jurisdiction is complementary to that of territorial waters in that it begins where the latter ends. Another dimension to the problem of the high seas and their exploitation was the discovery of the valuable potential resources of the ocean bed minerals. One possible solution is to set up an autonomous licensing agency, or United Nations agency, to allocate leases and permission to exploit in return for royalties to be used for the benefit of all states, but especially the developing nations, including those that are landlocked. The control of fisheries has traditionally been on a regional basis, and rests on the acknowledged needs to conserve stock, and has worked reasonably well. The problems have partly arisen through the development of large, efficient, distant-water fishing fleets, particularly by the USSR, German Democratic Republic, Poland, Korea and Japan. This threat is partly the cause of Latin American countries' claim to 200 n. miles of territorial waters. It also led to problems concerning Iceland's fisheries, on which the country is very dependent.

International cooperation. The challenge of the oceans brings out both the worst and the best in man. The massacre of the blue whales for short-term gain, latterly mainly by the USSR and Japan, is both stupid and self-defeating, as a controlled fishery could have yielded a permanent harvest, and preserved the largest animal ever to inhabit the earth. It also revealed the difficulty of obtaining international cooperation among competing nations.

On the other hand the oceans have fostered the spirit of exploration and achievement, and lured intrepid navigators to discover all parts of the globe. The curiosity of scientists, whetted by the intriguing problems posed by the seas and their abundant life, has been fully stretched by marine research. It is in this field that international cooperation has been most fruitful, and some examples have already been cited. Not all the research work has been for pure academic interest only, but it is more and more directed to providing man with essential food, raw materials and energy. In order to make the most of the resources, however, it is essential to understand the complex oceanic processes and interactions. This information is valuable in a wide range of spheres of knowledge, such as an understanding of the functioning of the atmosphere–ocean system, and its attendant droughts and floods, that cause such great human difficulties. The oceans are essentially one very large and intricate natural ecosystem, with many different subsystems in all of which man can intrude and cause unforeseen or purposeful alterations. It behoves us to study all aspects of the oceans so that our interaction with them can be beneficial to all concerned both at sea and on land.

REFERENCES

Bellamy, D., 1975, *The Life Giving Sea.* Hamish Hamilton, London.

Canby, T. Y., 1984, El Nino's ill wind, *National Geographic*, **165**(2), 144–184.

Chesher, R. H., 1969, 'Destruction of Pacific corals by the sea star, *Acanthaster planci*, *Science*, **165**, 280–282.

Clare, P., 1971, *The Struggle for the Great Barrier Reef.* Collins, London.

Cox, S., 1984, 'Ancient trade in a modern form', *Geographical Magazine*, **Oct.**, 521–527.

Deacon, M., 1971, *Scientists and the Sea*, Academic Press, London.

Drake, C. L., Imbrie, J., Knauss, J. A., and Turekian, K. K., 1978, *Oceanography*, Holt, Rhinehart & Winston, New York.

Gold, E., 1975, 'Law of the sea and ocean resources', in King, 1975, pp. 323–344.

Hardy, A., 1956, *The Open Sea. I. The World of Plankton*, Collins, London.

Hardy, A., 1959, *The Open Sea. II. Fish and fisheries*, Collins, London.

Hickling, C. F., and Brown, P. L., 1973, *The Seas and Oceans in Colour*, Blandford Press, London.

King, C. A. M., 1974, *Introduction to Marine Geology and Geomorphology*, Edward Arnold, London.

King, C. A. M., 1975, *Introduction to Physical and Biological Oceanography*, Edward Arnold, London.

McKee, A., 1967, *Farming the Sea*, Souvenir Press, London.

Russell, R. C. H., and Macmillan, D. H., 1952, *Waves and Tides*, Hutchinson, London.

Shepard, F. P., 1973, *Submarine Morphology*, 3rd edn, Harper & Row, New York.

Sverdrup, H. V., Johnson, M. W., and Fleming, R. H., 1942, *The Oceans*, Prentice-Hall, Englewood Cliffs, N.J.

Thorson, G., 1971, *Life in the Sea*, McGraw-Hill, New York.

Turekian, K. K., 1965, Riley, J. P., and Skirrow, G. (Eds.), *Chemical Oceanography*, Academic Press, New York.

Human Activity and Environmental Processes
Edited by K. J. Gregory and D. E. Walling
© 1987 John Wiley & Sons Ltd.

7

Disturbance of Lacustrine Systems

P. G. SLY

7.1 LAKES IN RELATION TO SURFACE WATER SUPPLY

In a penetrating study of environmental problems, Vallentyne (1972) placed human demands on freshwater in sharp perspective; in particular he defined two principal problems, associated with population growth and technological demand. Vallentyne stated that, although the total global annual precipitation exceeds $1 \times 10^5 \, km^3$, only about one-third of this or $3.8 \times 10^4 \, km^3$ becomes incorporated in surface drainage. These figures suggest that, with a present world population of about 4 billion, per capita surface water supplies amount to about $25 \times 10^3 \, l \, day^{-1}$. Even with world population growth to about 7 billion by A.D. 2000, that amount of water would seem more than sufficient. Vallentyne cites water demand as $2 \, l \, caput^{-1} \, day^{-1}$ for human physiological needs, $250 \, l \, caput^{-1} \, day^{-1}$ for domestic needs, $1500 \, l \, caput^{-1} \, day^{-1}$ for industrial needs, and agricultural needs of several thousand $l \, caput^{-1} \, day^{-1}$ in arid climates. If supplies were freely available to meet present needs they would generally exceed demand by about 3–4-fold. However, much of the world's freshwater supply is not where it is most needed. In Table 7.1 (data derived from Szestay, 1982), the 1970 and projected A.D. 2000 per capita river flows are presented for large river basins in five continents. Using a cut-off demand of about $4 \times 10^3 \, l \, caput^{-1} \, day^{-1}$ and based on these examples, only South America appears to have adequate water supplies for the period to A.D. 2000. Further, in all continents (and especially in the Americas) there are great disparities between the supply and demand situations in different basins. Although recycling of water, the use of groundwater recharge and water transfers can alleviate some problems of supply in more highly developed countries, the economies of other countries and the lack (or irregularity) of rainfall mean that water supply is an extremely acute problem in many parts of the world.

Lakes, of course, are a component of river basins and, although their annual outflow is recorded as part of the surface discharge from precipitation, they contain a major additional supply of water in the form of storage capacity.

TABLE 7.1   Per capita river flow in major river basins of the world (*modified after Szestay (1982); reproduced by permission of* Water Quality Bulletin)

Continent	River basin	Per capita river flow ($l\,day^{-1}$) A.D. 1970	A.D. 2000
Africa	Congo	186.3×10^3	81.0×10^3
	Zambezi	104.5	48.6
	Niger	30.2	12.4
	Senegal	24.6	12.2
	Orange	6.2	2.7
	Nile	4.6	2.2
Asia	Irrawaddy	56.7	27.8
	Brahmaputra	30.2	15.1
	Ob-Irtysh	30.0	23.8
	MeKong	20.3	8.9
	Yangtze	9.2	6.2
	Indus	6.6	2.8
	Ganges	5.2	2.6
	Tigris–Euphrates	4.8	1.8
	Hwang-Ho	2.5	1.7
Europe	Rhone	20.0	16.3
	Po	8.7	7.6
	Danube	6.9	6.1
	Rhine	4.6	4.3
	Vistula	4.6	3.8
N. America	Yukon	4266.6	2835.0
	Mississippi	25.9	20.3
	Colorado	17.8	13.7
	Rio Grande	2.0	1.6
S. America	Amazon	4374.0	1998.0
	Tocantins	391.0	175.5
	Orinoco	313.0	143.0
	Magadalena	34.6	14.9
	San Francisco	18.9	8.4
	Parana	18.4	11.3

On a volumetric basis Lake Baikal (USSR), with a volume of 23×10^3 km^3 and the North American Great Lakes, with a volume of 25×10^3 km^3, contain about 40 per cent of the world's surface freshwater supply (Wetzel, 1975). At about 120×10^3 km^3, the storage capacity of all the world's lakes represents about 3 years' surface drainage to the oceans. Lakes, therefore, in addition to their particular value as a special form of habitat, represent an extremely important resource of freshwater which will be used to augment both present and future human needs.

Lakes have played an important role in human development from earliest times, but it is only in the past 50–100 years that the combined effects of population growth and technological development have created overwhelming changes. Because of differences in the physical and chemical characteristics of lakes, and changes in the nature of human activities, the disturbance of natural lake systems has taken many forms.

7.2 GENERAL CHARACTERISTICS OF LAKES

(a) Origins and forms

Lakes have been categorized by origin (Hutchinson, 1957) and their sizes, depths and shapes are often closely related to common geological controls. Most very large inland waters (Caspian Sea, Black Sea, Lake Baikal in USSR; Lakes Victoria and Tanganyika, in Africa) have formed in regional tectonic traps or depressions caused by crustal uplift or downwarping. Waters held in areas of downwarping tend to be somewhat circular in form, whereas others are more linear and of great depth, (e.g. Victoria 79 m, Black Sea 2240 m, Baikal 1610 m, Tanganyika 1470 m, Caspian Sea 945 m). In the northern hemisphere, where the effects of Pleistocene glaciation were most intense, glacial scour has produced a number of long and deep 'fjord lakes' which have exploited previous drainage or lines of structural weakness or lithological difference (e.g. Lago Maggiore, Lac Como, and Lac Constance in Europe, lakes of the English Lake District). In North America, the edge of the Precambrian Canadian Shield is characterized by a series of glacially scoured lake basins (Great Slave, Winnipeg, and the Laurentian Great Lakes) which extend almost from the Arctic Ocean to the St Lawrence River. Small lakes are more numerous than large lakes and are of more diverse origins.

(b) Physical characteristics

The size, depth and shape of lakes is very important in determining their response to various forms of physical input (Fig. 7.1). Very large lakes respond to wind, atmospheric heating, barometric pressure and tidal forces and, to a lesser extent, river inflow (Sly, 1978). As lakes become smaller, the effects of wind stress are less important and differential barometric pressures and tidal forces become insignificant; the influence of atmospheric heating and river inflow, however, increase relatively to dominate very small lake systems. In Fig. 7.2 (based largely on Canadian data), an estimate of the relative numbers of large and small lakes is provided and this, in turn, is related to the effective depth of wave disturbance of the lake bed. According to these data, lake size has little significant effect on the maximum depth of bottom disturbance by waves, for lakes less than about 1 km² (100 ha) in area, and such lakes typically represent more than 95 per cent of all lake basins. However, in the remaining large lakes, the shape of the basin (or component sub-basins) is of additional and particular importance, since it may considerably modify the length of fetch. Further,

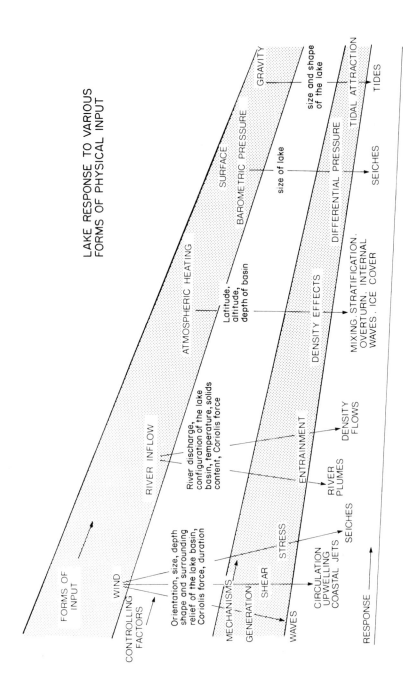

FIGURE 7.1   Lake response to various forms of physical input (*after Sly (1978); reproduced by permission of Springer-Verlag*)

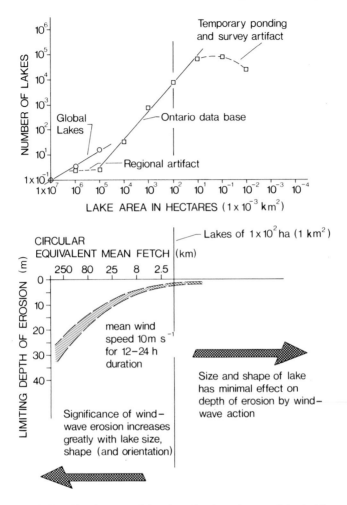

FIGURE 7.2   Relationships between lake size : depth and potential wind/wave erosion
of the lake bed

although there is a generally linear relationship between mean depth and basin
size, there can be major morphometric and shoreline differences between basins.
Thus, the effects of wave motion on bottom sediments will differ markedly
between lakes of comparable mean depth but different morphology.

   In most lakes, temperature structure is the single most important control;
not only upon the physical processes, but also upon associated chemical and
biological processes (Wetzel, 1975). The density of freshwater varies with
temperature, density being greatest at about 4°C. As a result of summer heating,
most lakes develop thermal stratification with warm surface waters (epilimnion)

separated from cold deep waters (hypolimnion) by a sharp temperature gradient (thermocline). Mass vertical circulation (mixing) takes place before lakes stratify or after they de-stratify. In very cold areas, which are subject to continual ice-cover, amictic lakes (Hutchinson and Löffler, 1956) have a stable temperature regime with coldest waters near the surface and there is no mixing. Cold (monomictic) lakes never exceed 4°C but mix once a year, during summer. In temperate regions, dimictic lakes are characterized by two periods of thermal mixing (spring and fall). Warm (monomictic) lakes are characterized by

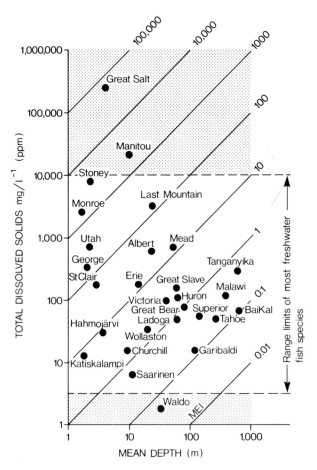

FIGURE 7.3   Range of total dissolved solids, mean depth and the morphoedaphic index (*after Ryder* et al. *(1974); reproduced by permission of the* Journal of the Fisheries Research Board of Canada)

temperatures which remain above 4°C and in very warm climates, where water temperatures are much above 4°C (oligomictic lakes), thermal mixing is occasional and irregular. Lakes that mix frequently, because of large daily temperature ranges or because of frequent wind stress, are termed polymictic. The replacement of nutrients used in biological processes is ensured in lakes which mix; in lakes which do not mix, chemical gradients and stagnation inhibit most forms of biota from all but near-surface waters.

Lakes are also subject to flushing; the displacement of lake waters (as outflow) by input contributions. Rates of lacustrine flushing may range from several times a year to once a decade or longer, whereas rates for riverine systems typically range from hours to a few days. If the densities of input waters are equivalent, the water masses will mix. If densities are different (owing to temperature, suspended solid content, etc.), input waters can pass through a lake with little or no mixing (entrained flow).

(c)   Chemical and biological characteristics

Lake water chemistry largely reflects the geology of the watershed (Golterman *et al.*, 1983) and, as an indication of the range of concentration, Fig. 7.3 shows

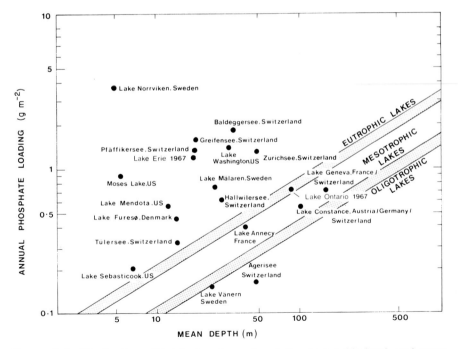

FIGURE 7.4   The loading of total phosphorus in relation to trophic level, and mean depth (*modified after Vollenweider (1968); reproduced by permission of the author*)

the relationship between total dissolved solids content (TDS) and mean depth (known as the morphoedaphic index for a representative sample of lakes). The limits for most freshwater fish species (associated with osmoregulation by salts and low level ionic in-balance) proposed by Ryder *et al.* (1974), are shown on this figure. Biological production in lakes is closely related to TDS content and to the concentration of major nutrients, particularly phosphorus. In Fig. 7.4, (modified after Vollenweider, 1968) mean total P concentrations, as indexed by trophic state, are related to the relationship between input load and mean depth for a sample of lakes. Although significant changes in TDS resulting from cultural activities in watersheds have been reported (Beeton, 1969), changes in loadings of total P result in a much more dramatic response. Increased P usually results in both a species shift (often to less desirable forms) and higher productivity. In extreme situations, P may be in such plentiful supply that it is replaced by nitrogen as a limiting nutrient (lakes characterized by N-fixing blue-green algae). The response to similar nutrient loads can differ markedly from lake to lake.

7.3   EVIDENCE OF CHANGE

(a)   Deep and shallow basin examples of geological timescales

Geologically speaking, lakes are transient features; although large deep basins have considerable longevity. The Black Sea, for example, probably formed as a marginal basin adjacent to a Cretaceous island arc, in which sediment accumulation may now exceed 10 000 m (Hsü and Kelts, 1978). Based on borehole data, sediments dating from Upper Miocene to Recent (Fig. 7.5) demonstrate changes in salinity in response to periodic links to ocean waters and/or periods of high evaporation and warm climate (based on the percentage of steppe pollen in the total pollen assemblage). With the exception of a coarse gravel facies formed at the Upper Miocene–Lower Pliocene boundary, sediments are characteristic of deep water conditions. Chemical sedimentation began during the warm climate of the Upper Miocene and probably reached its maximum with the chalk deposits of the early Pliocene, in which varve-like layers reflect summer calcite precipitation and winter detrital carbonates. The presence of siderite and associated diatoms implies deep weathering of the watershed, yielding large quantities of iron and silica to the lake which was probably quite eutrophic in its surface waters and anoxic at depth. The formation of $FeCO_3$ rather than sulphate deposits implies that the basin was largely freshwater and 'saline' species probably reflect the effects of surface evaporation-concentration. Terrigenous muds indicate the overriding influence of late and postglacial erosion in the surrounding watershed.

Deep, trough-shaped basins subject to variations in climate and water level tend to show less change in their limnological conditions than broad shallow

CORE PARAMETERS

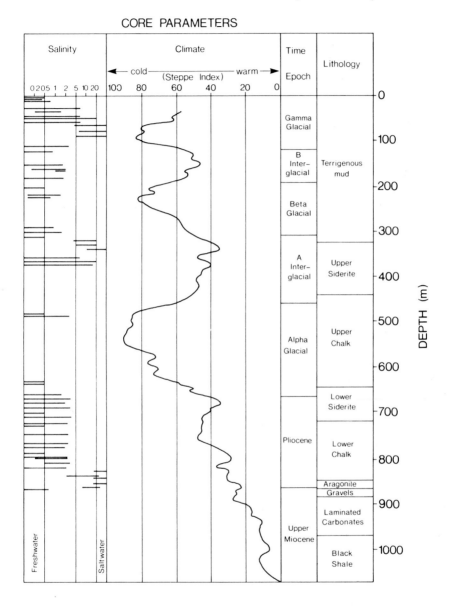

FIGURE 7.5 Stratigraphy and related core data from the Black Sea (*modified after Hsü and Kelts (1978); reproduced by permission of Blackwell Scientific Publications Ltd*)

lakes, similarly influenced. Thus Lake Balaton, which is a large $(600\,km^2)$ shallow (mean depth 3.3 m) lake in Hungary, shows considerable change in its sediment/water chemistry as a result of only minor changes of surface elevation (Müller and Wagner, 1978). Balaton is an alkaline lake (pH 8.6) in which the concentrations of most major ions are greater in the outflow than inflow, owing to the effects of evaporation-concentration. As a result of in-lake precipitation, however, Ca^{2+} and HCO_3^- ions are depleted in the outflow. Sediment core data record the presence of three periods of high magnesium/calcium ratio carbonate sediments, which coincide with dry climatic conditions or particularly high evaporation rates (Fig. 7.6). At about 5000 and 3000 years B.P. (before present) the lake acted as a closed system, with no significant outflow at times of maximum evaporation. At about 2000 years B.P. Roman engineers regulated the outflow (Sio' canal) at the eastern end of the lake, thus maintaining the lake as an open system. Because of this (and climatic variation), the effects of evaporation-concentration have been much less evident in modern sediments.

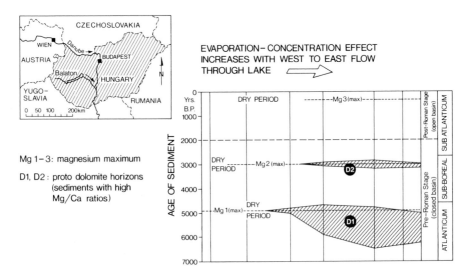

FIGURE 7.6　Relationship between Mg maxima and climate in Lake Balaton sediments (*modified after Müller and Wagner (1978); reproduced by permission of Blackwell Scientific Publications Ltd*)

(b)　Examples of recent climatic and cultural impacts

The combined impact of climate and cultural changes resulted in different conditions in three lakes (Vernet and Favarger, 1982) on the Rhone drainage in Switzerland and France (Fig. 7.7). Glacial ice retreated first from Lake Bourget and from Lake Leman between 14 000 and 12 000 years B.P. In Lake Bourget (maximum depth 46 m) the base of a sediment core exceeds 1600 years B.P. and

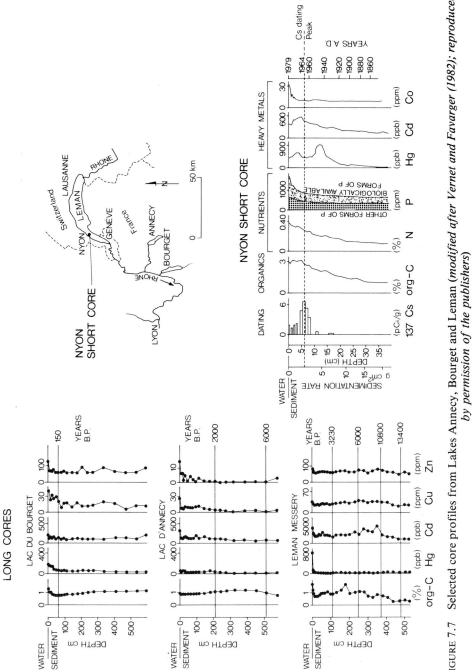

FIGURE 7.7   Selected core profiles from Lakes Annecy, Bourget and Leman (*modified after Vernet and Favarger (1982); reproduced by permission of the publishers*)

concentrations of heavy metals change little until after the turn of the twentieth century; their steep rise at that time, however, reflects a rapid expansion of local industry. The low concentration of organic carbon may represent the cool temperature effects of the 'little ice-age' prior to the 1800s. In Lake Annecy (maximum depth 35 m) there is little evidence of change prior to 2000 years B.P. (Roman introduction of chestnut trees: *juglans* pollen marker horizon), when a 25 per cent increase in the sedimentation rate was probably caused by increasing land use activities in the watershed. In Lake Leman (maximum depth 310 m), increasing organic carbon correlates particularly well with the climatic optimum of the Atlantic Period (4–5000 years B.P.) and with the modern increase in trophic status related to high nutrient inputs from industrial and domestic wastes, and farming practices. Studies on short cores from this lake demonstrate the effects of cultural development in more detail, summarized as follows (after Vernet and Favarger, 1982):

Mercury	prior to 1880	natural background levels
	1940–50	contaminant maximum, corresponding to large importation into Switzerland during wartime
	1970	secondary peak concentration related to development of chemical industry
	1970–79	decrease in waste
Copper	1900	start of vineyard treatments with $CuSO_4$
Cadmium	prior to 1920	background levels
Zinc	1970	contaminant maximum
	1970–79	decrease in waste
Cobalt	1970	start of contamination from unknown source

The progress of cultural eutrophication is also reflected by the concentrations of nutrient elements in these cores. Both organic carbon and nitrogen have increased since the turn of the last century, and the increase in total phosphorus at about 1950 is most notable. By analysing for different forms of phosphorus it was found that concentrations of apatite-P (detrital) had changed little throughout core depths. Concentrations of biologically bound organic-P had increased steadily since about 1900, and are now about 4–5 times background levels. The most biologically available forms of phosphorus, however, had increased particularly since 1964, reflecting contributions from domestic waste which are most easily recycled.

7.4   CONTAMINANT LEVELS

(a)   Enrichment

While it is possible to compare elemental compositions in sediments against global mean values, as an indication of cultural impact, the use of enrichment

TABLE 7.2   Average concentrations of the enriched elements in sedimentary materials from the Great Lakes area (*after Kemp and Thomas (1976); reproduced by permission of the Geological Association of Canada*)

| No. of samples | Lake Huron sediments | | Lake Erie sediments | | Lake Ontario sediments | | All rivers (28) | Niagara River (1) | Bluffs (16) |
	Recent (3)	Pre-colonial (3)	Recent (6)	Pre-colonial (6)	Recent (5)	Pre-colonial (5)			
Hg (ppb)	210	150	855	78	2350	78	285	560	44
Pb (ppm)	129	39	106	28	220	29	192	110	28
Zn (ppm)	197	94	279	98	475	104	411	317	39
Cd (ppm)	2	1	4	1	5	1	3	6	1
Cu (ppm)	58	38	57	29	98	44	118	309	24

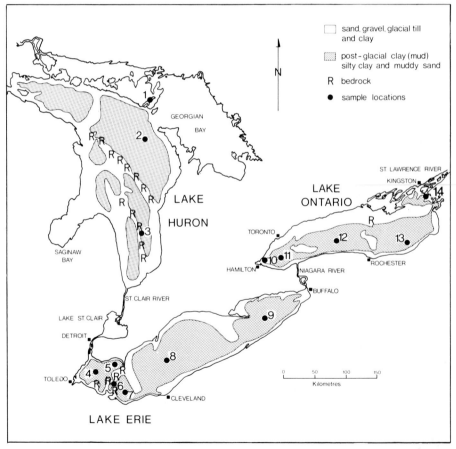

FIGURE 7.8   Core locations (SEF study) from the Great Lakes (*after Kemp and Thomas (1976); reproduced by permission of the Geological Association of Canada*).

TABLE 7.3   Sediment enrichment factors (SEF) for various elements at Great Lakes coring stations (*modified after Kemp and Thomas (1976); reproduced by permission of the Geological Association of Canada*)

Coring station	Huron					Erie				Ontario				
	1	2	3	4	5	6	7	8	9	10	11	12	13	14
Water depth (m)	56	245	91	11	12	11	.14	24	58	33	101	186	225	26
Present sedimentation rate (g m^{-2} year^{-1})	147	157	325	3850	3465	847	1109	1190	5049	452	420	423	366	1156
Mean annual sedimentation accumulation (mm)	0.7	0.9	1.4	7.6	6.7	1.1	2.7	5.1	13.4	1.0	2.0	2.1	2.0	5.4
Conservative elements														
Si	0	0	0	0	0	0	0	0	0	0	0	0	0	0
K	0	0	0	0	0	0	0	0	0	0	0	0	0	0
Ti	0	0	0	0	0	0	0	0	0	0	0	0.3	0.3	0
Na	0	0	0.7	0	0	0	0	0	0	0	0.3	0.3	0	0.5
Mg	0	0	0	0	0	0	0	0	0	0	3.0	0.7	0.4	0.3
Enriched elements														
Hg	0	0.9	2.3	22.8	4.3	13.2	9.7	11.4	7.3	6.7	6.4	131.5	64.8	13.1
Pb	1.1	3.1	6.2	3.2	0.9	3.4	2.6	6.7	3.7	6.7	8.4	7.3	8.7	14.0
Zn	0.5	1.7	3.4	2.6	0.8	1.4	1.4	3.2	2.2	3.1	4.5	6.2	6.2	4.5
Cd	0	1.9	0.4	1.4	1.9	4.0	0.8	4.6	2.7	1.1	3.6	8.3	6.8	11.3
Cu	0.4	0.8	1.9	1.4	0.7	1.3	1.1	1.2	0.8	2.2	1.7	1.7	2.1	1.9
Nutrient elements														
Org-C	0.4	0.3	0	0.8	0.5	1.1	1.3	2.8	1.7	1.0	2.5	2.3	2.7	2.0
N	0.5	0.3	0.3	1.3	1.1	1.2	2.1	3.0	2.0	0.6	2.4	2.8	3.1	2.3
P	0.8	0.4	0.4	0.4	0	0.5	0.3	0.7	0.8	0.7	1.4	2.2	0.8	1.8
Carbonate elements														
CO$_3$-C	0	0.3	0.8	-0.5	0	-0.4	-0.4	58.0	0	1.7	4.7	18.7	14.4	55.7
Ca	0	0.5	0.4	0	0	-0.3	-0.4	4.3	-0.3	1.2	10.8	10.7	13.3	7.9
Mobile elements														
Fe	0	0	0	0	0.3	0	0.3	0.3	0	0	0	0	0.3	0
Mn	2.3	2.5	1.8	0.4	0	0.4	0.4	2.7	1.1	0.4	4.5	4.2	9.3	1.0
S	-0.6	0	0	0.3	0.7	0.3	0	8.2	1.0	0.6	0.5	1.7	0.8	2.3

factors which compare modern and pre-cultural (nominally 100–150 years B.P.) compositions at the same site provide a more useful indication of local and regional changes. The sediment enrichment factor (SEF) is defined as:

$$\frac{(E_s - E_a)}{(Al_s - Al_a)} \Bigg/ \frac{E_a}{Al_a}$$

where E is the concentration of the selected element, Al is the concentration of aluminum, subscript 's' denotes surface material and subscript 'a' denotes pre-cultural material (Kemp *et al.*, 1976). The SEF is usually based on an aluminum standard but other conservative elements may be substituted. Table 7.2 based on data from the Great Lakes, gives the actual concentration of selected heavy metals in pre-cultural and recent lake sediments, modern suspended river

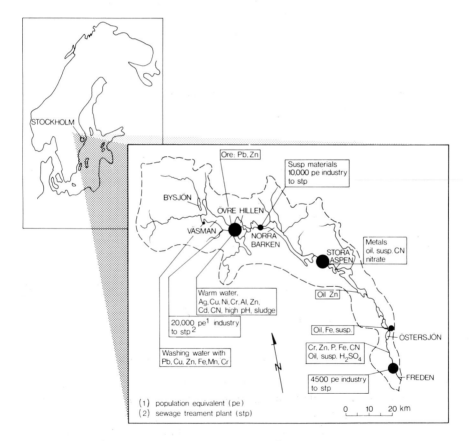

FIGURE 7.9   Industrial/urban contamination of the River Kolbäcksån (*modified after Håkanson and Jansson (1983); reproduced by permission of Springer Verlag*)

sediments, and old shoreline bluff materials. The SEF values from 14 Great Lakes cores (Fig. 7.8) are shown in Table 7.3. Enrichment in Lake Huron is slight but in Lake Erie and Lake Ontario both heavy metals and nutrient elements are considerably enriched (values of 2–3 or greater), with mercury representing an extreme case.

Using a similar approach Håkanson and Jansson (1983) showed that lakes in the River Kolbäcksån system, Sweden (Fig. 7.9) were also contaminated by various heavy metals, notably chromium and mercury. The degree of sediment contamination (contamination factors greater than 1) of Lake Stora Aspen was significantly greater than that of the other lakes, as shown in Table 7.4. The relationships between contaminants in sediments and biota are complex. Comparison between equivalent contamination factors in sediments and fish (pike liver), as shown in Table 7.5, indicate that in these lakes there is a strong positive correlation between mercury and cadmium in sediments and fish. The strong negative correlations for both copper and zinc, however, indicate that entirely different factors limit the biological availability and uptake of these elements.

TABLE 7.4 Degrees of sediment contamination in seven lakes from the River Kolbäcksån system, and the sequence of contamination factor (*modified after Håkanson and Jansson (1983); reproduced by permission of Springer Verlag*)

Lake (ranking order by C_d values	Degree of contamination C_d	Contamination factor			
		Very high $C_f > 6$	Considerable $C_f = 3–6$	Moderate $C_f = 1–3$	Low $C_f = <1$
Stora Aspen	134.1	Cr > Pb > Zn > Cd	Hg > Cu	—	—
Väsman	65.7	Hg > Pb > Cd > Zn	Cu	Cr	—
Norra Barken	59.1	Pb > Zn > Hg = Cd	Cu	Cr	—
Östersjön	54.4	Cr > Hg > Cu = Zn	Pb	Cd	—
Övre Hillen	44.3	Hg > Zn > Cd	Pb > Cu	Cr	—
Freden	34.4	Cr	Zn > Hg > Cd = Cu	Pb	—
Bysjön	11.1	—	Hg	Cd > Pb > Zn	Cu = Cr

C_f = relative enrichment of individual elements in modern sediment
C_d = sum of all C_f values greater than 1

(b) Contaminant availability and biological accumulation

As a result of normal physical and chemical sorption processes, metal ions become associated with clay mineral particulates and organics, and may form precipitates such as hydroxides, carbonates and sulphides, thus binding the metals and generally reducing their biological availability. However, if the pH of sediments or water is lowered (becomes more acidic) dissolution (hydroxides and carbonates) and desorption of cations occurs, which thereby remobilizes

TABLE 7.5 Comparison between sediment contamination and fish contamination (pike liver) for six metals in lakes from the River Kolbäcksån system (*modified data from Håkanson and Jansson (1983); reproduced by permission of Springer Verlag*)

Element	Hg		Cd		Pb		Cu		Cr		Zn		r
	C_f	PI_F	C_f	PI_F	C_f	PI_F	C_f	PI_F	C_f	PI_F	C_f	PI_F	$(n=6)$
Bysjön	3.9	10.2	2.5	4.6	1.8	3.5	0.8	7.6	0.8	4.5[a]	1.3	3.2	0.57
Väsman	28.4	9.8	6.5	14.7	18.3	2.2	4.8	1.4	1.6	1.8[a]	6.1	3.1	0.29
Ö. Hillen	20.0	11.8	6.1	22.5	5.0	3.9	4.2	3.0	1.9	2.3[a]	7.1	1.9	0.33
N. Barken	10.7	6.4	10.7	5.3	16.7	2.8	5.2	2.2	2.0	3.4[a]	13.8	1.8	−0.001
St. Aspen	5.4	7.6[a]	6.1	2.0[a]	18.0	1.9[a]	5.1	1.7[a]	90.0	5.4	9.7	0.7[a]	0.35
Österjön	6.9	4.4	2.3	1.4	3.0	1.1	6.0	3.3	30.2	34.3	6.0	3.9	0.997
Freden	4.0	7.8[a]	3.4	1.5[a]	1.8	0.9[a]	3.4	2.3[a]	16.4	13.0[a]	5.4	3.6	0.89
r $(n=7)$	0.48		0.29		0.03		−0.81		0.19		−0.65		0.24 $(n=42)$

[a] = data based on less than three analyses

C_f = relative enrichment of individual elements in modern sediment

PI_F = relative enrichment of individual elements in pike liver (Kolbäcksån system: other background sites)

the metals. In lakes in which oxygen deficits occur there may be release of reduced forms of iron and manganese. Further, as a result of microbial processes, zinc, copper and cadmium may be released from anoxic sediments; under some conditions, methylation of metals may also occur. Thus, because of different limonological characteristics (e.g. water chemistry, trophic state, thermal structure and flushing rates) factors affecting biological availability of contaminants in water and sediments (for individual elements or groups) will also vary from lake to lake. Whereas heavy metal contaminants tend to be present in lake systems at concentration levels at the part per million range, man-made organic contaminants are usually present at several orders of magnitude less. Those compounds which have a relatively high solubility tend not to bio-accumulate, and usually degrade fairly quickly. The principal concern relates to persistent organic contaminants which are characterized by low solubility and a strong tendency to be bio-accumulated. The effects of lethal toxicity from heavy metals tends to be greater than that of the persistent organic contaminants, at environmental concentrations, but suppression of growth, deformation, loss of fertility and other sub-lethal and carcinogenic effects are known to occur. The uptake of organic contaminants by biota is generally related to equilibrium partitioning between water and the lipid-pool of the biota (represented by *n*-octanol as a surrogate, Fig. 7.10). The polychlorinated biphenyls and similar persistent organic compounds are usually associated with particulates in water. However, many PCBs are quite volatile and this increases their rate of uptake by phytoplankton, and accumulation by fish is possible both as uptake during respiration and by ingestion of food particles. For organic comtaminants which are less volatile, e.g. DDT, the principal pathway for biological uptake is via the food-web. Although most artificial organic compounds are biogradable, rates of degradation under natural environmental conditions vary greatly.

(c)   Effects of remedial measures

The rise of contaminant levels in water, sediments and biota as a result of direct and indirect loadings is well known from many lake studies, but the changes which have taken place following control programmes are less well documented. The loss of contaminants as a result of flushing and downstream transport can be very rapid, especially in river–lake systems. In Lake St Clair, situated below Detroit on the connecting channel between Lake Huron and Lake Erie (Great Lakes), the shallow depth (mean depth 3.6 m) and a very short flushing time (about 3 days) have allowed natural processes to rapidly reduce contaminant levels in sediments, following legislative controls on contaminant inputs. This is demonstrated, dramatically in Table 7.6 which shows a 2–4-fold decrease in contaminant concentrations between 1970 and 1974 (Thomas *et al.*, 1975).

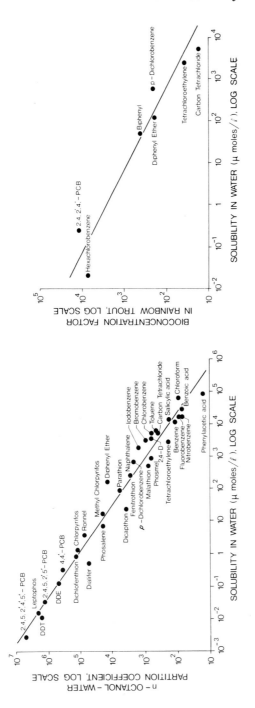

FIGURE 7.10 Lipophily and bio-accumulation. Both *n*-octanol–water partition coefficient and bioconcentration factor in rainbow trout are inversely proportional to aqueous solubility (*after Chiou* et al. (1977); reproduced with permission; Copyright 1977 American Chemical Society)

TABLE 7.6   Comparative mean concentration values of contaminants in Lake St Clair sediments, 1970 and 1974 in $\mu g \, kg^{-1}$ (ppb) (*after Thomas* et al. *(1975); reproduced by permission of the authors*)

Year		Hg	DDE	TDE	DDT	PCBs
1970	Mean	1549	2.1	3.6	2.0	19.1
	Standard deviation	2340	1.5	2.2	1.9	8.9
	Number of samples	50	49	49	23	49
1974	Mean	568	0.7	1.5	0.5	10.0
	Standard deviation	777	0.6	1.6	0.4	6.3
	Number of samples	54	53	54	19	54
t-value		2.8[a]	6.1[a]	5.5[a]	3.7[a]	6.0[a]

[a] = significant at 0.01 level

TABLE 7.7   Polychlorinated biphenyl (PCB) residues in Lake Michigan fish, 1972–80 (*modified after Rodgers and Swain (1983); reproduced by permission of the International Association for Great Lakes Research*)

Year	Number of samples	Average length (mm)	Total PCB[b] $\mu g \, g^{-1}$ (ppm)
Polychlorinated biphenyl residues in bloater chubs			
(Coregonus hoyi)			
1972	120[a]	255	5.7(1.0)
1973	160	250	5.2(0.4)
1974	110	257	5.6(0.3)
1975	170	249	4.5(0.4)
1976	110	253	4.1(0.2)
1978	100	254	3.1(0.3)
1980	110	245	2.2(0.3)
Polychlorinated biphenyl residues in coho salmon			
(Oncorhynchus kisutch)			
1972	10	693	10.9(2.1)
1973	29	620	12.2(0.8)
1974	30	665	10.5(0.9)
1975	30	645	10.8(0.6)
1976	30	635	9.2(0.5)
Polychlorinated biphenyl residues in lake trout			
(Salvelinus namaycush)			
1972	9	648	12.9(4.8)
1973	30	602	18.9(2.1)
1974	30	616	22.9(3.7)
1975	29	613	22.3(2.9)
1976	30	606	18.7(2.7)

Data from Contaminant Monitoring Program, Great Lakes Fishery Laboratory, US Fish and Wildlife Service, 1451 Green Road, Ann Arbor, MI 48105 (Willford, 1982, unpublished)
[a] = each sample of bloater chubs represents a composite of ten fish
[b] = concentrations in whole fish, wet weight with 95 per cent confidence interval in parentheses

Because of their great depth, very low flushing rates and generally low suspended sediment concentrations, most of the Great Lakes are particularly sensitive to the biological effects of PCB-type compounds, which first appeared as significant contaminants during the mid-1950s. Since 1972, however, inputs have been greatly reduced. As a result, contaminant levels in the biota have also declined (Murphy and Rzeszutko, 1977; Rodgers and Swain, 1983), although, because of analytical difficulties, it has not been possible to demonstrate comparable changes of ambient levels of PCBs in lake water. Table 7.7 shows whole fish concentration data for three species of fish from Lake Michigan. The salmonids (coho and lake trout) are at the top of a relatively large and complex food-chain, which results in a delayed response to contaminant reductions. The bloater, however, is a pelagic feeder (consuming large zooplankton), which responds much more rapidly to changing levels of PCB contamination. If loadings of PCBs to Lake Michigan continue to decrease, the salmonids will reach compliance with consumption guidelines in the 1980s, but if significant PCB loads continue (particularly from atmospheric sources) these fish may remain contaminated (above US guidelines) indefinitely (largely because of the greater degree of bio-accumulation in the salmonids).

7.5 SPECIAL PROBLEMS

(a) Example of mercury point-source contamination

The Wabigoon River in Ontario (Canada) flows northwest to the confluence with the English River, and thence to the Winnipeg River and Lake Winnipeg. This 420 km waterway connects numerous small lakes and reservoirs in a largely unpopulated area of boreal forest, on the Canadian shield. Between 1962 and 1970 approximately 10 tonnes of inorganic mercury were released into the river from a chlor-alkali plant/paper-mill complex at the town of Dryden (Fig. 7.11). Outputs were reduced between 1970 and 1975, when use of mercury was discontinued, but small amounts of residual mercury continue to leak into the river. Large quantities of wood chips and other paper-mill waste have also been discharged into the river. Anomalously high levels of mercury have been found in bottom sediments, as much as 240 km downstream, and in fish, as far as 380 km below Dryden (Jackson *et al.*, 1982). The legal limit of mercury for edible fish in Canada is 0.5 ppm and the presence of levels, considerably above this, has injured the economy, and possibly the health, of the Ojibway Indian communities in the area. For many years, the presence of mercury in the system gave little concern because it was thought that the heavy metal would move only a short distance and sink into the bottom sediments, and that there would be little effect on biota. Unfortunately, however, mercury can be transformed by bacteria into biologically available (and highly toxic) methyl mercury, and the rate of methylation is closely related to microbial activity. Sampling has

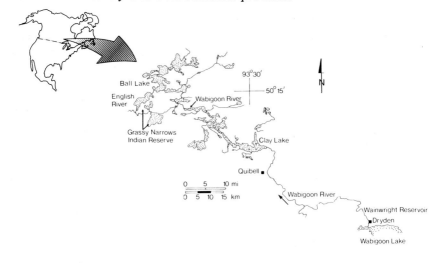

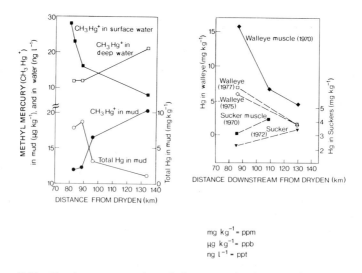

mg kg^{-1} = ppm
μg kg^{-1} = ppb
ng l^{-1} = ppt

FIGURE 7.11   Total mercury and methyl mercury in the aquatic environment of the Wabigoon–English River System (*modified after Jackson and Woychuk (1980); reproduced by permission of T. A. Jackson*)

shown that concentrations of total mercury (associated with both dissolved and particulate phases) are greatest at the time of spring runoff, owing to accelerated resuspension and transport. However, concentrations of dissolved methyl mercury are dominant during the summer period when it is most biologically available and when growth and production are most rapid. Seasonal changes in the chemistry of the mercury make it least available to biota during the winter–spring period. Methyl mercury becomes incorporated into fish tissues both as a result of feeding on benthic and planktonic organisms and by uptake through respiration. To demonstrate the effects of contamination and uptake by benthic organisms and fish, Jackson and Woychuk (1980) assembled data on pelagic fish (walleye, pike, cisco, whitefish and sauger), bottom feeders (sucker) and crayfish. These data, together with water and sediment data, are summarized in Fig. 7.11. The data were evaluated in terms of allochthonous (river-borne) and autochthonous (locally generated in lakes) methyl mercury. The mean mercury concentrations in pelagic fish decreased between Clay Lake and Ball Lake, and in further downstream lakes; these parallel the trend of allochthonous methyl mercury. In contrast, the mean mercury concentrations of suckers increased and follow the trends shown by sediment methyl mercury. The data imply that mercury contamination of pelagic fish is due to the availability of methyl mercury in the waters of both rivers and lakes of the Wabigoon–English River system, whereas bottom feeders are mostly influenced by conditions in the lacustrine parts of the system. Regional variations in the chemical speciation of mercury indicate that it becomes more readily solubilized and more easily methylated downstream from Dryden. Because of dilution, microbial activity is less likely to be inhibited by toxic concentrations of mercury, and mercury in mud is more readily bio-available than from wood chips. Thus, shallow downstream lakes, in which there are accumulations of fine mud and a high microbial activity associated with degradation of organic particles, are particularly important production sites for methyl mercury.

(b) Mercury in reservoirs

Mercury contamination of lakes occurs in many parts of the world and especially in the more highly industrialized countries, where the problems are greatest in shallow eutrophic lakes. Surprisingly, however, mercury contamination of fish has also been a problem in newly constructed reservoirs, often at sites with no apparent source of contamination. The reasons for this were not understood for several years but, in a recent study on the Churchill–Rat River system in northern Manitoba, Jackson (1984) has shown how flooding of adjacent land (to generate hydroelectric power) has adversely affected commercial fishing on South Indian Lake and lakes of the Notigi Reservoir complex. The effect has been greatest where inundation was most extensive. It appears that microbial activity has been stimulated by nutrients from freshly submerged soils and that mercury has been derived from natural organic and terrestrial sources and from erosion of silt and clay bank materials. Thus the cause and effects of this form of contamination bears much in common with that of the Dryden case, cited previously. Fortunately, nutrient levels in new reservoirs tend to drop after a few years, as the ecosystem adjusts from a terrestrial to an aquatic system. Thus the effect of methylation in new reservoirs is usually of a temporary nature.

(c)   Acidification: diffuse source inputs

Acid precipitation is characterized by high concentrations of sulphur (SO_2, SO_4^{2-}), nitrogen (NO, NO_2, HNO_3, NO_3^-, NH_3, NH_4^+) and hydrogen (H^+) ions. The major sources of contamination are usually associated with combustion of fossil fuels (power plants, smelters, vehicle emissions) the products of which become incorporated into air mass transport and are variously modified by chemical processes in the atmosphere. Atmospheric deposition of nitrate is generally one-third to one-half as great as sulphate deposition although, in the western United States, nitrate may represent nearly two-thirds of the total acid fraction in rainfall (Impact Assessment Work Group, 1983). On a very general basis, wet precipitation (rain, snow, hail) accounts for about 75 per cent of the total acid deposition. In areas of thin soil and bedrock lacking in neutralizing (carbonate) minerals, the normal bicarbonate (HCO_3^-) buffering system breaks down and sulphate (SO_4^{2-}) becomes the dominant anion in acidified waters. Cations (including H^+ ions) can be exchanged between sediments and water; at high concentrations H^+ ions tend to exchange with Ca, Mg and Al, lowering H^+ concentrations in the water. Net adsorption of H^+ may lead to a more acid soil water solution unless H^+ ions are consumed in weathering. In watersheds, therefore, less H^+ is adsorbed and more is transported in surface waters as the amount of basic cation in the soil profile is reduced (Tollan, 1980). The transport of cations between soil and water depends upon the availability of anions, and sulphate is a particularly important carrier for H^+. Acidification of lakes may be thought of as a large-scale titration process. Normal weathering is characterized by bicarbonate (HCO_3^-) and Ca^{2+} and Mg^{2+}. Waters with high bicarbonate levels resist acidification (and are said to be well-buffered). At a pH of about 5, sulphate replaces bicarbonate, as the major anion, and such lakes are considered to be strongly acidified.

Surface waters in areas of geologically similar bedrock may or may not be acidified, depending upon the characteristics of atmospheric deposition; trends in acid precipitation and surface water acidification tend to parallel one another (Fig. 7.12). Lakes and rivers in Norway and Sweden have been particularly hard hit by acid precipitation (Tollan, 1980). Comparison of data from south Norway during the 1920s–70s and since then shows that, in 128 lakes, 63 per cent of the sites showed a 0.25 pH unit decrease vs. an 0.25 pH unit increase at only 12 per cent of the sites. Only 4 per cent of these lakes had a pH < 5.0 before 1950, but 25 per cent of them had dropped to this level by 1977. In 1950, none of the 180 lakes from southern Sweden was below pH 5.5; by 1977, 28 per cent were less than pH 5.5 and 15 per cent were less than pH 5.0. Similar observations have been reported in eastern Canada and the United States (Galloway *et al.*, 1980; Thompson and Bennett, 1980).

Increased acidification of lakes is often associated with dissolution of carbonate and metals, which can be demonstrated in short sediment cores.

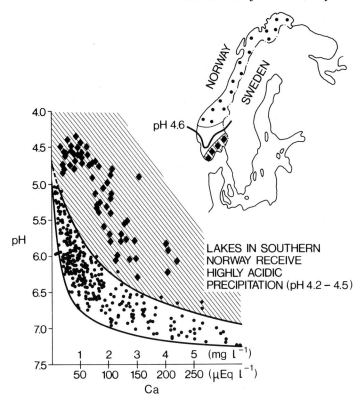

When calcium concentrations are assumed to be well
correlated with pre–acidification bicarbonate alkalinity,
the empirically drawn curve will distinguish between
acidified and nonacidified waters.

FIGURE 7.12   Calcium and pH concentrations in Norwegian lakes, 1974–77 (*modified
after Tollan (1980); reproduced by permission of the author*)

In Fig. 7.13 (based on Reuther *et al.*, 1981) data are presented from Lake
Hovvatin in southern Norway (pH 4.4). All three elements increase significantly
in the lower part of the core but the concentrations of both zinc and cadmium
decrease dramatically at the top. A comparison of the associations for these
and other metals indicates that the Zn decrease (also Co and Ni) is associated
with the easily reducible fraction, whereas remobilization of Cd occurs mostly
in the organic fraction. Lead (and Cu) is affected to a very much lesser extent.
The fact that similar changes were not seen in a nearby lake at pH 4.9 suggests
that mobility of these elements may be very sensitive to pH conditions. Although
the effects of acidification are compounded by high concentrations of Zn (and

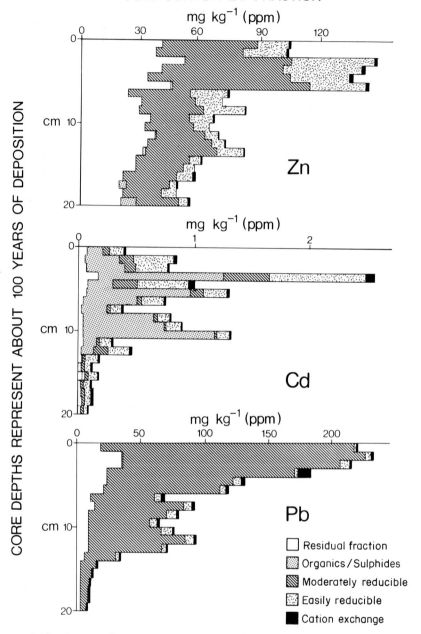

FIGURE 7.13   Core profiles of Zn, Cd and Pb, and their chemical forms from Lake Hovvatin, Norway (*modified after Reuther* et al. (1981); reproduced by permission of CEP Consultants Ltd)

other heavy metals) which can be toxic, it is the release of aluminium which causes most concern. In lakes with a low organic carbon content, Al is present as dissolved inorganic compounds which are extremely toxic to fish at about pH 5.0 (Tollan, 1980). Acidification affects lakes at all trophic levels. Typically, numbers of algal species are reduced and even tolerant species may be photosynthetically less active. However, heavier growths of filamentous algae may be due, in part, to reduced invertebrate feeding and lower rates of decomposition. Moss beds (sphagnum) are characteristic of more severely acidified waters. At pH 5.0–6.0, phosphorus precipitation by aluminium may also reduce nutrient availability. Diatoms show a community shift with pH, and there is a shift from bacteria to fungi (as decomposers) with a commensurate decrease in the rate of decomposition of organic materials. In Norway, there are three to four times as many mayfly species in waters of pH 6.5–7.0 than at pH 4.0–4.5; *Gammarus lacustris* (an important amphipod species) is lost between pH 6.5–5.5, and few freshwater snails with calcareous shells can survive at pH less than 6.0 (Tollan, 1980).

However, the regional decline of freshwater fisheries in eastern North America and Scandinavia probably represent the most dramatic effects of lake (and river) acidification (Tollan, 1980). During the period 1929–37, only 4 per cent of 217 Adirondack Mountain lakes (United States) were below pH 5.0 and devoid of fish; by 1975, more than half the lakes had levels less than pH 5.0, and 90 per cent were devoid of fish. (In the Adirondack region as a whole, however, 4 per cent of the total area of lakes has pH below 5.0 and about 10 per cent has pH 5.0–6.0. This is because the effects of acidification are greatest in the higher lakes (above 500 m) which also have the smallest surface and watershed areas (SFI, 1984).) In Ontario (Canada), there are several hundred lakes devoid of fish and numerous rivers in Nova Scotia (pH 4.5–5.0) are now devoid of salmon. In Scandinavia the effects of acid precipitation were felt early this century. Roach disappeared from many lakes in the southwest of Sweden in the 1920s and 1930s and, in this very sensitive area, about half the lakes are now less than pH 6.0 (there are perhaps 10 000 lakes in a similar condition throughout Sweden). Similarly, in southern Norway, huge areas (about 13 000 km²) of lake water are now devoid of fish and perhaps 20 000 km² of lake water are highly endangered. At the present rate of acidification, Norway will have lost 80 per cent of its trout population by 1990. Regional correlations between acidity and fish loss are very strong, and small lakes at high altitude are usually most susceptible. Increased acidity causes high egg and fry mortality; in adult fish, deformation, gill-clogging and stunted growth are common and massive fish-kills often result from pH-events (snowmelt, and runoff). Generally, salmonids are most sensitive to acid stress, with brook trout a little more resistant than lake trout or atlantic salmon.

The effects of lake acidification are usually associated with the impact of acid precipitation but there are situations where acidification has been caused by other forms of input, such as drainage of mine waste dumps or industrial discharge. The forms of impact from these input sources are very similar to those produced by acid rainfall; their extent, however, is very much more localized.

(d)   Mass balance and the importance of atmospheric loads

In addition to the effects of acidification, atmospheric precipitation (both wet and dry) contributes significantly to the mass balance of materials entering lakes from all sources, and this is particularly important for large lakes situated in relatively small watersheds (such as the Great Lakes of North America). In the upper lakes region (Lakes Superior, Michigan and Huron) direct atmospheric contributions of major nutrient elements represent a significant part of the total loading budget (total phosphorus 10–20 per cent; total nitrogen 30–60 per cent). Atmospheric sulphate (as SO_4) represents 15–30 per cent of the mass balance and atmospheric inputs of copper and lead represent about 30–40 per cent of their total loads (Upper Lakes Reference Group, 1976). Atmospheric inputs of PCBs prior to 1975 (Lake Michigan) represented about 15 per cent of the total PCBs input. However, although there has been little decrease in the atmospheric input of PCBs, there has been a major decrease in tributary loadings of these contaminants and presently the atmospheric load represents about 55 per cent of the total PCBs input to the lake. Atmospheric inputs contribute materials directly to the surface waters where primary production is greatest and, thus, may have an immediate effect on plankton. Atmospheric inputs are seasonally variable and around Lake Michigan about 85 per cent of the phosphorus is associated with wind-blown soil and dust from the mid-west, which has a spring–summer maximum. On the other hand, sulphate is strongly associated with power plant emissions which tend to peak in the winter period.

7.6   LAKE LEVELS

Lake levels have been regulated to avoid periodic flooding of surrounding shorelines or to provide navigable depths, and (more often) to limit the effects of downstream flooding by using the storage capacity of lakes to reduce the range of discharge flow. Mostly, however, regulation has been used to store water and to manipulate its supply for drinking, agricultural and industrial uses, and for power generation. Regulation modifies natural regimes in two ways, it changes the water levels within lake basins and it alters the seasonal pattern of fluctuations.

If lake levels are raised or the range of their natural fluctuations increased, rates of erosion are increased; usually as a result of wave action on soft shoreline materials, but also as a result of changes in groundwater saturation or melting of permafrost which cause slumping. The effect of maintaining lake levels within narrow limits may reduce the width of the littoral zone and lessen productivity, especially in small lakes. In lakes where water supply requires a major draw-down (usually on an annual cycle), disruptions of chemical and biological characteristics occur, limiting species range or making them dependent upon some form of artificial support (such as hatchery propagation). In several lakes and newly formed reservoirs it has been found that, as a result of fresh inundation, nutrient levels increase and that there is an early increase in fish productivity. Usually, however, such increases are not maintained and, as a result of flushing and exhaustion of immediately available nutrient sources, there follows a considerable period during which biological productivity is much reduced; recovery to equilibrium conditions can take as long as 20–25 years.

The presence of regulatory structures, themselves, also add to the perturbation of natural lake systems. Atlantic salmon no longer enter Lake Ontario and American eels can do so only because of the construction of eel-ladders in the power-dams of the St Lawrence River (Christie, 1973). The construction of connecting channels may also allow invasion of lakes by exotic species and the invasion of the parasitic sea lamprey into the Great Lakes is a prime example of the massive destruction of indigenous fish communities which is possible (Smith and Tibbles, 1980). Further, regulation of lake levels can have the effect of changing the relative proportions of deep and shallow water, and wetland areas within a basin.

Natural fluctuations of lake levels occur over different time scales, which may be superimposed on one another. Seasonal fluctuations reflect changes in the annual precipitation regime, and in large lakes or lakes in large drainage basins there may be a lag in response to precipitation/evaporation changes so that excessively high or low lake levels may predominate over several years. Short-period changes in relative surface elevation due to barometric pressure are only possible over very large lakes, but almost all lakes will show periodic changes in surface level due to wind stress. In extreme situations such as Lake Eyre, in central Australia, the lake fills completely during some years, only to dry out completely (under high evaporation) during the following years. Lakes subject to high evaporation rates usually concentrate salts and little perturbation of natural systems is required if commercial extraction of salts involves evaporation-concentration in ponded areas adjacent to main lake bodies.

(a) Examples of shore erosion problems

Lakes in the more developed countries of Europe and North America, which often lie within temperate climates, have frequently become the focus of population centres and agricultural concentrations. Surface levels of these lakes are generally stabilized to minimize fluctuations and to enhance adjacent land values (in contrast, reservoirs are designed so that levels can change significantly without interference of shoreline use). High water level events in these lakes can result in property damage due to flooding and shore erosion (large lakes),

TABLE 7.8   Economic, evaluations of lowering or raising the mean levels of the Great Lakes, in millions of dollars (*modified after International Great Lakes Levels Board (1973); reproduced by permission of the International Joint Commission*)

		Lowering the mean levels			
	0.00	− 0.25	− 0.50	− 0.75	− 1.00 ft
		− 0.08	− 0.15	− 0.23	− 0.3 m
Approximate annual benefits					
All Power	0.00	− 0.7	− 1.4	− 2.1	− 2.9
lakes Navigation	0.0	− 1.6	− 3.4	− 5.4	− 7.6
Shore property	0.0	+ 5.6	+ 10.1	+ 13.6	+ 16.1
Total	0.0	+ 3.3	+ 5.3	+ 6.1	+ 5.6
Geographical breakdown of shore property benefits					
Superior	0.0	+ 1.3	+ 2.2	+ 2.7	+ 3.0
Michigan–Huron	0.0	+ 2.7	+ 4.9	+ 6.8	+ 8.1
Erie	0.0	+ 0.8	+ 1.5	+ 2.0	+ 2.4
Ontario	0.0	+ 0.8	+ 1.5	+ 2.1	+ 2.6
Total	0.0	+ 5.6	+ 10.1	+ 13.6	+ 16.1
		Raising the mean levels			
	0.00	+ 0.25	+ 0.50	+ 0.75	+ 1.00 ft
		+ 0.08	+ 0.15	+ 0.23	+ 0.3 m
Approximate annual benefits					
All Power	0.0	+ 0.6	+ 1.2	+ 1.7	+ 2.2
lakes Navigation	0.0	+ 1.4	+ 2.7	+ 3.9	+ 5.0
Shore property	0.0	− 7.0	− 15.2	− 24.8	− 35.6
Total	0.0	− 5.0	− 11.3	− 19.2	− 28.4
Geographical breakdown of shore property benefits					
Superior	0.0	− 1.8	− 4.1	− 6.9	− 10.3
Michigan–Huron	0.0	− 3.2	− 6.9	− 10.9	− 15.4
Erie	0.0	− 1.0	− 2.2	− 3.7	− 5.2
Ontario	0.0	− 1.0	− 2.0	− 3.3	− 4.7
Total	0.0	− 7.0	− 15.2	− 24.8	− 35.6

and low water events result in loss of recreation facilities, navigation and water supply problems. In Table 7.8 examples of estimated costs, associated with possible changes to existing water levels of the Great Lakes, give an indication of the socioeconomic forces which influence lake level regulation that is generally designed to minimize seasonal variations within an annual cycle.

(b)   Example of wetland problems

The Peace–Athabasca delta system (Fig. 7.14) is one of the largest freshwater deltas in the world and covers an area of about 3800 km² in northern Alberta (Canada). It lies in the southern part of the Mackenzie drainage system and receives much of its water as spring runoff from the eastern slopes of the

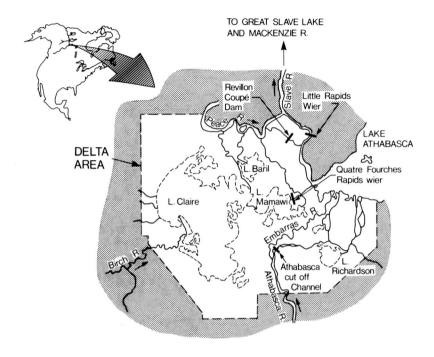

LAKE ATHABASCA LEVELS (1960–71)

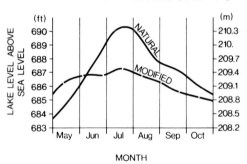

FIGURE 7.14 Water level changes in the Peace–Athabasca Delta area *(modified after Canada Water Year Book (1976); reproduced by permission of the Minister of Supply and Services)*

Rocky Mountains. Because of limited discharge in the normal outflow channels of Lake Athabasca (into the Slave River), the level of Lake Athabasca rises substantially during spring and early summer, and floods the shallow delta area surrounding Lakes Claire, Mamawi and Baril *(Canada Water Year Book*, 1976).

In 1967 Williston Lake was formed behind a dam on the headwaters of the Peace River and, as a result of continued filling, water levels in the Peace–Athabasca delta (more than 1100 km downstream) were generally lowered. The delta lies within an area of boreal vegetation and almost immediately there was a major successional change. The groundwater table dropped in the vast meadowland area and there was an explosive growth of new willows which marked a shift away from an aquatic system dominated by sedges and bullrushes. Fig. 7.14 shows that the natural water levels in relation to the man-modified regime and it is important to note that the change in mean water level amounts to less than 1 m. This seemingly small change in elevation, however, reduced the shorelines of the perched basins (seasonally refilled by over-topping) by an equivalent amount. The muskrat population crashed (a principal base for the trapping industry of Indian Bands) and it became clear that, without a very quick return to high water levels in the delta, scrub willow would become permanently established. The vegetation shift was a threat to about 120 000 ha of grazing area used by 10 000 bison of the Wood Buffalo National Park. There were also major problems associated with low water levels in Lake Athabasca, including access to docks, decreased available draught on vessels used for local transport and an increased need for channel dredging. Effects of water level changes on fish communities in the delta and Lake Athabasca were less certain but it was clear that a long-term reduction of the delta shoreline (by about 50 per cent) would have a major deleterious effect on the staging and nesting area of waterfowl on a major North America flyway (*Canada Water Year Book*, 1976).

As a temporary measure, a weir was constructed on the outlet of Lake Mamawi (Quatre Fourches Rapids) to raise levels in much of the delta, although allowing some overflow into Lake Athabasca. As a result of this work and a combination of other fortunate events (including a spring ice-jam which raised levels of the Peace River to overflow into the delta) spring–summer water levels reached 210 m (a.s.l.) in 1972; thus limiting the effect of colonization by scrub willow. Subsequently the level of Lake Athabasca was raised by constructing a weir on the outlet at Little Rapids, and a dam was constructed on the bypass channel (Revillon Coupé dam). The temporary weir at Quatre Fourches was removed, to return the variation between Lake Athabasca and delta levels to a more normal regime. The total cost of these generally successful remedial actions was about $4 million. Further, as a result of the studies on the impact of Peace River regulation on the delta, it was realized that the area was extremely sensitive to even minor changes in water level regime. Additional construction work was completed on the Athabasca cut-off channel to ensure that natural bank erosion would not threaten the future water balance. Although natural year to year flow variations have caused water levels to fluctuate greatly in the Peace–Athabasca Delta, to an extent greater than the effect of regulation of the upper Peace River, the effect of variation itself has inhibited permanent colonization of the wetlands by scrub willow. Thus it was the fact that long-term mean water levels had been reduced (though only slightly) which was of such importance in this extremely sensitive area (*Canada Water Year Book*, 1976).

7.7   PERSPECTIVE

Shoreline modifications, diversions and water level control were often some of the earliest forms of disturbance to lakes that have taken place; subsequently, as population centres developed and agriculture intensified, additional problems associated with nutrient enrichment (so-called 'cultural eutrophication') and over-fishing have occurred. Most recently, problems of contamination from toxic substances have developed. Further, whereas the earliest forms of impact were often slight or very localized, they have grown in size to become lakewide or greater (as a result of downstream transport or atmospheric contributions). Globally, there are very few lakes that remain truly pristine and devoid of man's influence. In the foreseeable future, demands for use of water resources can only increase, and effective mechanisms are urgently required to maintain a balance between competing uses. Undoubtedly many more problems will arise as a result of diversions, storage, contamination, and destruction of indigenous populations; but it should be possible to identify and address most new problems before they actually develop. Further, although remedial actions can be extremely costly, it is clear that many are effective. Therefore management to maintain lake ecosystems is clearly more a matter of social will, than of technical ability.

REFERENCES

Beeton, A. M., 1969, 'Changes in the environment and biota of the Great Lakes', In *Eutrophication: Causes, Consequences, Correctives*, National Academy of Science, Washington, D.C., pp. 150–187.

Canada Water Year Book, 1976, Fisheries and Environment Canada, Ottawa, pp. 49–55.

Chiou, C. T., Freed, V. H., Schmedding, D. W., and Kohnert, R. L., 1977, 'Partition coefficient and bio-accumulation of selected organic chemicals', *Environ. Sci. Technol.*, **11**, 475–478.

Christie, W. J., 1973, *A Review of the Changes in the Fish Species Composition of Lake Ontario*. Technical Report No. 23, Great Lakes Fishery Commission, Ann Arbor, Michigan, pp. 1–65.

Galloway, J. N., Norton, S. A., Hanson, D. W., and Williams, J. W., 1980, 'Changing pH and metal levels in streams and lakes in the eastern United States caused by acidic precipitation', *Proc. Internat. Symp. on Restoration of Lakes and Inland Waters, September 1980, Portland, Maine*, US EPA 440/5-81-010, Washington, D.C., pp. 446–452.

Golterman, H. L., Sly, P. G., and Thomas, R. L., 1983, '*Study of the Relationship between Water Quality and Sediment Transport*', Technical Papers in Hydrology No. 26, UNESCO, Paris, 231 pp.

Håkanson, L., and Jansson, M., 1983, *Lake Sedimentology*, Springer-Verlag, New York, 316 pp.

Hsü, K. J., and Kelts, K., 1978, 'Late Neogene sedimentation in the Black Sea', in Matter, A. and Tucker, M. E. (Eds), *Modern and Ancient Lake Sediments*, Internat. Assoc. Sedimentol., Special publ. No. 2., Blackwell Scientific, Oxford, pp. 129–145.

Hutchinson, G. E., 1957, *A treatise on Limnology. I—Geography, Physics and Chemistry*, John Wiley, New York, 1015 pp.

Hutchinson, G. E. and Löffler, H., 1956, 'The thermal classification of lakes', *Proc. National Academy Science*, **42**, 84–86.

Impact Assessment Work Group, 1983, *U.S.–Canada Memorandum of Intent on Transboundary Air Pollution Final Report*, Vol. 1, Environment Canada, Ottawa.

International Great Lakes Levels Board, 1973, *Regulation of Great Lakes Water Levels*, Report to the Internat. Joint Comm., IJC, Regional Office, Windsor, Ontario, 294 pp.

Jackson, T. A., 1984, 'Speciation, mobilization and biological accumulation of mercury in recently formed reservoirs of northern Manitoba', *Abstr. 27th Conf. Internat. Assoc. Great Lakes Res., May 1984, St. Catharines, Ontario.*

Jackson, T. A., and Woychuk, R. N., 1980, 'Mercury speciation and distribution in a polluted river-lake system as related to the problems of lake restoration', *Proc. Internat. Symp. on Restoration of Lakes and Inland Waters, September 1980, Portland, Maine*, U.S. EPA, 440/5-81-010, Washington, D.C., pp. 93–101.

Jackson, T. A., Parks, J. W., Jones, P. D., Woychuk, R. N., Sutton, J. A., and Hollinger, J. D., 1982, 'Dissolved and suspended mercury species in the Wabigoon River (Ontario, Canada): seasonal and regional variations', *Hydrobiologia*, **91/92**, 473–487.

Kemp, A. L. W., and Thomas, R. L., 1976, 'Sediment geochemistry: evidence of cultural impact', *Geoscience Canada*, **3** 191–207.

Kemp, A. L. W., Thomas, R. L., Dell, C. I., and Jaquet, J.-M., 1976, 'Cultural impact on the geochemistry of sediments in Lake Erie', *Jour. Fish. Res. Board Can.*, **33**, 440–462.

Müller, G., and Wagner, F., 1978, 'Holocene carbonate evolution in Lake Balaton (Hungary): a response to climate and impact of man', in Matter, A., and Tucker, M. E., (Eds), *Modern and Ancient Lake Sediments*, Internat. Assoc. Sedimentol. Special Publ. No. 2, Blackwell Scientific, Oxford, pp. 57–81.

Murphy, T. J., and Rzeszutko, C. P., 1977, 'Precipitation inputs of PCBs to Lake Michigan', *Jour. Great Lakes Res.*, **3**, 305–312.

Reuther, R., Wright, R. F., and Förstner, U., 1981, 'Distribution and chemical forms of heavy metals in sediment cores from two Norwegian lakes affected by acid precipitation', *Proc. Internat. Symp. on Heavy Metals in the Environment, Amsterdam*, CEP Consultants, Edinburgh, pp. 318–321.

Rodgers, P. W., and Swain, W. R., 1983, 'Analysis of polychlorinated biphenyl (PCB) loading trends in Lake Michigan', *Jour. Great Lakes Res.*, **9**, 548–558.

Ryder, R. A., Kerr, S. R., Loftus, K. H., and Regier, H. A., 1974, 'The Morphoedaphic Index, a fish yield estimator—review and evaluation', *Jour. Fish. Res. Board Canada*, **31**, 663–688.

Sly, P. G., 1978, Sedimentary processes in lakes, in Leman, A. (Ed.), *Lakes, Chemistry Geology and Physics*, Springer-Verlag, New York, pp. 65–89.

Smith, B. R., and Tibbles, J. J., 1980. 'Sea lamprey (*Petromyzon marinus*) in Lakes Huron, Michigan, and Superior: history of invasion and control, 1936–78', *Can. Jour. Fish. Aquat. Sci*, **37**, 1780–1801.

SFI, (Sport Fishing Institute), 1984, 'Adirondack acid rain update', *SFI Bulletin*, **360**, 5–6.

Szestay, K., 1982, 'River basin development and water management', *WHO/OMS, Water Quality Bull.*, Environment Canada, **7**, 155–162.

Thomas, R. L., Jaquet, J.-M., and Mudroch, A., 1975, 'Sedimentation processes and associated changes in surface sediment trace metal concentrations in Lake St. Clair 1970–1974', *Proc. Internat. Conf. Heavy Metals in the Environment, Toronto, Ontario,* October 1975, pp. 691–708.

Thompson, M. E., and Bennett, E. B., 1980, 'Variations in the degree of acidification of river waters observed in Atlantic Canada', *Proc. Internat. Symp. on Restoration of Lakes and Inland Waters, September 1980, Portland, Maine*, U.S. EPA 440/5-81-010, Washington, D.C., pp. 453–456.

Tollan, A., 1980, 'Effects of acid precipitation on aquatic and terrestrial ecosystems', *Proc. Internat. Symp. on Restoration of Lakes and Inland Waters, September 1980, Portland, Maine,* U.S. EPA 440/5-81-010, Washington, D.C., pp. 438–445.

Upper Lakes Reference Group, 1976, *The Waters of Lake Huron and Lake Superior,* Vol. I, Report to the Internat. Joint comm., IJC Regional Office, Windsor, Ontario, 236pp

Vallentyne, J. R., 1972, 'Freshwater supplies and pollution: effects of the demophoric explosion on water and man', in Polunin, N. (Ed.), *The Environmental Future,* Macmillan Press, Basingstoke, pp. 181–211.

Vernet, J. P., and Favarger, P.-Y., 1982, Climatic and anthrophogenic effects on the sedimentation and geochemistry of Lakes Bourget, Annecy and Leman', *Hydrobiologia,* **91/92,** 643–650.

Vollenweider, R. A., 1968, *Scientific Fundamentals of the Eutrophication of Lakes and Flowing Waters, with Particular Reference to Nitrogen and Phosphorus as Factors of Eutrophication,* OECD, Paris, 159 pp.

Wetzel, R. G., 1975, *Limnology,* W. B. Saunders Co., Philadelphia, PA., 743 pp.

PART III
GEOSPHERE

INTRODUCTION

The geosphere has not been used as a term as extensively as other spheres such as the hydrosphere, the atmosphere, and the biosphere. It refers to the sphere of the earth's land surface but may be closely associated with the cryosphere which is the zone embracing glacial and ground ice. Perhaps in the areas of the geosphere and the cryosphere the significance of human activity in relation to environmental processes has been appreciated much more recently than is the case in relation to other spheres of the earth's environment. It is only since 1960 that there has been extensive acknowledgement of the effects that human activity can have in relation to the exogenetic and the endogenetic processes of the earth's surface.

Exogenetic processes include weathering, slope, river, ice, wind and coastal processes. The last two are considered in other sections but the first four are particularly pertinent here. In each case the rates at which processes operate has been changed, new processes have been instigated in some locations or processes have been arrested in others, and all such changes have very considerable implications. The fields which have emerged include studies which have identified the consequences of changes in process, studies which have mapped the extent of such changes, and also investigations which have suggested the ways in which remedial action may be undertaken to minimize the most adverse effects which human activity can have. It has therefore been appreciated that environmental impact statements are needed especially in relation to the geomorphological implications which arise because of the effect that human activity has upon exogenetic processes.

However, it is perhaps in relation to endogenetic processes that we have had to adapt an approach which was originally very historically based. It was necessary not only to investigate the ways in which the geosphere had changed using enhanced chronological indicators, but it was also necessary to investigate the ways in which process modifications had occurred. In the investigation of the effects of human activity upon endogenetic as well as upon exogenetic processes, one of the important reasons for future research is to establish the

181

extent to which there are spatial variations and spatial patterns in the consequences of human activity. Indeed in a general context W. L. Graf (1984) has argued that such a spatial approach is a very necessary attribute of the training of a landscape manager.

The next three chapters relate to processes on slopes, in river channels and involving ground ice. In each case the emphasis is upon the meso-scale but leads to an appreciation of the spatial variations in the way in which human impact is registered. A similar theme emerges from the fourth chapter where D. R. Coates surveys the endogenetic processes that can be modified by human activity. In the course of research developments concerned with the geosphere in the last two decades there has been considerable progress towards a more relevant geomorphology and some have styled this applied geomorphology. However, whereas past attention has been focused upon the impact of human activity upon the processes of the geosphere, what is now required is that we focus upon the development and design of ways of controlling processes to minimize the feedback effects which follow from particular courses of human action.

REFERENCE

Graf, W. L., 1984, *The Colorado River—Instability and Basin Management*, Association of American Geographers, Washington DC.

Human Activity and Environmental Processes
Edited by K. J. Gregory and D. E. Walling
©1987 John Wiley & Sons Ltd.

8

Slopes and Weathering

M. J. SELBY

8.1 INTRODUCTION

Most of the land surface is composed of hillslopes and, except in areas of great aridity, low temperature and extreme steepness, they have a mantle of weathered rock and soil. By comparison with the underlying rock, soil is a weak and permeable material which can be readily modified by processes of erosion and by direct human interference. The processes of weathering which convert resistant rock to weaker soil are thus essential forerunners of erosion.

The results of weathering can be appreciated from a consideration of the characteristics of a deep profile (Table 8.1). The depth of weathering profiles and the completeness of the horizons are extremely variable. On ancient surfaces of stable continents profile depths may exceed 100 m in the humid tropics, but such depths are rare and 30 to 50 m is a more common maximum depth. Profile depths decrease as slope angles increase so that in many humid temperate environments slopes steeper than about 45° have either bare rock or very thin soil profiles. In some parts of the humid tropics, where weathering rates are very high, slopes with a vegetation cover may have angles of up to 70°, although they are then subject to periodic landsliding. Such high angles are relatively common in Tahiti and Papua New Guinea.

The description of weathering profile features, given in Table 8.1 assumes that the soil is developed *in situ*; where it is derived from transported materials the profile characteristics may be far more variable and the trend of strength and permeability with depth may be highly irregular.

On hard rocks the depth of weathering sets a limit above which human activity can modify the landscape with ease and with little expense. Soft rocks, such as unconsolidated sands or clays may have strengths no greater than those of soils and they also are easily modified by human activity—whether this is by deliberate intervention with machinery or from the, usually unwanted, results of induced erosion.

TABLE 8.1   An idealized weathering profile with some characteristic properties

VI	Residual soil at the surface	No trace of the original rock is preserved; strength is lowest and permeability is greatest unless clay minerals have been formed
V	Completely weathered	Original rock fabric may be preserved but colour and mineral composition are those of soil not rock
IV	Highly weathered	More than 50 per cent of the material is soil. Corestones are common and rock fabrics and structure are dominant. Joints may be open and are then zones of weathering
III	Moderately weathered	More than 50 per cent of the material is rock. Corestones are fitting; joints are partly open
II	Slightly weathered	Discolouration is largely along joints
I	Fresh rock	Joints are closed; rock is not discoloured

Rock compressive strength, that is resistance to crushing, is measured in newtons per square metre and ranges from about 200 MN m^{-2} for a hard igneous rock to about 20 kN m^{-2} for a soft moist clay.
Source: Dearman (1974).

8.2   FACTORS CONTROLLING EROSION

The factors controlling the rate and type of erosion occurring on hillslopes include climate, topography, rock type, vegetation and soil character (Fig. 8.1). They are linked in a web of relationships in which climate and geological factors are the most independent, with soil character and vegetation cover being dependent upon them, and closely related to each other. Vegetation, for example is closely controlled by the amount, duration and intensity of rainfall, by the heat available during the growing season, and also by the capacity of soil to supply nutrients and an anchorage for plant roots. In its turn vegetation influences soil through the action of roots, the provision of organic matter, the uptake of nutrients and by providing strength which resists the impact of raindrops and the shear stresses imposed by running water and landsliding.

The extent to which human interference can modify the impact of the factors of erosion is varied. Rock type and topography cannot be altered on a large scale except at enormous cost; but on a minor scale the formation of agricultural terraces can reduce slope angles and shorten the effective length of slopes. The effect of climate can be most readily modified by reducing the impact of raindrops upon soil (see also Chapters 12 and 13).

A vegetation cover or a mulch of plant material, or plastic sheeting can reduce splash erosion to negligible amounts. By far the easiest way to modify erosion

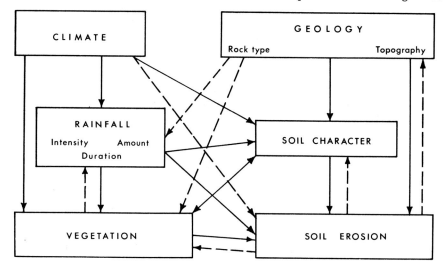

FIGURE 8.1   Relationships amongst the factors influencing soil erosion

is by manipulating plant cover. Thus man can greatly increase potential erosion by exposing soil, or reduce the erosion hazard to very low values by keeping a complete vegetation cover on the soil.

Erosion is a function of the eroding power, that is the erosivity, of raindrops, running water and sliding or flowing soil masses, and the erodibility of the soil or:

$$\text{Erosion} = f(\text{erosivity, erodibility})$$

For given soil and vegetation conditions one storm can be compared quantitatively with another and a numerical scale of values of erosivity can be created. Erodibility is similarly quantifiable because the physical and chemical composition of soil controls its resistance to erosion, because measured soil properties can be related statistically to the sediment removed by a particular process, and because the shear strength of soil can be determined for various moisture contents.

(a)   Erosivity

The effectiveness of erosion processes on a slope is closely related to the disposition of water on or in the soil. A number of alternative routes for the water can be visualized as follows:

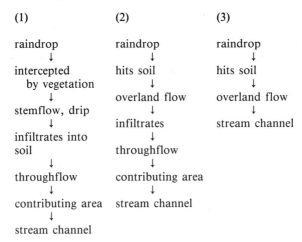

(1)	(2)	(3)
raindrop	raindrop	raindrop
↓	↓	↓
intercepted	hits soil	hits soil
by vegetation	↓	↓
↓	overland flow	overland flow
stemflow, drip	↓	↓
↓	infiltrates	stream channel
infiltrates into	↓	
soil	throughflow	
↓	↓	
throughflow	contributing area	
↓	↓	
contributing area	stream channel	
↓		
stream channel		

In a fully vegetated catchment it is rare for overland flow to occur except along the edges of stream channels where water, which has infiltrated and then moved vertically and laterally through the soil profile, reaches the saturated zone and emerges at the surface as stormflow (Fig. 8.2). The number of quantitative studies which have verified this model are few, but those of Betson (1964), Whipkey (1965), Ragan (1968) and Dunne (1970) support it for humid temperate forests. The implications of this partial-area model are very great, for it follows that surface wash is likely to be relatively slight and the supply

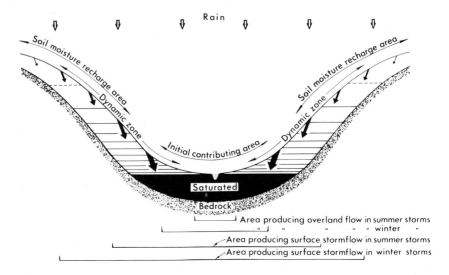

FIGURE 8.2   The dynamic area concept of runoff production as developed by the Tennessee Valley Authority (1964)

of sediment to a stream must come from channel bank collapse and, more rarely, from minor sources such as subsurface pipes, animal burrows or, where they occur, landslide scars. The routing of most drainage waters through the soil helps to explain why in excess of 50 per cent of the denudation of most forested catchments is by solution (Meade, 1969).

The infiltration capacity of many soils, especially those on steep slopes, is very variable, and a raindrop or thread of water which has moved down a plant stem may reach an impermeable patch of soil and move over the soil or litter layer before reaching a more permeable patch or crack into which it can infiltrate. Reported rainfall intensities reach little more than 225 mm h^{-1} but infiltration capacities range from 2 to 2500 mm h^{-1} (Hudson, 1971; Morgan, 1969; Selby, 1970). It is, therefore, only under extreme storm events that overland flow is likely to occur under natural conditions. Bare arable soils frequently

FIGURE 8.3   Rill and gully erosion on long slopes in the South Island, New Zealand

have relatively low infiltration rates because of compaction of the soil by machinery, so that overland flow is relatively common on croplands when the soils are exposed.

The overland flow model of runoff was developed by Horton (1933, 1938, 1945) who postulated that overland flow would occur as soon as rainfall intensity exceeded the infiltration rate. This model appears to be particularly applicable to areas of bare rock, to arid and semiarid climates, to some arable soils, to frozen soils, and to periods of snowmelt. It is not generally applicable to runoff from vegetated surfaces under humid climates. Where it does occur overland flow has relatively low kinetic energy for the entrainment of sediment compared with the energy of large falling raindrops (Hudson, 1971, p. 62). Most of the erosion occurring on bare soil surfaces results from impacting drops breaking down soil aggregates, splashing them into the air, and causing turbulence in surface runoff.

Erosivity under a vegetation cover is therefore low, and from catchments with a full forest or grass cover erosion takes place largely in solution and from the few areas of exposed soil in banks, terracettes and animal burrows. There flow can become channelled, as for example when vegetation cover is removed from slopes, then the erosion rate can be vastly increased as rills and gullies cut through the protective root network and entrain the less cohesive subsoil (Fig. 8.3).

(b)   Erodibility

The soil characteristics which control erodibility are particle size distribution, structure, organic matter content, permeability, and root content (Wischmeier, *et al.*, 1971). These variables are of importance because they control the infiltration capacity and the cohesive and frictional strength of the soil. Root content and organic matter content are contributors to apparent soil cohesion. Roots alone can increase effective soil shear strength by two or three times (O'Loughlin, 1974; Waldron, 1977) and this is significant when considering the implications of man-induced changes of vegetation and land use.

8.3   LANDSLIDING

The mechanisms and causes of landsliding have been discussed at considerable length in the literature (e.g. Terzaghi, 1950; Carson and Kirkby, 1972). Discussion here will consequently be confined to the forces resisting and promoting landslides, to the implications for rural and urban land use, to the periodicity of landsliding, and to its effects on valley floors.

(a)   Shallow translational landslides

By far the most common forms of landslides are shallow features which are classified as debris slides, debris avalanches or debris flows depending upon the degree of remoulding of the soil material and the water content of the sliding soil. The stability of a hillslope against landsliding may be expressed as a factor of safety (*F*) where

$$F = \frac{\text{the sum of resisting forces}}{\text{the sum of driving forces}}$$

When the slope is at equilibrium $F = 1$, when *F* is greater than unity then the slope is potentially stable under existing conditions with the probability of stability increasing with high values of *F*. When the value of *F* is less than unity the slope is in a condition for failure. A stable slope may become less stable in time, as weathering decreases soil strength, because of higher water table levels or because vegetation is removed. The state of stability and the effects of changes may be analysed by taking into account the factors contributing to resisting and driving forces.

Rotational landslides, usually called slumps, are most common in deep and uniform strata of clays, clay shales and highly shattered rocks. The stability of slopes subject to such failures is analysed by dividing the landslide, or potentially failing slope unit, into 'slices' of relatively uniform depth and inclination. For each slice, the resisting and driving forces are calculated. The total of all forces is then derived to form a summation of the stability of the whole slope or feature. The analysis for each slice is similar to that undertaken for a slope which is likely to produce a shallow planar slide of great length, and considerable width compared with the depth. Such a slide is analysed by considering the forces acting on a single representative slice. The analysis assumes that the landslide is not affected by friction along its margins, and that the slope does not have a toe of soil resisting the slide. The slope is, therefore, considered as though it has infinite length and the method of analysis is called an infinite slope analysis (Skempton and De Lory, 1957).

The resisting forces are the effective shear strength of the soil and the driving forces are the gravitational forces tending to move the soil down the hillslope, together with the tendency of water to reduce soil strength and to cause a buoyancy effect within the soil. The significance of the gravitational forces will increase as the slope angle increases. Effective shear strength at any point in the soil is given by the Coulomb equation:

$$s = c' + (\sigma - u) \tan \phi'$$

where *s* is the effective shear strength at any point in the soil; c' is the effective cohesion as reduced by loss of surface tension; σ is the normal force imposed by the weight of solids and water above the point in the soil; *u* is the pore water pressure derived from the unit weight of water (γ_w) and the hydrostatic head ($\gamma_w \, mz$) (see Fig. 8.4); and ϕ is the angle of friction with respect to effective stresses.

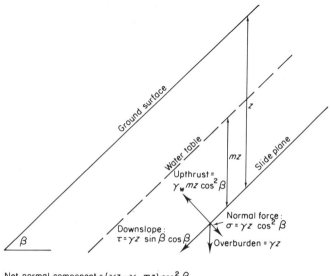

Net normal component = $(\gamma z - \gamma_w mz) \cos^2 \beta$

FIGURE 8.4   Stresses within a soil mass on a slope

Substituting for σ, the Coulomb equation can be rewritten as:

$$s = c' + (\gamma z \cos^2 \beta - u) \tan \phi'$$

where γ is the unit weight of soil; z is vertical depth of soil above the failure plane; β is the angle of inclination of the slope.

As the driving forces are $\gamma z \cos \beta \sin \beta$ then:

$$\text{Factor of safety} = F = \frac{c' + (\gamma z \cos^2 \beta - u) \tan \phi'}{\gamma z \sin \beta \cos \beta}$$

It is convenient to express the vertical height of the water table above the slide plane as a fraction of the soil thickness above the plane, and this is denoted by m. Then if the water table is at the ground surface $m = 1.0$, and if it is just below the slide plane $m = 0$. Pore water pressure on the slide plane assuming seepage parallel to the slope, is given by:

$$u = \gamma_w mz \cos^2 \beta$$

thus

$$F = \frac{c' + (\gamma - m\gamma_w)z \cos^2 \beta \tan \phi'}{\gamma z \sin \beta \cos \beta}$$

Given that

$$\phi = 12°; \ c' = 11.9 \ \mathrm{kN \ m^{-2}}; \ \gamma = 17 \ \mathrm{kN \ m^{-3}}; \ \beta = 15°;$$
$$z = 6 \mathit{metres}; \ m = 0.8; \ \gamma_w = 9.81 \ \mathrm{kN \ m^{-3}}$$

then

$$F = \frac{11.9 + \ (17 - 0.8 \times 9.81) \ 6 \times 0.92 \times 0.2}{17 \times 6 \times 0.25 \times 0.96} = 0.9$$

Thus the slope is prone to long-term failure when the water table is about a metre below the ground surface. If the water table can be lowered by drainage to just below the slide plane (when $m = 0$) then $F = 1.3$ and the slope will be stable against shallow landsliding.

In extremely dry summer periods soils frequently crack and, depending upon the shrinkage limits of the soil, these cracks may extend to depths of a metre or more. In both California and the North Island of New Zealand severe summer rainstorms frequently fall upon dry catchments with deeply cracked soils. As a result water rapidly fills the cracks and seeps downslope under the hydrostatic pressure of water filling the crack. The effective piezometric surface is then well above the soil surface ($m = 1.3$ to 1.6). Such storms frequently give rise to landsliding where rainfall intensities are very high and are in excess of the infiltration capacity of the soil at the base of the tension cracks (Selby, 1967, 1976; Campbell, 1975).

The components of the infinite slope analysis show that only two variables can be readily modified by human influence to increase slope stability, namely water content of the soil and apparent cohesion. Water content can be modified by drainage programmes and apparent cohesion can be modified in so far as it is influenced by the root content of the soil and therefore by the vegetation or land use cover.

(b) Effects of vegetation upon slope stability

The root networks of all plants provide mechanical reinforcement of the soil, and in shear box testing of soil strength this is evident as an increase in apparent cohesion (Table 8.2). Some roots grow downwards through potential failure zones of shallow landslides into underlying soil or rock. One study in Ohio, USA (Riestenberg and Sovonick-Dunford, 1983), found that the average shear strength contributed by tree roots, at the potential shear plane, was about

TABLE 8.2   Strength added to soil by plant roots

Plant	Soil	Increases in apparent cohesion (kPa)
Conifers (pine, fir)	glacial till	0.9–4.4
Alder	silt loam	2.0–12.0
Birch	silt loam	1.5–9.0
Podocarps	silty gravel	6.0–12.0
Alfalfa (lucerne)	silty clay loam	4.9–9.8
Barley	silty clay loam	1.0–2.5

Note: in any soil the strengthening effect of roots varies greatly vertically and laterally.
Source: Unpublished material; and O'Loughlin, 1974; and Waldron, 1977.

5.7 kN m^{-2} and the frictional strength of the soil was only about 0.7 kN m^{-2}. The tree roots thus increased the factor of safety against landsliding nine-fold. Root strength allows forested slopes in that area to remain stable at angles as high as 35°, whereas similar slopes devoid of trees are subject to landsliding at inclinations of 12° to 14°. In some forests the tree roots do not penetrate the underlying rock or pass through the failure zone, but they may still have an effect on soil strength by interlocking in a network which binds together the upper soil horizons. This interlocking strength is very variable, although it usually falls in the range of 1.0 to 12.0 kPa. Trees, but not smaller plants, have three other effects on slope stability (Brown and Sheu, 1975; Gray, 1970). First, wind throwing and root wedging occurs as trees are overthrown by strong winds or under heavy snowfalls. Secondly, trees increase the surcharge on a soil mantle and have the effect of increasing the normal load upon the failure plane and thus of increasing slope stability—but this is only a significant effect when large, closely spaced trees occur on thin soils and provide an increase in normal stress of up to 5 kPa, while increasing shearing stress by only half that (Bishop and Stevens, 1964). Thirdly, trees may also increase evapotranspiration from the soil and so lower water tables.

The importance of trees for slope stability is demonstrated in many areas where deforestation has been followed, after an interval of a few years, by periods of severe shallow landsliding. Examples of this sequence are known for New Zealand, Alaska, British Columbia, the Himalayan foothills and Japan. The interval which occurs between clearance and landsliding may be due to the lack of a storm with sufficient intensity and duration to raise porewater pressures to critical levels, but it is more commonly attributable to the slow decay of the tree roots which are left in the soil. The thinnest roots decay first so that over the first year or so after clearing, root tensile strength per unit volume of soil

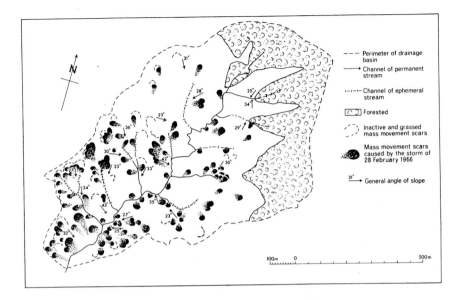

Perimeter of drainage basin

Channel of permanent stream

Channel of ephemeral stream

Forested

Inactive and grassed mass movement scars

Mass movement scars caused by the storm of 28 February 1966

General angle of slope

FIGURE 8.5   The distribution of shallow translational landslides in one small catchment in the Hapuakohe Range, New Zealand

decreases rapidly and then over the following two to ten years the larger roots decay until they contribute nothing to soil apparent strength. The actual rate of decay depends upon the tree species, the soil climate and the rooting pattern of the trees.

The effect of trees upon slope stability is particularly noticeable where adjacent tree-covered and grass-covered slopes are subjected to the same storm. The incidence of landslides is nearly always far greater on the grass-covered slopes (Selby, 1967; Pain, 1971; Swanston, 1970) as illustrated in the basin shown in Fig. 8.5.

(c)   Effects of landsliding on valley floors

The debris from very large individual landslides or that from many small slides occurring during a single storm over a catchment can have effects on valley floors which may be evident for hundreds of years (Figs 8.6 and 8.7). Large slumps and flows in rocks, such as mudstones, may infill valleys to depths of many metres and, once the vegetation cover has been disrupted by a slide, continued erosion by gullies may provide so much sediment that streams are incapable of removing it. This happened in parts of Japan (Machida, 1966) and in northern New Zealand where a series of severe rainstorms in the 1930s triggered numerous landslides. In the Waipaoa River

FIGURE 8.6　A massive landslide in mudstones

catchment some valley floors have been infilled to depths of over 30 m since then and aggradation is still continuing in spite of attempts to afforest the slopes.

The northern Ruahine Range (Fig. 8.7), where the underlying bedrock is severely shattered sandstones and argillites, has been subjected to a series of severe storm events over the last 400 years which have left behind a sequence of terraces (Grant, 1965). The approximate age of each terrace has been determined from the oldest trees on it and from the known age of volcanic ash deposits which have fallen onto each terrace surface; the basic assumption being made that the terrace cannot be younger than the trees or ashes on it. Terraces are dated as being A.D. pre-1300, 1450, 1650 and 1840. The more recent terraces suggest that very large catastrophic events may occur every 200 years or so, but it is certain that more frequent severe storms also cause much erosion. The debris of lesser events may, however, be largely removed from catchments either in a series of pulses as large floods carry waves of debris down the channels, or by minor floods which occur several times a year.

It can be seen from Fig. 8.7 that near the headwaters of the streams large influxes of debris bury the channel forms and low terraces. The debris in this channel was all deposited in one major event and the minor terraces alongside the channel were formed in the two years afterwards. The datable evidence

FIGURE 8.7   A valley floor infilled with rock debris derived from numerous shallow
landslides in the Northern Ruahine Range, New Zealand

for severe storm periods is consequently preserved not on the slopes, nor
in the channels of the headwaters, but in the middle reaches of the rivers where
terraces are not necessarily buried by the subsequent wave of debris, and terrace
remnants are preserved, at least discontinuously, along reaches of migrating
channels.

Catastrophic slope erosion and infilling of valley floors is not confined to
Japan and New Zealand. It has been described from the Caucasus and from
California (Starkel, 1976; Lamarche, 1968). In the Caucasus the term 'sjel' is
applied to rapid mudflows and debris-laden streams with very high densities.
Mudflows with densities of 1.7 to 2.3 t m^{-3}, are capable of transporting very
large boulders because the boulders have densities of 2.5 to 2.6 t m^{-3} and
virtually float in the mud. Several sjels may occur in one valley in a single year.

The effects of landslides on valley floors can be of even greater economic
importance than the landslides themselves. In steep hill country, roads, railways,
and power and telephone cables frequently follow the valley floors and may
be buried by debris. Perhaps of wider significance is the general raising of stream
beds so that channels become unstable, flood protection banks are overtopped,
bridges are buried or their clearance above flood levels is reduced, and the
agricultural production from floodplains is lost or impaired.

Very large landslides of the rock slide and rock avalanche type which are reported from high mountain ranges such as the Canadian Rockies (Cruden, 1976), the Peruvian Andes (Browning, 1973) and the Himalayas (Kingdon-Ward, 1955) have an even more devastating effect for they are capable of burying towns and blocking valleys. The lakes impounded behind the landslide debris may eventually burst through the temporary dam to release a devastating flood wave upon the valley downstream.

8.4 PERIODICITY OF EXTREME EVENTS ON SLOPES

Severe storms with heavy rainfall are the most effective agents of change on hillslopes. The greater the magnitude, that is the amount and intensity, of the rainfall, the lower is its frequency. Very severe storms, therefore occur separated by long intervals of time but the frequency with which land-forming processes are effective depends partly upon the type of process and partly upon the climate of a region. In any area it must also be affected by the permeability of the regolith and the closeness of the hillslopes to limiting angles of stability.

It has been contended that in rivers most of the work of transportation is carried out by floods which are of such magnitude that they recur with a frequency of at least once in five years (Wolman and Miller, 1960). On hillslopes, processes such as solution occur during every prolonged rainfall but the processes which remove large quantities of debris, such as gullying and landsliding, occur at infrequent intervals. The frequency of these catastrophic events has a major effect upon the rate at which the surface of the earth changes.

Hillslope processes which are active frequently do not greatly disturb the approximate balance which usually exists between the rate of weathering and the removal of regolith. Except in deserts, therefore, there is an approximate equilibrium between weathering, soil development, the vegetation cover, and the rate of erosion on hillslopes. Extreme events destroy this equilibrium, breaking the vegetation cover, stripping away regolith along the track of gullies and landslides, and producing large influxes of sediment into valley floors. As it is uncommon for several extreme events to occur in a short interval, the scars and the deposits may then be slowly modified by lesser intensity processes such as creep, solution, and wash, while weathering, soil formation and vegetation gradually re-establish a surface which is in approximate equilibrium with the energy of the processes usually acting on the slope. After a period the equilibrium may be broken by another extreme storm. This type of episodic development of the land surface can be visualized as occurring in a stepwise manner (Fig. 8.8) in which storm events are following by gradual periods of adjustment to the normally active processes and then, when an approximate equilibrium is re-established, by a period of relative stability.

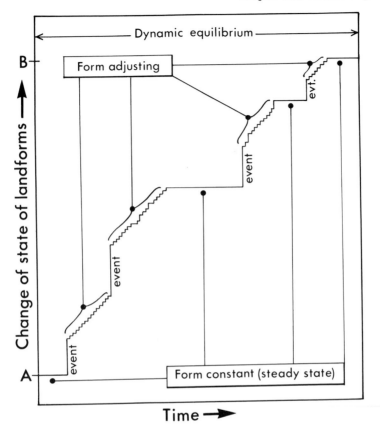

FIGURE 8.8   Implications of dynamic equilibrium

The importance of extreme events in different regions of the world is very variable. Starkel (1976) reviewed studies of extreme events and concluded that four classes of region may be distinguished.

(1) Regions with a frequency of extreme events of 5–10 per century and with events of such magnitude that in each the denudation greatly exceeds the denudation by all low-intensity processes during 100 years. Such areas are most common in tropical monsoon and mediterranean climates, and in steep uplands and farming lands of the temperate zone where human action has removed or changed the vegetation cover.
(2) Regions in which extreme events are rare and in which such events do not exceed the total denudation of 100 years of low-intensity processes. In these areas heavy rainfalls or snowmelts occur each year. Such regions are common in the semiarid zone.

(3)   Regions with very rare extreme events. Because of normally very low precipitation a storm may achieve much more denudation than usually occurs in 100 years. Arid zones and some parts of the boreal zone are in this category.

(4)   Stable regions showing little variation from normal denudation rates. Many Arctic, Antarctic, continental boreal and lowland zones of the temperate regions are in this group.

In any part of the world, mountains are far more subject to extreme events than lowlands, but even in them the effects of extreme denudation are frequently limited to quite small localities. Thus catastrophic events may occur in a mountain range nearly every year but each time hitting a different area so that overall the frequency is much less than one event a year. Similarly, the sediment produced by a storm event may range from a few tonnes per square kilometre, to hundreds of thousands of tonnes per square kilometre and the average downwearing of the affected areas from 0.01 to 200 mm. This range occurs because the area of the ground suffering extreme erosion during a storm may vary from less than 1 to more than 5 per cent.

Data on the frequency of extreme events are not available for many parts of the world, but some comparisons are presented in Fig. 8.9. In the hills and lowlands of Western Europe periods of landsliding appear to be related to intervals of wet climate since the last glacial (Fig. 8.10) (Starkel, 1966). These periods last for a few hundred years and occur at intervals of 2000 years. At the other end of the scale the frontal ranges of the Himalayas, north of the Ganges delta, receive prolonged and intense rainfalls nearly every year and extreme landforming events probably have a frequency of about once in five years (Starkel, 1972).

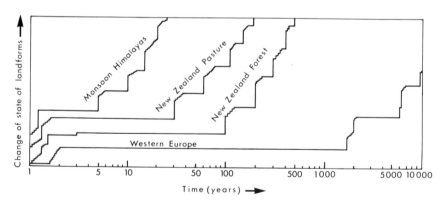

FIGURE 8.9   Curves showing the magnitude and frequency of landform change by landsliding for selected environments

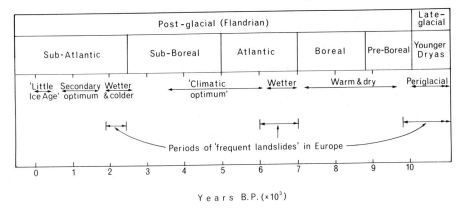

FIGURE 8.10 Periods of landsliding in lowland Western Europe during the last 11 000 years

Less frequent are the storms causing landslides in the North Island of New Zealand. Under original forest the hills are affected by shallow landsliding about once in every 100 years, but in the last century many uplands have been cleared of forest, and pastures now cover the hills. Pastures are less protective than the forests, and storms of lower intensity, which recur about once in thirty years, now cause landslides (Selby, 1967, 1976; Pain, 1969).

8.5 SLOPES IN URBAN AREAS AND ALONG ROUTEWAYS

In densely populated areas the spread of settlement on to steep slopes may be necessary, and it is often desired by people seeking extensive views or freedom from the air pollution of enclosed valleys. On a slope which has a factor of safety close to unity, almost any increase in the driving forces or decrease in resisting forces can be hazardous. When roads and buildings are placed on slopes, or sewers, water pipes and power cables are buried in them, there is nearly always an accompanying change in the stresses acting on the slope materials. With good planning the probability of instability developing, as a result of stress changes, can be diminished; but with bad planning or no planning, the probability may be increased.

Slope failure may be encouraged by many results of construction works, but three features are particularly important: first, the removal of the toes of slopes reduces support for the soil and rock above; secondly, the removal may expose joints, weak beds or other features of low strength which are dipping in the direction of the slope, and so make possible a large failure along the plane of weakness; thirdly, the presence of extra water is a common triggering event for landslides (see Selby, 1982, for a detailed review).

The removal of the toes of slopes is frequently necessary if roads are to be built across a slope. Where soils are thin and the road can be cut into stable rock there may be no hazard, but deep soils, with downslope water seepage, may be severely affected. Where trees are removed to provide space for buildings the potentially unstable area may be greatly enlarged. Protective measures involve improving slope drainage by ditches, and by porous pipes driven into the slope, or almost horizontally across it. Ditches may be filled with coarse gravel to provide a buttressing effect and prevent undercutting of the upper

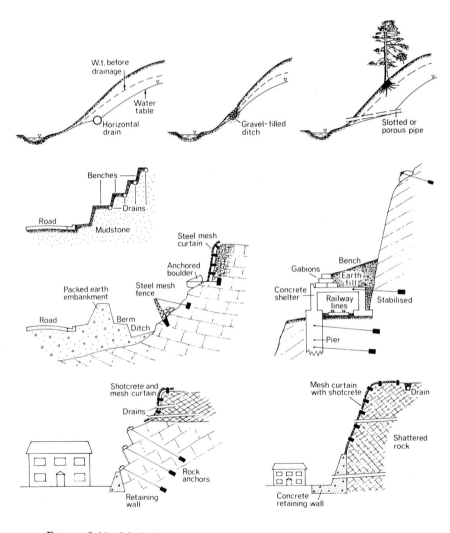

FIGURE 8.11   Methods of stabilizing slopes and protective structures

slopes. Replanting trees in critical areas, and placing reinforcing walls or gabions around the base of undercut slopes, are both common protective measures. Gabions are rectangular baskets of wire mesh which are filled with large rock fragments; they have the advantage of being freely draining, they can be filled with any 'hard' rock or brick materials, and, where cheap local labour is available, they can be constructed easily, quickly and at a low price. For these reasons, gabions are commonly used in many developing and developed countries for small-scale bridge abutments, supports for road shoulders, training and protecting walls in stream channels and as walls protecting structures from rock falls (Fig. 8.11).

Exposure of weak rock units may require expensive and difficult engineering remedies. The simplest situations are those in which zones of fractured rock can be protected from erosion by being covered with wire nets, which in very hazardous places may be sprayed with shotcrete (a cement mixture used as a strong sealer), and larger rock blocks may be anchored to the slope individually. Exposure of beds which dip down the slope may give rise to conditions in which large rock masses could slide. Remedial measures involve draining the slope, and drilling through the potentially unstable mass into firm rock and placing rock or cable anchors, often surrounded by concrete, to bolt the potentially unstable mass in place. In a few extreme conditions in steep mountains, routeways are protected by being placed in tunnels or roofed galleries. Some tunnel portals and galleries have thick fills of soil on them to absorb the shock of falling rock debris.

The presence of excess water at a site may result from badly designed storm water drains, it may be seepage from leaking swimming pools, the irrigation of lawns, runoff from roofs and sealed areas, or from the diversion of natural drainage.

A detailed slope stability analysis may indicate whether the slope is close to limiting equilibrium, but before detailed planning commences many potential hazards can be recognized by careful study of aerial photographs and soil conditions. The best indications of future instabilities are their existence in the past, for many landslides occur within the site of old slides simply because landslide debris usually has a significantly lower shear strength than undisturbed soil, and the drainage of debris from old slides is usually impeded. Signs of instability include cracked or hummocky surfaces, crescent-shaped depressions, crooked fences, trees or lamp-posts leaning uphill or downhill, uneven road surfaces, swamps or wet ground in elevated positions, rushes or other wet-site vegetation growing on slopes, and water seeping from the ground.

All surface investigations should include measurement of slope angles of adjacent stable and unstable areas with similar geology, groundwater conditions and vegetation cover to that of the area which is to be developed. This information provides a useful guide to the likely failure of slopes before and after construction. Such preliminary investigations then have to be followed by detailed investigations of rock, soil and groundwater conditions.

From suitable information it is possible to determine which slopes can be safely traversed by roads or be cut into for house sites, and which areas need special reinforcement of cut slopes and diversions of drainage.

Slope stability investigations and protection works are often expensive but they are almost invariably cheaper than the remedial works which are necessary if a slope does fail in an urban area. Precautionary measures taken before development also have the advantage that they can be charged to the intending occupier, but remedial works are often charged to the community, either through local taxes or increased insurance premiums.

8.6   DENUDATION AND LAND USE PLANNING

The four main variables which influence the rates of downwearing of the land surface—climate, rock type, vegetation and slope angle—themselves contain such wide variations that it is probably premature to suggest that we have anything better than an order of magnitude estimate and understanding of the rate of denudation. Knowledge of actual rates is, however, essential for informed planning of land use. In most parts of the world it is probably true that lowland and gentle hillslope surfaces are worn down at rates of the order of 50 mm per 1000 years, and mountain slopes are worn down at rates of 500 mm per 1000 years (see Selby (1974b) and Saunders and Young (1983) for reviews of the evidence). Agricultural, deforestation and construction activities by man have probably increased world denudation rates two or three times above the long-term geological rate and locally the accelerated rates may be one thousand times as great.

An appreciation of the effect of vegetation cover on erosion rates has been obtained from the work of the United States Soil Conservation Service (see Chapter 13) but it is still unclear how the partial-area concept of contributing areas for storm flow, and also understanding of the periodicity of landsliding, affect currently available estimates of denudation. Many measurements of processes have been made on slope units which are not part of the dynamic zone of water and sediment supply to streams and hence these measurements may not be representative of slope processes.

The partial-area model has provided a valuable indication of where soil conservation works need to be located for maximum effect. Conservation works such as tree planting and the formation of thick grass strips are clearly best positioned in the storm flow contributing areas around a channel. In areas where overland flow can occur the primary aim must be to reduce the depth and length of flow over the slope by employing strips of closely spaced vegetation or shallow grassed channels along the contour (see Chapter 13).

Landslide hazards are far more difficult to prevent. Extremely vulnerable areas can be planted, or even contained by retaining walls if this is warranted, but appreciation of the probable frequency of landsliding allows a comparison to be made between an economic assessment of the cost of repair works and the cost of protection works. Whether or not expensive protection works are erected depends, of course, not only on cost comparisons but on the social and political acceptance of dangers to life and property. Whether or not the knowledge of slope processes now available is used for better planning in rural and urban areas depends partly upon supplying expertise to decision makers but, at least as importantly, it depends upon the political will to use that information.

REFERENCES

Betson, R. P., 1964, 'What is watershed runoff?' *Journal of Geophysical Research*, **69**, 1541–1551.

Bishop, D. M., and Stevens, M. E., 1964, 'Landslides in logged areas in SE Alaska', *US Department of Agriculture Forest Service Research Paper, NOR-1*, pp. 1–18.

Brown, C. B., and Sheu, M. A., 1975, 'Effects of deforestation on slopes', *Journal of the Geotechnical Engineering Division, American Society of Civil Engineers*, **GT2**, 147–65.

Browning, J. M., 1973, 'Catastrophic rock slide, Mount Huascaran, North-central Peru, May 31, 1970', *American Association of Petroleum Geologists Bulletin*, **57**, 1335–1341.

Campbell, R. H., 1975, 'Soil slips, debris flows and rainstorms in the Santa Monica Mountains and vicinity, southern California', *US Geological Survey Professional Paper, 851*, pp. 1–51.

Carson, M. A., and Kirkby, M. J., 1972, *Hillslope Form and Process*, Cambridge University Press, Cambridge.

Cruden, D. M., 1976, 'Major rock slides in the Rockies', *Canadian Geotechnical Journal*, **13**, 8–20.

Dearman, W. R., 1974, 'Weathering classification in the characterisation of rock for engineering purposes in British practice', *Bulletin International Association of Engineering Geology*, **9**, 33–42.

Dunne, T., 1970, 'Runoff production in a humid area', *US Dept of Agriculture ARS 41-160*, pp. 1–108.

Grant, P. J., 1965, 'Major regime changes of the Tukituki River, Hawke's Bay, since about 1650 A.D.', *Journal of Hydrology (N.Z.)*, **4**, 17–30.

Gray, D. H., 1970, 'Effects of forest clear cutting on the stability of natural slopes', *Bulletin Association of Engineering Geologists*, **7**, 45–66.

Horton, R. E., 1933, 'The role of infiltration in the hydrologic cycle', *American Geophysical Union Transactions*, **14**, 446–60.

Horton, R. E., 1938, 'The interpretation and application of runoff plot experiments with reference to soil erosion problems', *Soil Science Society of America Proceedings*, **3**, 340–349.

Horton, R. E., 1945, 'Erosional development of streams and their drainage basins', *Geological Society of America Bulletin*, **56**, 275–370.

Hudson, N., 1971, *Soil Conservation*, Batsford, London.

Kingdon-Ward, F., 1955, 'Aftermath of the great Assam earthquake of 1950', *Geographical Journal*, **121**, 290–303.

Lamarche, V. G., 1968, 'Rates of slope degradation as determined from botanical evidence, White Mountains, California', *US Geological Survey Professional Paper 3521*, pp. 341–377.

Machida, H., 1966, 'Rapid erosional development of mountain slopes and valleys caused by large landslides in Japan', *Geographical Reports of Tokyo Metropolitan University*, **1**, 55–78.

Meade, R. H., 1969, 'Errors in using modern stream-load data to estimate natural rates of denudation', *Geological Society of America Bulletin*, **80**, 1265–1274.

Morgan, M. A., 1969, 'Overland flow and man', in Chorley, R. J. (Ed.), *Water, Earth and Man*, Methuen, London, pp. 239–255.

O'Loughlin, C., 1974, 'The effects of timber removal on the stability of forest soils', *Journal of Hydrology (N.Z.)*, **13**, 121–34.

Pain, C. F., 1969, 'The effect of some environmental factors on rapid mass movement in the Hunua Ranges, New Zealand', *Earth Science Journal*, **3**, 101–107.

Pain, C. F., 1971, 'Rapid mass movement under forest and grass in the Hunua Ranges, New Zealand', *Australian Geographical Studies*, **9**, 77–84.

Ragan, R. M., 1968, 'An experimental investigation of partial area contributions', *International Association for Scientific Hydrology Publication*, **76**, pp. 241–51.

Riestenberg, M. M., and Sovonick-Dunford, S., 1983, 'The role of woody vegetation in stabilizing slopes in the Cincinnati area, Ohio', *Geological Society of America Bulletin*, **94**, 506–518.

Saunders, I., and Young, A., 1983, 'Rates of surface processes on slopes, slope retreat and denudation', *Earth Surface Processes and Landforms*, **8**, 473–501.

Selby, M. J., 1967, 'Geomorphology of the greywacke ranges bordering the Lower Waikato Basin', *Earth Science Journal*, **1**, 37–58.

Selby, M. J., 1970, 'Design of a hand-portable rainfall-simulating infiltrometer with trail results from Otutira Catchment', *Journal of Hydrology (N.Z.)*, **9**, 117–132.

Selby, M. J., 1974a, 'Dominant geomorphic events in landform evolution', *Bulletin International Association of Engineering Geology*, **9**, 85–89.

Selby, M. J., 1974b, 'Rates of denudation', *New Zealand Journal of Geography*, **56**, 1–13.

Selby, M. J., 1976, 'Slope erosion due to extreme rainfall: a case study from New Zealand', *Geografiska Annaler*, **58A**, 131–138.

Selby, M. J., 1982, *Hillslope Materials and Processes*, Oxford University Press, Oxford.

Skempton, A. W., and De Lory, F. A., 1957, 'Stability of natural slopes in London Clay', *Proceedings Fourth International Conference on Soil Mechanics and Foundation Engineering, London*, Vol. 2, pp. 378–381.

Starkel, L., 1966, 'Post-Glacial climate and the moulding of European relief', *Proceedings International Symposium on World Climate 8000 to 0 B.C.*, Royal Meteorological Society, London, pp. 15–31.

Starkel, L., 1972, 'The role of catastrophic rainfall in the shaping of the relief of the Lower Himalaya (Darjeeling hills)', *Geographica Polonica*, **21**, 103–147.

Starkel, L., 1976, 'The role of extreme (catastrophic) meteorological events in contemporary evolution of slopes', in Derbyshire, E. (Ed.), *Geomorphology and Climate*, Wiley, London, pp. 203–246.

Swanston, D. N., 1970, 'Mechanics of debris avalanching in shallow till soils of southeast Alaska', *USDA Forest Service Research Paper PNW-103*, pp. 1–17.

Tennessee Valley Authority, 1964, 'Bradshaw Creek-Elk River, a pilot study in area-stream factor correlation', *Research Paper 4 (TVA)*, pp. 1–64.

Terzaghi, K., 1950, 'Mechanism of landslides', *Bulletin Geological Society of America*, Berkey Volume, pp. 83–122.

Waldron, L. J., 1977, 'The shear resistance of root-permeated homogeneous and stratified soil', *Soil Science Society of America Journal*, **41**, 843–849.

Whipkey, R. Z., 1965, 'Subsurface stormflow from forested slopes', *International Association for Scientific Hydrology Bulletin*, **10**(2), 74–85.

Wischmeier, W. H., Johnson, C. B., and Cross, B. V., 1971, 'A soil erodibility nomograph for farmland and construction sites', *Journal of Soil and Water Conservation*, **26**, 189–193.

Wolman, M. G., and Miller, J. P., 1960, 'Magnitude and frequency of forces in geomorphic processes', *Journal of Geology*, **68**, 54–74.

Human Activity and Environmental Processes
Edited by K. J. Gregory and D. E. Walling
©1987 John Wiley & Sons Ltd.

9

River Channels

K. J. GREGORY

It is with rivers as it is with people: the greatest are not always the most agreeable nor the best to live with.

Henry van Dyne, 1852–1933

9.1 INTRODUCTION

The impact of human activity on river channels can be found in the extent to which the river has been *agreeable* and the way in which rivers have been *lived with* because these two things have inspired the deliberate changes of river channels by human action. In turn such changes have produced feedback effects in terms of channel adjustments and have encouraged the development of better ways of engineering a river channel to ensure that the river is more agreeable and easy to live with.

The impact of human activity on river channels is not confined to the last few decades. This is underlined in China where rivers are of great significance and where the Chiang Jiang is regarded as 'China's main street', because of its potential for navigation, and the Huang He, or Yellow River, is thought of as 'China's sorrow' because of its history of flooding, its shifting channels, and its high silt load. The dimensions of the river problem in China are indicated when it is recalled that a quarter of the Huang He's drainage basin is able to provide loess to contribute to the high silt load which represents up to 46 per cent, by weight, of the flow in the river channel; that in a century the Hwang He carries sediment equal to a layer 3 cm thick from the entire basin; that the river has flooded and broken through levees many times during the last 4000 years; and that during the last 4262 years there have been twenty changes of the river's course. Some of these changes have been very substantial and the major shifts are illustrated in Fig. 9.1. Because of the high population of 110 million within the drainage basin with a quarter of a million square kilometres that were susceptible to flooding, the impact of major events like

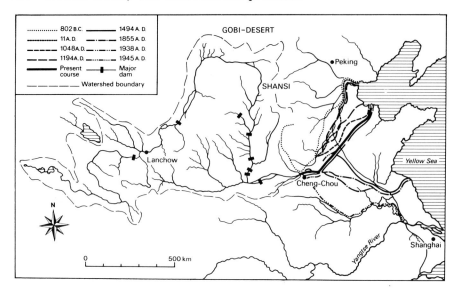

FIGURE 9.1   Major changes of the course of the Yellow River (Huang He) China. The major changes of course illustrate the major shifts that can occur and the diversion to the southeast was deliberately effected during wartime conditions. Some major dams are shown. (Based on Chou, 1976.)

1855 when 250 000 people died, and in addition the characteristics of the river itself, the river channel has been the subject of control, particularly since Liberation. Since then irrigation work and flood prevention have been further developed so that 1800 km of restraining dyke have been built, flow velocities have been reduced by the construction of training walls, piers, and baffles, conservation methods have been extended in the loess areas, and many large dams have been constructed. Six large dams exist on the main river (Fig. 9.1) and the Sanmen gorge dam built between 1956 and 1960 could control 92 per cent of the water; but siltation in the first four years of operation reduced the reservoir capacity by 50 per cent.

Although developed more recently, control of the Mississippi has also been very extensive. The first levees were built in 1699 by individual landowners; by 1844 levees were almost continuous from 30 km below New Orleans to the mouth of the Arkansas river on the west bank and to Baton Rouge on the east bank of the river; and in 1879 Congress founded the Mississippi River Commission. Regulation over the succeeding century led to the gradual control of floods. Up to 1931 there was a flood every 2.8 years. Flooding occurred every 2.2 years between 1932 and 1943, but only once every 5.5 years between 1944 and 1955, and there was only one overbank condition between 1956 and 1969 (Stevens *et al*, 1975).

These two examples illustrate the need to control rivers as man's activities have increasingly extended within the river's domain, and so we need to establish the ways in which man has modified processes and river channels (Section 9.2), to understand the effects that these modified processes and modified channels have had (Section 9.3), and to know of the alternative strategies for river channel development (Section 9.4).

9.2   DOWN THE RIVER

R. L. Sherlock, who wrote *Man as a Geological Agent* in 1922, noted that when Queen Victoria came to the throne, the River Tyne in northeast England was a tortuous, shallow river, full of sand banks and eddies, and it was fordable, whereas in 1922 the river was severely affected by industry and was regulated. Fifty-six years later in a paper on world dams, Beaumont (1978) noted that 20 per cent of the total runoff in Africa and North America is regulated by reservoirs, 15 per cent in Europe, and 14 per cent in Asia. There was a surge of dam-building between 1900 and 1940 and a marked rise occurred after 1950. In a summary of impounded rivers (Petts, 1984) it was shown that, whereas about 20 per cent of river runoff in Africa and North America was contributed by impoundments in the early 1980s, it was estimated that by the year 2000 some 66 per cent of the world's river flow would be controlled by dams. Between 1922 and 1984, many changes of river channel processes occurred and this was accompanied by a growing awareness of the effects which these changes of process (Chapter 4) could have upon river channels.

(a)   Direct modifications of processes

Processes in river channels have been modified directly and indirectly as a result of human influence. Direct modifications involve those cases where the river channel is modified in some way. Mrowka (1974) cited dam and reservoir construction, channelization, bank manipulation and levee construction, and irrigation diversions as examples of direct channel manipulation. Such changes can be resolved into two major categories. The first category consists of those changes which arise consequent upon a change of the river channel at a specific location. This includes: the construction of a dam to impound a reservoir for water supply, for flow regulation, or for the generation of hydroelectric power; the abstraction of river flow for water supply or for diversion to another river system; the return of water at a specific outfall point from industrial use or from water supply systems; and also the building of structures which intersect the river channel at a specific point, such as the bridges needed by road and rail transport networks. In all these cases the water and sediment in the river may be modified. Thus, the construction of a dam provides a storage reservoir upstream and this will modify the pattern of downstream discharges (Rutter and Engstrom, 1964). If the reservoir is not full then inflow may be contained within the reservoir so that peak discharges downstream may be reduced by 98 per cent; but even if the reservoir is near capacity level the routing of a flood peak through the reservoir can lead to a peak discharge downstream reduced by more than 50 per cent (Moore, 1969). Because storage of water in the reservoir may increase the residence time of water in the river system, losses by seepage and evaporation can be increased, giving a reduced volume of river flow. In addition to the major dams of the world there are many small ones used for local water supplies and conservation measures, so that it is evident that flow regulation is of widespread occurrence. In general the magnitude and frequency of peak flows is reduced downstream from a dam, and below Hoover dam on the Colorado the magnitude of the mean annual flood has been reduced by 60 per cent (Dolan *et al.*, 1974). Also, on the Colorado, which has been described as the world's largest plumbing system (Fradkin (1981), quoted by Graf (1985)), it has been shown that some 9.87 million m^3 of bottom sediment were scoured from the channel downstream of Glen Canyon dam (constructed between 1956 and 1963), although it subsequently became quite stable between 1965 and 1975 (Pemberton, 1976). Reservoirs provide a sediment trap in which up to 95 per cent of the bedload and suspended sediment carried by the river can be retained; this encourages scour downstream of the dam. At the upstream end of the reservoir, sediment accumulation may be encouraged since the velocity of the water is reduced on reaching the impounded water body. Some effects on processes are obvious but others are less immediately apparent. Petts (1984) has used three orders of impact: first order effects occur simultaneously with dam closure and affect the transfer of energy and material into the dam and downstream from it; second order impacts are the changes of channel structure and of primary production which follow; and third order impacts arise because of the effects of these two impacts on the fish and invertebrate communities.

Whereas other point modifications may have less dramatic effects on water and sediment discharge, the consequences will depend on the way in which abstraction of water and sediment relates to peak flow rates. If a constant amount of water is abstracted from river flow this will not significantly affect high discharges, so that any influence on river regime may be apparent for only a short distance downstream. Similar local effects are likely where bridges are built (Gregory and Brookes, 1983), because water velocity may be increased and sediment supply inhibited at the bridge site.

A second category of directly induced change of river regime arises where a reach of river channel has been modified. This includes: where the pattern has been modified by a series of cutoffs; where the channel has been modified by dredging to maintain navigation or to supply gravel; and where the channel has been regulated to reduce flooding, to drain wet areas, to control bank erosion or to improve river alignment (Section 9.4). The effect of creating a larger channel, or one which is reduced in roughness, is to increase water velocity so that the frequency of peak discharges may be increased downstream of the modified reach. Thus the Missouri river from Sioux City, Iowa, to its confluence with the Mississipi has been reduced in length from 269 km in 1890 to 217 km in 1960 as a result of flood control and navigation improvement (Schumm, 1971; Richardson and Christian, 1976).

The general ways in which river channels can be modified are indicated below (Section 9.4(a)).

Modifications at specific points along a river, and along river reaches, can therefore induce adjustments of the frequency of peak discharges, of the total runoff amount, and of the level and persistence of low flows. In addition, the amount of sediment available for transport may be reduced, as for example, where a river channel is regulated to inhibit bank erosion, or increased where gravel extraction makes more bed material available or where mine spoil is provided to the channel. In studies in central Wales it has been shown (Lewin *et al.*, 1977) how the wastes from nineteenth-century mining were indiscriminately fed into local streams such as the Ystwyth until preventative legislation was enacted. Further examples of man-induced changes of channel processes are included in Table 9.1.

(b) Indirect modifications of channel processes

Indirect changes include all those instances where a change of the character of the catchment are is responsible for an alteration of the magnitude and frequency

TABLE 9.1   Examples of man-induced changes relevant to river channel processes

Area	Runoff regulated by reservoirs (% of total runoff) (after Beaumont, 1978)	Percentage of world large dams (after Petts, 1984)
Europe	15.1	20
Asia	14.0	29
Africa	21.0	3
North America	20.6	38
South America	4.1	7
Australasia	6.1	3

USA downstream flood control works to 1966 (after Todd, 1970)

Number of projects	Type	Extent
263	Reservoir	Total storage 124 310 acre feet
426	Levee and floodwall	14 115 km
193	Channel improvement	8578 km

USA channelization by embanking, resectioning and lining had reached 26 550 km with a further 16 090 km predicted according to Leopold (1977).

of water and sediment discharge. A variety of landuse effects is now known to have occasioned changes in discharge, in sediment hydrographs, and in bedload transport (Chapter 5) and these changes involve either an increase or a decrease in the activity of river channel processes. Whereas deforestation may induce greater activity because of increased sediment availability and higher runoff rates, the converse may obtain where runoff is decreased and sediment transport reduced following afforestation or conservation measures. Land use changes can be responsible for adjustments of river channel processes which in turn affect channel landforms (Gregory, 1977a). Thus the removal of forest in a basin on South Island, New Zealand, was shown by O'Loughlin (1970) to have induced more frequent peak discharges and similar increases were found in the Beskid Sadecki Mountains of Poland following land use change as potato cultivation was increased (Klimek and Trafas, 1972).

Where the temporal sequence of water flow and sediment transported by rivers is altered, then changes in the processes operating within the river channel should occur as a direct consequence. Such changes can be anticipated from equations such as that proposed in 1955 by E. W. Lane to relate bed material load (Q_s), particle diameter (D), water discharge (Q_w), and stream slope (S) in the form:

$$Q_s D \simeq Q_w S$$

He deduced (Lane, 1955) the possible implications of six types of change from this equality. Thus, if water was diverted into the system then Q_w would increase so that the equality could be maintained only if either Q_s or D increased or if S decreased. A similar approach has been available for many years from regime theory. This is based upon the notion that a channel carrying a definite long-term pattern of water and sediment discharge, and subjected to a definite set of constraints that can adjust to the action of the flow, will acquire a definite regime (Blench, 1972). River morphometrics is then defined as the routine of measuring quantities relevant to river regime, and morphometric changes may be interpreted from the present regime (Blench, 1972). The principal changes of channel process that may arise from a change in regime will be expressed either in increased scour by erosion of the banks or bed, or in increased aggradation which is reflected in channel sedimentation. The consequences of man-induced changes of river channel processes, should be substantial and Beckinsale (1972) has argued that man has been responsible for many of the changes in sediment load of rivers since the Quaternary. However, he reminds us that the effects of such changes are never simple; for example, deposition on flood plains in western Europe since Roman times must be viewed in relation to a small rise in sea level. Furthermore, initially greater sediment transport following deforestation could have been succeeded by scour and channel incision as sediment sources were exhausted. An additional complication is afforded by the relative significance of climatic fluctuations and the influence of man and these two factors have been compared in the Mediterranean basin by Vita-Finzi (1969). Crampton (1969) argued that the sediment yield of rivers in south east Wales was increased following the deforestation of Iron Age times but that this was succeeded by a phase when the resulting fan deposits were dissected because sediment yields decreased again and scour of channels occurred.

In addition to the sequence of temporal changes of processes it is also important to remember the significance of threshold levels and of negative feedback mechanisms. Thus a change in process will induce a result which will be expressed in scour or in aggradation only if the change exceeds a critical threshold level. If a channel is in regime then the sediment load may have to be decreased below a critical level before scour will take place and the threshold level will depend upon the shear strength of the bed and bank materials. The time required to achieve a new steady state after disruption, the relaxation time, has been modelled by a rate law using half-life values analogous to those used for radioactive materials and chemical mixtures (Graf, 1977). Because the change to a new steady state may depend upon the occurrence of a large event or flood then a process change may not immediately prompt a response. A further complication of the expected sequence of change may be the incidence of negative feedback. If the sediment transported by a river is decreased substantially then it would be expected that the river bed may be scoured in accordance with the notion embodied in Lane's equation (p. 212). However, a layer of coarse material may develop on the channel bed (effectively an increase of D) and so resist scouring. This is illustrated by the Colorado river channel downstream of Glen Canyon Dam, because degradation was controlled by the development of an armouring of cobble-sized materials on the bed (Pemberton, 1976); and below the Aswan dam, Hammad (1972) deduced that an armoured condition may develop before an appreciable change of bed slope could occur on the Nile.

9.3 RIVER CHANNEL REACTION

Changes in river channel processes should induce adjustments of river channel and flood plain morphology. A number of studies have explored the potential adjustments which can occur and Strahler (1956) indicated how gullying in the headwaters of a basin could have aggradation downstream as its corollary (Fig. 9.2). The general model has been exemplified specifically by a study of the sedimentation which averaged 1 m in the Alcovey River swamps in Georgia (Trimble, 1970), and of the alluvial fills which accumulated in the Orange and Vaal basins of South Africa following over-grazing and burning of the degraded grassveld after 1880 (Butzer, 1971). An equally important study by Wolman (1967a) identified the possible changes in channels after reservoir construction and urbanization and a general model (Wolman, 1967b) was developed for the sequence of changes that would be expected with land use changes in the north east USA since 1700 (Fig. 9.2).

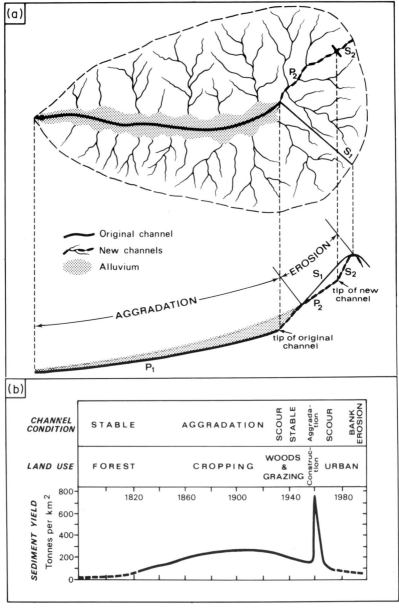

FIGURE 9.2   Models of river channel adjustment. Model (a) demonstrates the way in which a change of land use can induce channel extension by gullying and downstream aggradation marked by floodplain growth. (Source: A. N. Strahler, 1956.) Model (b) depicts the way in which changes of sediment yield since 1700 in the northeast USA (see also Fig.–5.7(c)) are associated with man-induced alterations of land use and induce alterations in river channel stability. (Source: Wolman, 1967b.)

Changes of river channels consequent upon changes in river channel processes have been termed river metamorphosis by Schumm (1969, 1971, 1977) and he has provided many illustrations of such metamorphoses. Similarly Richardson and Simons (1976) argued that changes in flow, in either magnitude or time distribution, should be expressed in changes in channel geometry, channel planform and sediment transport. They argued further that river characteristics, including the height of the water surface, depth, and channel geometry, could all respond to changes in sediment transport. There is obviously a range of parameters that can respond to change and Hey (1982) contended that rivers possess seven degrees of freedom and concluded that it is possible to define bankfull hydraulic geometry by the simultaneous solution of the appropriate process equations (Table 9.2) although many of the necessary equations are not yet adequately defined. This approach demonstrates that changes may occur in process variables (water velocity), in variables which express the cross-sectional geometry of the river channel (maximum flow depth, wetted perimeter, hydraulic radius), and in expressions for river channel length (slope) and planform (sinuosity, meander arc length). Many of these are interrelated and geometry and planform provide the main variables. To focus upon these morphological consequences of channel adjustment we can visualize a series of four spatial scales at which adjustment may be represented (Gregory, 1976) and these range from the sedimentary bedforms in the channel, through the river channel cross-section, and the river channel planform, to the drainage network. There can be adjustment of size, shape, and composition at any of these scales and examples of each type are illustrated in Table 9.3. It must be remembered that an adjustment of a specific river channel may involve changes indicated in any one of the nine categories (Table 9.3). Thus an increase in peak discharges due to the effects of urbanization could, for example, lead to an increase of channel cross-section size, an alteration of channel shape, an increase in the size of meanders and metamorphosis of planform along selected reaches from single thread to multithread as sedimentary bars accumulated. Although changes of each of the individual components included (Table 9.3) have been recorded associated with changes in river channel process, there is an outstanding need

TABLE 9.2 Degrees of freedom of a river channel

Adjustment of	may occur in response to input of
width, velocity	water
depth, hydraulic radius	sediment
slope	bed and bank material
velocity, wetted perimeter	valley slope
plan shape, maximum flow depth	
sinuosity	
meander arc length	

Source: Hey (1982).

TABLE 9.3   Potential river channel adjustments

Potential adjustments of:	Fluvial landform		
	River channel cross-section	River channel pattern	Drainage network
Size	INCREASE OR DECREASE OF RIVER CHANNEL CAPACITY Erosion of bed and banks can produce a larger channel which maintains the same shape. Sedimentation can produce a smaller channel which maintains the same shape.	INCREASE OR DECREASE OF SIZE OF PATTERN Increase or decrease of meander wavelength whilst preserving the same planform shape.	INCREASE OR DECREASE OF NETWORK EXTENT AND DENSITY Extension of channels or shrinkage of perennial, intermittent and ephemeral streams.
Shape	ADJUSTMENT OF SHAPE Width/depth ratio may be increased or decreased.	ALTERATION OF SHAPE OF PATTERN A change from regular to irregular meanders after deforestation.	DRAINAGE PATTERN CHANGED IN SHAPE Inclusion of new stream channels
Composition	CHANGE IN CHANNEL SEDIMENTS Alteration of grain size of sediments in bed and banks possibly accompanied by development of berms or bars.	PLANFORM METAMORPHOSIS Change from single to multithread channel or converse.	NETWORK COMPOSITION CHANGED The replacement of channels with no definite stream channel (dambos in West Africa) by a clearly defined channel.

General changes are indicated in capitals and examples given in lower case letters.
Source: Gregory (1976).

to understand the exact way in which the several possible adjustments interrelate and interact during the river metamorphosis. To predict the future behaviour of rivers and river channels is not easy, however, as a result of uncertainty (Burkham, 1981) which arises because it is not easy to predict which of the possible variables or degrees of freedom will change in a particular situation. More generally Schumm (1985a) has argued that there are at least seven problems associated with reasoning from present landforms and geomorphic processes for interpretation of past, and prediction of future, geomorphic processes and forms and he relates these seven problems to scale, location, convergence, divergence, singularity, sensitivity and complexity.

A great range of studies have now been completed and 180 such studies in a variety of world areas were classified into two major groups (Gregory and Burkhard, 1986) largely corresponding to the direct (Section 9.2(a)) and the indirect changes of process (Section 9.2(b)) above. Basin effects include land use changes including urbanization and water transfers that modify river discharges and can lead to changes of river channels. A second group of effects are dominantly point changes including dam construction, channelization and bridge crossings. Although there is not always a clear distinction between the two groups it was possible to use a channel change ratio (CCR) to express the relationship of the final channel dimension (CDF) with the original or initial value (CDI) and this ratio CCR = CDF/CDI is analogous to the channel enlargement ratio used by Hammer (1972). Average values of such ratios have to be interpreted cautiously but channels downstream of urban runoff have average values of 3.38 and the ratio varies from 0.13 to 15.0, showing that in some areas urban channels have reduced in capacity whereas others have increased to as much as 15 times their original values. Downstream of reservoirs the results from 49 studies show that capacities can be changed to as little as 0.16 of the former size or to 2.38 times the former value with an average of 0.79. The summary of results from 180 studies (Gregory and Burkhard, 1986) indicates that a great range of amounts of river channel change can occur, that such changes can occur along rivers with basin areas as much as 40 000 km^2, and that there is no universally agreed type of reaction that will occur in a particular area. The categories of adjustment proposed (Table 9.3) can be used to elucidate changes of river channel cross-section, of channel planform or pattern, and of drainage networks.

(a) Adjustments of river channel cross-sections

Downstream of dams and reservoirs the magnitude and frequency of peak discharges is often decreased by the effect of water storage, and sediment is trapped in the reservoir. Water released from the dam may scour the channel immediately below the dam and this has to be allowed for when the dam is designed in order to avoid the possibility of erosion undermining the dam

(Komura and Simons, 1967). Maximum degradation below the dam can be up to 15 cm per year but it may occur along a channel length equal to 69 channel widths (Wolman, 1967a). Adjustment to bed degradation was apparent for nearly 250 km below Elephant Butte dam in the USA (Stabler, 1925). The importance of a reduced peak discharge downstream of this dam on the Rio Grande has induced changes in channel size after storage operations were initiated in 1915 (Sonderreger, 1935). A number of studies have now been undertaken to show how channel capacities are smaller than expected downstream of dams. Downstream of Clatworthy reservoir on the Tone, Somerset, the channel capacities were shown (Gregory and Park, 1974) to be less than half the expected size (Fig. 9.3(a)), and the effect of the reservoir was apparent downstream until the drainage area was 2.5 times that draining to the reservoir. Studies have also investigated the way in which channel size reduction is achieved, and in the Willow Creek Basin of Montana detention reservoirs were shown to be the cause of a reduction in channel width (Frickel, 1972). The regulated discharge downstream of dams can induce changes in vegetation limits and sometimes in the encroachment of riparian vegetation. Downstream of the dams constructed between 1899 and 1943 in the headwaters of the river Derwent in the southern Pennines, a river channel capacity reduced to 40 per cent of its former level occurred as a result of the formation of a bench along certain reaches and this bench has been dated by reference to large-scale maps and plans and by using ring dating of trees growing on the bench compared with those on the margin of the flood plain proper (Petts, 1977, 1984). In managing river channels downstream of dams and reservoirs it has been argued (Petts, 1984) that changes in the ecology and in the morphology must be considered in their interrelationships. An excellent illustration of the way in which the results of specific studies may be combined for predictive purposes was provided by Williams and Wolman (1984) who analysed changes downstream from dams on 21 alluvial rivers in the western USA to provide a series of equations which could be used to indicate the likely changes of alluvial rivers downstream from a dam, although channel width can increase, decrease, or remain constant in the reach downstream from the dam. Using results from 45 previously published studies it was shown (Gregory and Burkhard, 1986) that, downstream of dams and reservoirs, channels can be reduced to about one-tenth of their original width or they can sometimes be enlarged to three times their original depth for example (Fig. 9.4(a)).

Downstream of urban areas, channel capacities may increase as a result of the increased frequency of flood discharges. In Philadelphia, Pennsylvania, comparison of channel capacities in relation to age and character of urban area indicated (Hammer, 1972) that channels within and downstream of urban areas can be up to four times the size of their rural counterparts. If river discharge is the only variable to change, then the size of stream channels is increased as expected, but sediment availability is greatly increased during building operations,

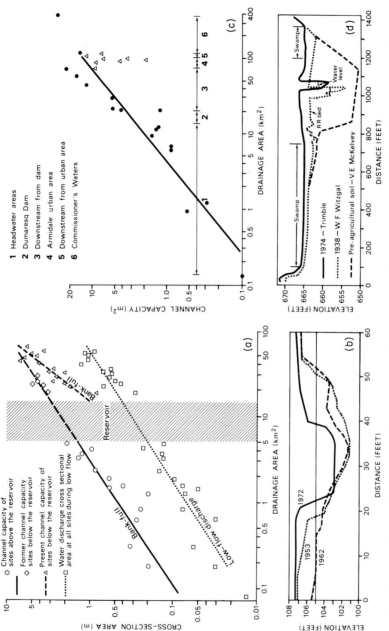

FIGURE 9.3 Changes of channel geometry. May be analysed by comparing the relation between channel capacity and drainage area above and below a reservoir as illustrated by data from the basin of the Tone, Somerset, UK, in (a) (Gregory and Park, 1974), and above and below an urban area as exemplified by the basin of Dumaresq Creek in New South Wales, Australia, in (c) (Gregory, 1977b). Alternatively successive surveys may be employed to show how channels change as illustrated in (b) by sections surveyed by Leopold (1973) to reveal the changes along the Watts Branch near Rockville, Maryland, during urbanization. Such successive surveys may be utilized to demonstrate flood plain aggradation shown in (d) for Coon Creek, Wisconsin (Trimble, 1976).

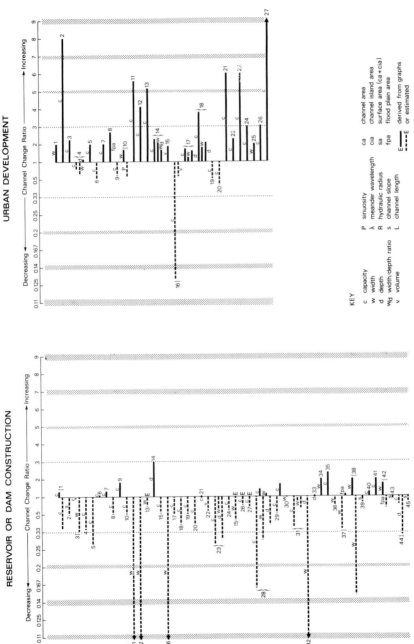

FIGURE 9.4   The magnitude of river channel change. In (a) each line represents a research study of a particular area affected by flows modified by reservoir or dam construction. Channel change ratio is final dimensions/original dimension. In (b) each study relates to an area downstream of urban development. The numbers in both diagrams refer to a specific research paper but the complete list of references is not given. Further details are given in Gregory and Burkhard (1986).

then drastically reduced when urbanization is complete. Thus along the Watts Branch, Maryland, it was shown by repeated surveys of river channel cross-sections (Fig. 9.3(b)) that capacities declined for twelve years during building operations but later began to increase (Leopold, 1973) (Fig. 9.3). If an urban area provides sediment to a major river channel then aggradation may occur and along Dumaresq Creek, New South Wales, channel capacities were shown to be as little as 0.13 (Fig. 9.3(c)) of the size expected (Gregory, 1977b). An important consideration in the erosion of urban channels is the impact of individual storms. This has been illustrated in the Patuxent Basin, Maryland: where catastrophic floods widened urban channels by 50 per cent more than rural ones; where the size and shape of urban channels change at rates at least three times greater than those in comparable rural areas; where urban channels hold about fifteen times as much sediment as rural channels; and where deposition and scour are more severe (Fox, 1976). Changes in channel dimensions downstream of areas of urban development are summarized from 27 previously published studies in Fig. 9.4(b) and this indicates that, whereas occasional reduction in diagnostic parameters may occur, in most cases there is channel enlargement which may be as much as a capacity increase of 15.3 times along the Estom Creek, Bathurst, Sydney, Australia (Hannam, 1979).

Associated with such changes in stream-channel size have been changes in channel shape and below urban areas it has been argued that width may change more easily than depth because of the armouring which may occur on the channel bed. Where discharges are increased the channel may be incised so that the width–depth ratio is decreased. Thus streams in a number of areas of New South Wales may have become incised since a modified runoff regime was occasioned by nineteenth-century settlement and colonization (Dury, 1968), and in Wisconsin, Trimble (1976) has analysed channel change since initial settlement (Fig. 9.3(d)).

Many of these changes in river-channel geometry have involved alterations of sediment transport, deposition, and temporary storage, and so the composition of the channel cross-section has changed. This was clearly illustrated in the Meadow Hills area of Denver, Colorado, where suburban development led first to an expansion of floodplains (Fig. 9.5(b)) which was later followed by downcutting of streams (Graf, 1975). Within, and downstream from, urban areas specific channel deposits may be produced, reflecting the urban sedimentology reviewed by Douglas (1985), who has cited the humorous characterization by Ager (1981) of the upper dustbinian (with plastic) and the lower dustbinian (without plastic). In urban areas the interrelationships of changing processes, are such that a distinctive urban ecosystem may be identified (Douglas, 1983).

The interrelation of changes of capacity, channel shape, and channel composition still requires further elucidation and the channels of a particular area at any one time may be at a particular point in their adjustment process. This is illustrated in the Ruahine Range of New Zealand where forest clearance, for example along the Tamaki river, saw channel width increased from 5 to 10 m in the early 1920s, to 54 m by 1942, and to 60 m by 1976 (Mosley, 1978). In this area the increase of sediment supply rates from the deforested valley throats and alluvial fans at the foot of the range has induced stream-channel trenching. Thus the influence of vegetation on the channel banks, the changes in discharge, and the increased sediment supply, all have to be considered in relation to interpretation of channel adjustment.

(b)   Changes of river channel planform

Two kinds of change of pattern may be distinguished (Lewin, 1977), namely autogenic, which are inherent in the river regime, and allogenic, which occur in response to changes in the system, including those induced by man. It is often difficult to completely differentiate between these two types, but change in the size of channel pattern is a consequence of increased frequency of flood events.

Six possibilities of metamorphosis of river channel planform have been outlined by Schumm (1985b) because a straight channel could become sinuous or braided; a meandering channel may become straight or braided; and a braided channel could become straight or meandering. Historical evidence has been of considerable use in reconstructing changes of channel pattern (Hooke and Kain, 1982) and along the Gila River in Arizona up to 15 historical sources were used (Graf, 1984) to identify reaches of stability and hazardous instability.

A detailed study of river planform in Devon was made (Hooke, 1977) by comparing patterns of tithe survey maps with those on several editions of large-scale Ordnance Survey maps. This revealed changes in planform between ca. 1840 and 1903, including a decrease in mean wavelength along 75 per cent of the sections studied and further changes along 69 per cent of the sections between 1903 and 1958. There was also an increase in sinuosity in this period. Thus the size of the pattern was changing and this was also accompanied by a change in regularity, although the influence of human interference is difficult to separate from other variables. The White River, Indiana, has been analysed by comparing reaches of the river from air photographs taken in 1937 and 1966–68 with a map published in 1880 (Brice, 1974). From this comparison it was concluded (Brice, 1974) that the pattern became less regular and sinuous after 1880 and this may have been the result of accelerated migration following nineteenth-century deforestation of the floodplain. Spectral analysis of the development of the pattern also confirms the decline in regularity (Ferguson, 1977). Examples have been described where a river shape changes following man's activity and along the Durajec river in the Carpathians of southern Poland comparison of surveys made in 1787, 1880 and 1955 showed how the channel was becoming more sinuous and branching; and this change of shape has been associated with deforestation and with the advent of potato cultivation in the early nineteenth century because this allowed high rates of slope runoff (Klimek

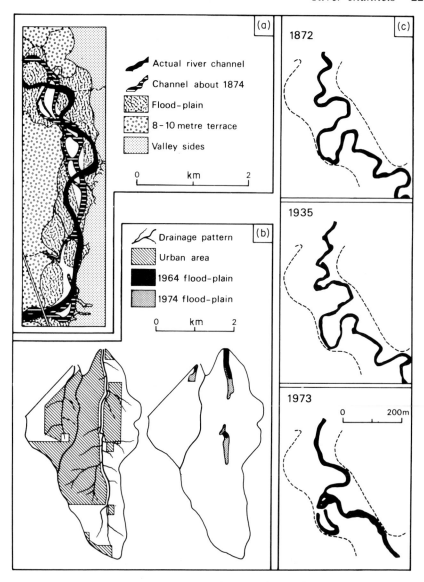

FIGURE 9.5 Changes of channel pattern. (a) illustrates the changes documented for a section of Wisłoka River, Polish Carpathians. Nineteenth-century changes in land use led to the development of a multi-thread channel in the nineteenth century. (Source: Klimek and Starkel, 1974.) (b) indicates the extension of floodplain identified by Graf (1975) in the Denver area of Colorado, USA, which followed suburban development and the production of large quantities of sediment. Changes in the course of a section of the River Bollin, Cheshire, UK, were analysed by Mosley (1975) as illustrated in (c) to reflect changes in land use and flood frequency

and Trafas, 1972). A similar change has been identified along the Wisłoka Valley (Fig. 9.5(a)). Changes of shape may result from the confinement of river channel patterns by man, and Lewin and Brindle (1977) have shown how the incidence of rail, road and other engineering works may retard the autogenic migration of meanders downstream and modify meander shapes, and how in other cases the loops become confined and cutoffs occur. A change in shape may often depend for its execution upon the occurrence of flood events which exceed the critical threshold value. Thus the channel of the River Bollin in Cheshire (Fig. 9.5(c)) was stable between 1872 and 1935, but then rates of channel shifting were increased and seven meander cutoffs contributed to a decline in sinuosity from 2.34 to 1.37 (Mosley, 1975). Although this change in shape was triggered by the large floods of the 1930s it may have been a response to increased runoff following agricultural land drainage and urban development in the area in the twentieth century (Mosley, 1975).

Planform metamorphosis is most dramatic when the channel changes from one type of pattern to another. The South Platte River, USA, has always been cited as a classic example of a braided stream, but whereas the channel was about 800 m wide in 1897 about 90 km above its junction with the North Platte River, it had narrowed to 60 m by 1959 (Schumm, 1977), and the tendency has been for a single thread channel to replace the former braided and less sinuous one. Along this river and along the North Platte and Arkansas rivers there have been flood control works and diversions for irrigation, and the narrowing of the North Platte river is associated with a decrease in the mean annual flood from $370 \, \mathrm{m^3 \, s^{-1}}$ to $85 \, \mathrm{m^3 \, s^{-1}}$ (Schumm, 1977).

(c)   Drainage network adjustments

Changes of the drainage network may accompany changes in channel geometry and channel pattern induced by man. The most dramatic changes are those which arise from extension of the network, and gully development has been well documented (e.g. Cooke and Reeves, 1976; Graf, 1983) and is discussed in Chapter 13. The metamorphosis of channel patterns described in the United States (Schumm, 1977) is often associated with the incision of stream channels and the development of gullies as illustrated by the Strahler model (Fig. 9.2(a)). Less widespread are the changes which result from contraction, but these occur where the water table has been lowered so that perennial or intermittent streams cease to flow. Metamorphosis of the drainage network must be considered in terms of the dynamic elements of the network. The relative significance of perennial, intermittent and ephemeral streams may change, so that it is not merely a question of considering the extent of the network, it is also desirable to consider the function of the streams composing the network as well.

Alterations of drainage pattern have occurred widely both directly and indirectly. Man's direct influence is shown where drainage ditches are dug so that the pattern of the stream network is effectively changed, and in England and Wales land-drainage channels have an estimated length of 128 000 km (Marshall *et al.*, 1978). In areas recently afforested the pattern may have been amplified by drainage channels, so that in the Coalburn catchment, Northumberland, the stream density was increased by about fifty times (Institute of Hydrology, 1973). Indirect changes of the drainage network arise when a change in land use produces new elements in the pattern and this is illustrated where a land use change has introduced new rills or streams to complement the existing network.

Such changes may be cases of adjustments in the composition of the drainage network. In the seasonal tropics, there are depressions without a stream channel but along which water flow occurs in the wet season, and these are called dambos in parts of Africa. If land use pressure induces changes in runoff rates this may be expressed in the production of a stream channel along such depressions and this may develop into a valley-floor gully. An illustration of a change in the composition of the drainage network is provided on the Southern Tablelands

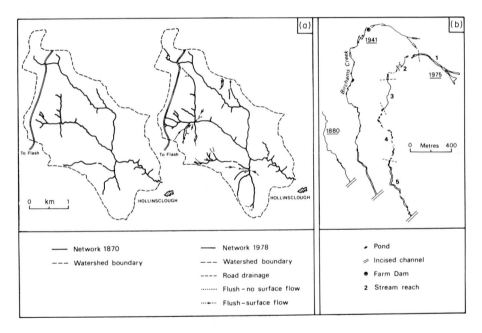

FIGURE 9.6   Examples of network change. (a) illustrates the change in the drainage network of the head of the Dove basin southern Pennines, UK. The 1870 network is based upon the first edition 1:10560 map and the network in 1978 is based upon the 1970 map and field survey in 1978 (see Gregory and Ovenden, 1979). (b) shows the changing channel form of Birchams Creek, a tributary of the Yass river north east of Canberra ACT, Australia. The creek in 1880 is based upon Surveyor Potter; aerial photographs were used for the 1941 pattern and field survey by R. J. Eyles produced the 1975 pattern. (Source: R. J. Eyles, 1977.)

of New South Wales. At the time of exploration and initial settlement, many drainage lines contained chains of ponds, but these were often destroyed by channel entrenchment between 1840 and 1950 as tree removal and grazing pressure affected the area (Eyles, 1977). This change involved a sequence of changes from chain of 'scour' ponds, to discontinuous gully, to continuously incised channel, to channel containing 'fixed bar' ponds, to permanently flowing stream (Fig. 9.6(b)). Only recently have such changes been identified but they may have been more widespread than previously envisaged. In upland Britain many headwater channels have juncus-lined depressions without a stream channel but comparison of nineteenth- and twentieth-century maps indicate that these depressions without a stream channel may have become less frequent during the last century (Fig. 9. 6(a)). As drainage from farms, tracks and metalled roads has become more prominent, stream channels have developed where none existed previously (Gregory and Ovenden, 1979). In the USA depressions called cienagas without clearly defined channels have sometimes been metamorphosed with the development of a stream channel.

It is now apparent that interpretation of the metamorphosis of fluvial land-forms must consider the interrelationships between channel geometry, channel planform and drainage network (Table 9.3). This is evident in south-western Wisconsin where many headwater and tributary channels now have relatively wide and shallow cross-sections compared with the pre-settlement channels (Knox, 1977), and also in New South Wales where the drainage system of Dumaresq Creek includes evidence of change from dambo to stream channel, indications of reduced channel capacities consequent upon reservoir construction, and clear signs of metamorphosed channels arising from urban runoff (Gregory, 1977b). Similar interrelationships may have prevailed in other areas, although they are more difficult to decipher if the time scale of man's activities has been greater and the controlling influences themselves more complex. In the analysis of changes in the fluvial system it is desirable to focus upon the interrelationship between force and resistance which Graf (1979) has argued to be central to geomorphology. The force exerted by flowing water in stream channels and over the land surface may have been increased by human activity as discharges and flood frequency has been increased for example, and resistance may have been altered either by removing riparian vegetation or by conservation measures which increase resistance. In future analyses it is desirable to express relationships in terms of force and resistance and this has been the basis for analysis of stream channel response to mining in Colorado (Graf, 1979). In the vicinity of Central City nineteenth century mining induced greater discharges and lower surface resistance which was estimated from old photographs (Graf, 1979).

9.4   RESTRAINING RIVERS

To control river flow and to prevent channel adjustment a variety of engineering

TABLE 9.4 Engineering methods of restraining rivers

(a) Channel control structures

Type	Method
Revetment	utilized primarily for bank protection and usually applied to bank previously sloped or shaped to a designed form.
blanket revetment	constructed of rock, concrete, asphalt or other materials
pervious revetment	open fence, baskets
solid fence	one or more rows of fencing backed by other materials
groins	short solid structures at right-angles to flow
Training structures	employed to guide flow so that effective channel will be scoured and maintained along required course
timber pile dikes	single or multiple rows of piling
jacks	
rock dikes	
rock-filled pile dikes	
Closure of chutes or secondary channels	where flow is precluded in a secondary channel or diverted from a major to a minor channel or a new channel is cut (as in the case of a cutoff)

(b) Channel modification

Type	Method
Emergency clearing	following catastrophic storms, bulldozing of channels to remove debris and sediment and to straighten channels
Change of channel geometry	widening, deepening or straightening, often involving removal of trees and bank vegetation
Clearing of channel	removal of debris, trees and obstructions, leading to increased flow velocities

Source: Vanoni (1975).

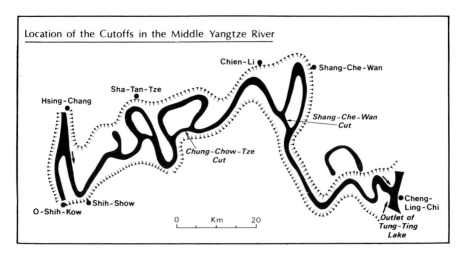

FIGURE 9.7   Cutoffs along the Yangtze (Chiang Jiang), China. The two meanders were cut off in 1968, and the effects upon sediment discharge and upon the river channel have been investigated by Ching-Shen *et al.*, 1977

methods can be employed (Table 9.4) and these include channel control structures which are designed primarily to prevent erosion of an existing bank, and training structures which are used to guide the flow and/or to promote deposition of sediment in specific areas (Vanoni, 1975). The channel pattern can also be deliberately modified (Table 9.4), as in the case of the Chiang Jiang where one reach of the river near Tung-t'ing Hu has been reduced from 240 to 160 km by artificial cutoffs (Fig. 9.7). In addition to these types of planned project there are also emergency works following catastrophic storms. Thus after Hurricane Camille in 1969 many reaches of stream channels in Virginia were cleared and straightened independently of any major planned programme (Keller, 1976). Other techniques of channel 'improvement' include channel clearing and deliberate alteration of channel shape and size (Table 9.4).

These various techniques of channel improvement, collectively referred to as channelization, have been applied to many locations along world rivers. The original purpose of the channelization of a specific reach was to prevent erosion or flooding at that location or to improve drainage or navigation but subsequently it has been realized that channelization procedures can have implications beyond the reach of river which is controlled and that integrated approaches to river planning and control should be further developed. In the USA channelization had affected 26 650 km of river channel by the 1970s and Leopold (1977) estimated that a further 16 090 would be modified by the year 2000, whereas in England and Wales there were 8504 km of major or capital works undertaken 1930–80 together with 35–500 km of river which is maintained

(Brookes *et al.*, 1983) so that the density of channelized river is $0.06 \, \text{km km}^{-2}$ in England and Wales compared with $0.003 \, \text{km}^{-2}$ in the USA.

(a) Consequences of channelization

Although the effects of channelization are immediately evident in the river reach which is protected, the ancillary consequences both upstream and downstream take longer to become apparent. Extensive reaches of the East and West Prairie rivers in Alberta, Canada, were channelized and straightened between 1953 and 1971 to provide flood relief. However, upstream of the channelized portions, nearly 5 m of degradation occurred between 1964 and 1974 and there has also been aggradation downstream which is reducing the capacity of the channels to cope with floods (Parker and Andres, 1976). A specific instance of adjustment consequent upon channel relocation was provided along the Peabody River in New Hampshire where the shortening of the river course gave a slope of $15.2 \, \text{m}$ km^{-1} replacing the former slope of $9.6 \, \text{m km}^{-1}$ and the river had reduced this to $13.3 \, \text{m km}^{-1}$ after seven years by degradation at the upstream end and aggradation downstream (Yearke, 1971).

A large literature has now developed about the implications of channelization but the main ones have been identified as downstream effects, damage to channel and floodplain, damage to fish and wildlife, and aesthetic degradation (Keller, 1976). One type of downstream effect is indicated by the examples from Alberta, but this will not always apply because flooding and sedimentation could be reduced downstream. Damage to channel and floodplain can occur because significant bank erosion may be induced by channelization procedures and this has been illustrated for the headwaters of the Blackwater River, in Missouri, where a new channel was dredged in 1910. The channel increased in cross-sectional area from $19 \, \text{m}^2$ to between 160 and $484 \, \text{m}^2$; channel widths increased from about 15 m to 60 m, and bridges had to be rebuilt (Emerson, 1971). Many changes have been inspired by changes of slope, because if these are too severe then the river proceeds to reduce slope by upstream degradation and sedimentation downstream. Channelization may remove the pools and destroy the vegetation and also increase the velocities such that habitats for fish and wildlife are dramatically changed. The general loss of the picturesque and the varied features of a river consequent upon improvement schemes has been termed aesthetic degradation (Keller, 1976).

(b) Integrated approaches

Channelization was first used for specific-problem reaches without concern for the effects on the channel system as a whole. At a time when concern for environmental quality has become very apparent it is evident that changes in a river system must be designed against the background of the character of the

TABLE 9.5   Methods of retaining rivers: examples of alternatives to channelization (afer Brookes *et al.*, 1983)

Guidelines	Area to which applied
Pools and riffles	
Construction of pools and riffles provides stable channel morphology that is biologically productive and aesthetically pleasing	General application in North America
Rest stops	
Corrected channels alternated with short lengths of natural channel which provide an acceptable habitat for fish. Permits fish migration in an otherwise impossible situation.	Hawaii
Stream restoration	
Minimal straightening of channels: retention of trees to promote bank stability; minimization of channel reshaping; use of bank stabilization techniques; emulation of the morphology of natural stream channels	Briar Creek, Charlotte, North Carolina, employed since 1975
Stream renovation	
Similar to restoration, but water-based methods of channel maintenance used or small-tracked vehicles. Hand-labour preferred.	Wolf River, Tennessee
Biotechnical engineering	
Emulation of the morphology of natural channels with meanders and asymmetrical profiles; preservation or creation of natural habitats for flora and fauna; bank stabilization with living vegetation	Bavaria, West Germany

References to specific articles are given in Brookes *et al.*, (1983).

river system and its basin as a functioning unit. This necessitates: a greater physical understanding of the present mechanics of rivers and of the way in which they react to change; use of an approach which is based upon integrated consideration of the river, channel, floodplain and basin as a system; and an assessment of water resource projects in relation to environmental values (Keller, 1976). It is therefore possible to envisage a number of ways in which the channelization procedures can be made less obtrusive in the landscape (Table 9.5). Designed methods should be based upon a clear understanding of river behaviour and on river regulation with the aid of nature (Winkley, 1972). It is desirable to apply an appropriate classification of the zones of a river system, to know the interactions of form and process in each zone, and to plan the future developments of the river in its basin against this zonal pattern. Thus Palmer (1976) advocated reference to geohydraulic river environment zones which are of four types along the rivers of Washington and Oregon: namely the boulder zone (headwater) and floodway zone, the pastoral zone, and the estuarine zone. A large number of techniques have been developed by ecologists to provide a discipline of bioengineering or biotechnical engineering which uses living vegetation rather than artificial materials for bank protection: using morphological modifications to achieve the objectives of channelization; using instream habitat devices which increase the habitat in a channelized river; using procedures specifically developed for particular areas (Brookes, 1985). Such specific techniques include techniques of stream restoration and renovation which involve minimal modification of the channel (Keller, 1976).

9.5 PROSPECTS FOR RIVERS AND HUMAN ACTIVITY

An article by L. B. Leopold (1977) who has studied rivers for many years was entitled 'A reverence for rivers'. Whereas rivers were originally conceived as an element of the landscape that could be altered at specific sites as required, research has now shown how the river functions as a system, and how it may react not only in the area affected directly by man but elsewhere as well. Studies have been made of the quantity and quality of the water flowing along the channel, of the quantitative aspects of the river channels and their changes, and it now remains to look more closely at the quality of riverscape. We can only maintain this quality, or enhance it, by working with the river and not against it and hence we must continue to have a reverence for rivers. In 1979 it was suggested that there were clear indications of further applications for hydrogeomorphology (Gregory 1979) and in 1985 W. L. Graf showed how geographers were taking an active role in dealing with the management of the Colorado River Basin where their major attribute, a spatial perspective, is very appropriate. Such a spatial perspective combined with a reverence for rivers may provide the basis for optimum interaction of rivers and human activity.

REFERENCES

Ager, D. V., 1981, *The Nature of the Stratigraphical Record*, London, Macmillan.
Beaumont, P., 1978, 'Man's impact on river systems: a world wide view', **Area**, **10**, 38–41.
Beckinsale, R. P., 1972, 'The effect upon river channels of sudden changes in load', *Acta Geographica Debrecina*, **10**, 181–186.
Blench, T., 1972, 'Morphometric changes', in Oglesby, R. T., Carlson, C. A., and McCann, J. A. (Eds), *River Ecology and Man*, Academic Press, New York, pp. 287–308.
Brice, J. C., 1974, 'Meander pattern of the White River in Indiana—an analysis', in Morisawa, M. E. (Ed.), *Fluvial Geomorphology*, State University of New York, Binghamton, pp. 178–200.
Brookes, A., 1985, 'River Channelization', *Progress in Physical Geography*, **9**, 44–73.
Brookes, A., Gregory, K. J., and Dawson, F. H., 1983, 'An assessment of river channelization in England and Wales', *The Science of the Total Environment*, **27**, 97–112.
Burkham, D. E., 1981, Uncertainties resulting from changes in river form, *Proceedings American Society of Civil Engineers Journal Hydraulics Division*, **107**, 593–610.
Butzer, K. W., 1971, 'Fine alluvial fills in the Orange and Vaal basins of South Africa', *Proceedings Association of American Geographers*, **3**, 41–48.
Ching-Shen, Pan, Shao-Chuan, Shi, and Wen-Chung, Tuam, 1977, *A Study of the Channel Development after the Completion of Artificial Cutoffs of the Middle Yangtze*, Peking.
Chou, Tsung-Lien, 1976, 'The Yellow River: its unique features and serious problems', *Rivers '76, American Society of Civil Engineers*, pp. 507–526.
Cooke, R. U., and Reeves, R. W., 1976, *Arroyos and Environmental Change in the American South-West*, Clarendon Press, Oxford.
Crampton, C. B., 1969, 'The chronology of certain terraced river deposits in the southeast Wales area', *Zeitschrift für Geomorphologie*, **13**, 245–259.
Dolan, R., Howard, A., and Gallenson, A., 1974, 'Man's impact on the Colorado River in the Grand Canyon,' *American Scientist*, **62**, 392–401.
Douglas, I., 1983, *The Urban Environment*, Edward Arnold, London.
Douglas, I., 1985, 'Urban sedimentology', *Progress in Physical Geography*, **9**, 255–280.
Dury, G. H., 1968, Footnote in Woodyer, K. D., 1968, 'Bankfull frequency in rivers', *Journal of Hydrology*, **6**, 114–142.
Emerson, J. W., 1971, 'Channelization: a case study', *Science*, **1973**, 325–326.
Eyles, R. J., 1977, 'Changes in drainage networks since 1820. Southern Tablelands, NSW,' *Australian Geographer*, **13**, 377–386.
Eyles, R. J., 1977, 'Birchams Creek: the transition from a chain of ponds to a gully', *Australian Geographical Studies*, **15**, 146–157.
Ferguson, R., 1977, 'Meander migration: equilibrium and change', in Gregory, K. J. (Ed.), *River Channel Changes*, Wiley, Chichester, 235–248.
Fox, H. L., 1976, 'The urbanizing river; a case study in the Maryland Piedmont', in Coates, D. R. (Ed.), *Geomorphology and Engineering*, State University of New York, Binghamton, pp. 245–271.
Fradkin, P. L., 1981, *A River No More: The Colorado River and the West*, Alfred A. Knopf, New York.
Frickel, D. G., 1972, 'Hydrology and effects of conservation structures, Willow Creek Basin, Valley County, Montana 1954–68', *US Geological Survey Water Supply Paper 1532-G*.
Graf, W. L., 1975, 'The impact of suburbanization on fluvial geomorphology, *Water Resources Research*, **11**, 690–692.

Graf, W. L., 1977, 'The rate law in fluvial geomorphology', *American Journal of Science*, **277**, 178–191.

Graf, W. L., 1979, 'Mining and channel response', *Annals Association American Geographers*, **69**, 262–275.

Graf, W. L., 1983, 'The arroyo problem—palaeohydrology and palaeohydraulics in the short term', in Gregory, K. J. (Ed.), *Background to Palaeohydrology*, Wiley, Chichester, pp. 279–302.

Graf, W. L., 1984, A probabilistic approach to the spatial assessment of river channel instability, *Water Resources Research*, **20**, 953–962.

Graf, W. L., 1985, *The Colorado River. Instability and Basin Management*. Association of American Geographers, Resource Publications in Geography, 86 pp.

Gregory, K. J., 1976, 'Changing drainage basins, *Geographical Journal*, **142**, 237–247.

Gregory, K. J., (1977a, 'The context of river channel changes', in Gregory, K. J. (Ed.), *River Channel Changes*, Wiley, Chichester, pp. 1–12.

Gregory, K. J., 1977b, 'Channel and network metamorphosis in northern New South Wales, River Channel Changes', Gregory, K. J. (Ed.), *River Channel Changes*, Wiley, Chichester, pp. 389–410.

Gregory, K. J., 1979, 'Hydrogeomorphology: how applied should we become?', *Progress in Physical Geography*, **3**, 85–101.

Gregory, K. J., and Brookes, A., 1983, 'Hydrogeomorphology downstream from bridges', *Applied Geography*, **3**, 145–159.

Gregory, K. J., and Burkhard, P., 1986, 'The global background to river channel adjustment'. Submitted for publication.

Gregory, K. J., and Ovenden, J. C., 1979, 'The permanence of stream networks in Britain', *Earth Surface Processes*, **5**, 47–60.

Gregory, K. J., and Park, C. C., 1974, 'Adjustment of river channel capacity downstream from a reservoir', *Water Resources Research*, **10**, 870–873.

Hammad, H. Y., 1972, 'Riverbed degradation after closure of dams', *Proc. Amer. Soc. Civ. Engrs., Jnl. Hyd. Div.*, **98**, 591–607.

Hammer, T. R., 1972, 'Stream channel enlargement due to urbanization', *Water Resources Research*, **8**, 1539–1540.

Hannam, I. D., 1979, 'Urban soil erosion: an extreme phase in the Stewart subdivision, west Bathurst', *Journal Soil Conservation Service NSW*, **35**, 19–25.

Hey, R. D., 1974, 'Prediction and effects of flooding in alluvial systems', in Funnell, B. M. (Ed.), *Prediction of Geological Hazards, Geol. Soc. Misc. Pap. 3*, pp. 42–56.

Hey, R. D., 1982, 'Gravel-bed rivers: form and process', in Hey, R. D., Bathurst, J. C., and Thorne, C. R. (Eds), *Gravel-bed Rivers*, Wiley, Chichester, pp. 5–13.

Hickin, E. J., 1983, 'River channel changes: retrospect and prospect', in Collinson, J. C., and Lewin, J. (Eds), *Modern and Ancient Fluvial Systems*, Blackwell, Oxford, pp. 61–83.

Hooke, J. M., 1977, 'The distribution and nature of changes in river channel patterns. The example of Devon', in Gregory, K. J. (Ed.), *River Channel Changes*, Wiley, Chichester, pp. 265–280.

Hooke, J. M., and Kain, R. J. P., 1982, *Historical Change in the Physical Environment: A Guide to Sources and Techniques*, Butterworth, London.

Institute of Hydrology, 1973, *Research 1972–3*, Natural Environment Research Council.

Keller, E. A., 1976, 'Channelization environmental, geomorphic and engineering aspects', in Coates, D. R. (Ed.), *Geomorphology and Engineering*, State University of New York, Binghamton, pp. 115–140.

Klimek, K., and Starkel, L., 1974, 'History and actual tendency of floodplain development at the border of the Polish Carpathians', *Abh. d. Akademie der Wiss in Gottingen, Math-Phys. Klass 3*, **29**, 185–196.

Klimek, K., and Trafas, K., 1972, 'Young Holocene changes in the course of the Durajec River in the Beskia Sadecki Mts (Western Carpathians)', *Studia Geomorphologica Carpatho-Balcanica*, **6**, 85–92.

Knox, J. C., 1977, 'Human impacts on Wisconsin stream channels', *Annals Association of American Geographers*, **67**, 323–342.

Komura, S., and Simons, D. B., 1967, 'River bed degradation below dams', *Proc. Amer. Soc. Civ. Engrs. Jnl. Hyd. Div.*, **93**, 1–14.

Lane, E. W., 1955, 'The importance of fluvial morphology in hydraulic engineering', *Proc. Amer. Soc. Civ. Engrs.*, **81**, 1–17.

Leopold, L. B., 1973, 'River channel change with time: an example', *Bulletin, Geological Society of America*, **84**, 1845–1860.

Leopold, L. B., 1977, 'A reverence for rivers', *Geology*, **5**, 429–430.

Lewin, J., 1977, 'Channel pattern changes', in Gregory, K. J. (Ed.), *River Channel Changes*, Wiley, Chichester, pp. 167–184.

Lewin, J., and Brindle, B. J., 1977, 'Confined meanders, in Gregory, K. J. (Ed.), *River Channel Changes*, Wiley, Chichester, pp. 221–234.

Lewin, J., Davies, B. E., and Wolfenden, P. J., 1977, 'Interactions between channel change and historic mining sediments', in Gregory, K. J. (Ed.), *River Channel Changes*, Wiley, Chichester, pp. 353–368.

Marshall, E. J. P., Wade, P. M., and Clare, P., 1978, 'Land drainage channels in England and Wales', *Geographical Journal*, **144**, 254–263.

Moore, C. M., 1969, 'Effects of small structures on peak flow', in Moore, C. M., and Morgan, C. W. (Eds), *Effects of Watershed Changes on Stream flow*, University of Texas Press, Austin, 101–117.

Mosley, M. P., 1975, 'Channel changes on the River Bollin, Cheshire 1872–1973', *East Midland Geographer*, **6**, 185–199.

Mosley, M. P., 1978, 'Erosion in the south-eastern Ruahine Range: its implications for downstream river control', *NZ Journal of Forestry*, **23**, 21–48.

Mrowka, J. P., 1974, 'Man's impact on stream regimen and quality', in Manners, I. R., and Mikesell, M. W. (Eds), *Perspectives in Environment*, Assoc. Amer. Geog., Publ., 12, pp. 79–104.

O'Loughlin, C. L., 1970, 'Streambed investigations in a small mountain catchment', *NZ Journal of Geology and Geophysics*, **12**, 684–706.

Painter, R. B., *et al.*, 1974, 'The effect of afforestation on erosion processes and sediment yield', *Proc. Symp. Effects of Man on the Interface of the Hydrological Cycle with the Physical Environment*, Int. Assoc. Sci. Hyd. Publ., 117, pp. 62–67.

Palmer, L., 1976, 'River management criteria for Oregon and Washington', in Coates, D. R. (Ed.), *Geomorphology and Engineering*, State University of New York, Binghamton, pp. 329–346.

Parker, G., and Andres, D., 1976, 'Detrimental effects of river channelization', *Rivers '76*, Amer. Soc. Civ. Engrs., pp. 1248–1266.

Pemberton, E. L., 1976, 'Channel changes in the Colorado river below Glen Canyon dam', *Proc. Fed. Third Inter Agency Sedimentation Conf.* pp. 5–61 to 5–73.

Petts, G. E., 1977, 'Channel response to flow regulation: the case of the River Derwent, Derbyshire', in Gregory, K. J. (Ed.), *River Channel Changes*, Wiley, Chichester, pp. 145–164.

Petts, G. E., 1984, *Impounded Rivers: Perspectives for Ecological Management*, Wiley, Chichester.

Richardson, E. V., and Christian, 1976, 'Channel improvements on the Missouri River', *Proc. Third Fed. Inter Agency Sedimentation Conf.* 5–113 to 5–124.

Richardson, E.V., and Simons, D. B., 1976, 'River response to development', *Rivers '76*, Amer. Soc. Civ. Engrs., pp. 1285–1300.

Rutter, E. J., and Engstrom, L. R., 1964, 'Reservoir regulation', in Chow, V. T. (Ed.), *Handbook of Applied Hydrology*, McGraw-Hill, New York, Section 25.

Schumm, S. A., 1969, 'River metamorphosis', *Proc. Amer. Soc. Civ. Engrs. Jnl. Hyd. Div.*, **95**, 251–273.

Schumm, S. A., 1971, 'Channel adjustment and river metamorphosis', in Shen, H. W. (Ed.), *River Mechanics*, I, Water Resources Publications, Fort Collins, pp. 5–1 to 5–22.

Schumm, S. A., 1977, *The Fluvial System*, Wiley, New York.

Schumm, S. A., 1985a, 'Explanation and extrapolation in geomorphology: seven reasons for geologic uncertainty', *Transactions Japanese Geomorphological Union*, **6**, 1–18.

Schumm, S. A., 1985b, 'Patterns of alluvial rivers', *Annual Reviews Planetary Sciences*, **13**, 5–27.

Sherlock, R. L., 1922, *Man as a Geological Agent*, Witherby, London.

Sonderreger, A. L., 1935, 'Modifying the physiographic balance by conservation measures', *Trans. Amer. Soc. Civ. Engrs. Paper No. 1897*, pp. 284–304.

Stabler, H., 1925, 'Does desilting affect cutting power of streams?', *Eng. News Records*, **95**, 960.

Stevens, M. A., Simons, D. B., and Schumm, S. A., 1975, 'Man-induced changes of Middle Mississippi River', *Proc. Amer. Soc. Engrs. Jnl. Waterways Harbors, Coast, Eng. Div.*, **101**, 119–133.

Strahler, A. N., 1956, 'The nature of induced erosion and aggradation', in Thomas, W. L. (Ed.), *Man's Role in Changing the Face of the Earth*, (University of Chicago Press), 621–638.

Todd, D. K., 1970, *The Water Encyclopedia*, Water Information Center, New York.

Trimble, S. W., 1970, 'The Alcovey River swamps. The result of culturally accelerated sedimentation', *Bull. Georgia Academy of Sci.*, **28**, 131–141.

Trimble, S. W., 1976, 'Modern stream and valley sedimentation in the Driftless area, Wisconsin, USA', *International Geography '76*, Vol. 1, pp. 228–231.

Vanoni, V. A. (Ed.), 1975, *Sedimentation Engineering*, Amer. Soc. Civ. Engrs. Task Committee Sedimentation Committee of Hydraulics Division, pp. 531–546.

Vita-Finzi, C., 1969, *The Mediterranean Valleys*, Cambridge University Press, Cambridge.

Warner, R. F., 1984, 'Man's impact on Australian drainage systems', *Australian Geographer*, **16**, 133–141.

Williams, G. P., and Wolman, M. G., 1984, 'Downstream effects of dams on alluvial rivers', *U.S. Geological Survey Professional Paper 1286*.

Winkley, B. R., 1972, 'River regulation with the aid of Nature', *Int. Comm. Irrigation and Drainage*, Eighth Congress, pp. 433–457.

Wolman, G., 1967a, 'Two problems involving river channel changes and background observations', *Quantitative Geography, Part II, North Western studies in Geography*, **14**, 67–107.

Wolman, G., 1967b, 'A cycle of sedimentation and erosion in urban river channels', *Geog. Annaler,* **49a**, 385–395.

Yearke, L. W., 1971, 'River channel erosion due to channel relocation', *Civil Engineering*, **August**, 39–40.

Human Activity and Environmental Processes
Edited by K. J. Gregory and D. E. Walling
©1987 John Wiley & Sons Ltd.

10

Permafrost and Ground Ice

H. M. FRENCH

10.1 INTRODUCTION

Permafrost, or perennially frozen ground, influences virtually all aspects of man's activities in the regions which it underlies. In North America, the importance of permafrost was realized only relatively recently, when attention focused on Alaska and other northern regions during and immediately after the Second World War. The building of the Alaskan Highway in 1942 from Dawson City to Fairbanks and the construction of the Canol pipeline from Norman Wells to Alaska in 1943 were major engineering undertakings which highlighted the inadequacies of traditional methods of construction. In addition, the difficulties of water and sewage provision, and the limitations placed on agriculture and mining by permafrost meant that large-scale permanent settlement of Arctic North America was unrealistic until permafrost problems were understood. Referring to Alaska, S. W. Muller, of the US Army Corps of Engineers, wrote in 1945, 'The destructive action of permafrost phenomena has materially impeded the colonization and development of extensive and potentially rich areas in the north. Roads, railways, bridges, houses and factories have suffered deformation, at times beyond repair, because the condition of permafrost ground was not examined beforehand, and because the behaviour of frozen ground was little, if at all, understood' (Muller, 1945, pp. 1–2).

The Soviet Union, by virtue of its longer history of settlement in permafrost regions, was aware of these problems earlier than other countries and by the late 1920s had established a Permafrost Institute of the Siberian Academy of Sciences at Yakutsk in Eastern Siberia. This institute now employs several hundred permafrost scientists. The nearest equivalent in North America is the Cold Regions Research and Engineering Laboratory (CRREL) of the US Army, located at Hanover, New Hampshire.

Permafrost underlies approximately one-fifth of the land surface of the earth. In the northern hemisphere, large areas of permafrost exist in the Soviet Union, Canada, The People's Republic of China, and Alaska (Fig. 10.1). Permafrost is

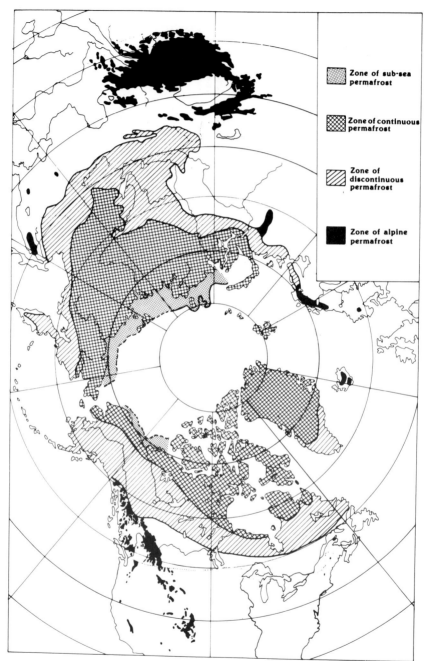

FIGURE 10.1   Distribution of permafrost in the northern hemisphere (modified from Péwé, 1983)

usually classified as being either continuous or discontinuous in nature. In areas of continuous permafrost, frozen ground is present at all localities except for localized thawed zones, or taliks, existing beneath lakes, river channels and other large water bodies which do not freeze to their bottoms in winter. In discontinuous permafrost, bodies of frozen ground are separated by areas of unfrozen ground. At the southern limit of this zone permafrost becomes restricted to isolated 'islands' occurring beneath peaty organic sediments. Permafrost may vary in thickness from a few centimetres to several hundreds of metres. In parts of Siberia and interior Alaska, permafrost has existed for several hundred thousand years; in other areas, such as the modern Mackenzie Delta, Canada, permafrost is young and currently forming under the existing cold climate.

10.2   MOISTURE IN PERMAFROST

Traditionally, permafrost has been defined on the basis of temperature (Brown and Kupsch, 1974): that is, ground (i.e. soil and/or rock) that remains at or below 0°C for at least two consecutive years. However, permafrost may not

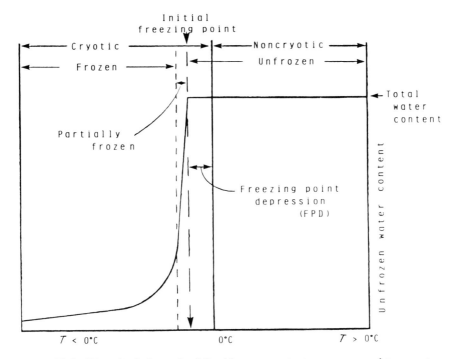

FIGURE 10.2   Hypothetical graph of liquid water content versus ground temperature showing relationships between temperature and phase conditions of permafrost (source: van Everdingen, 1976)

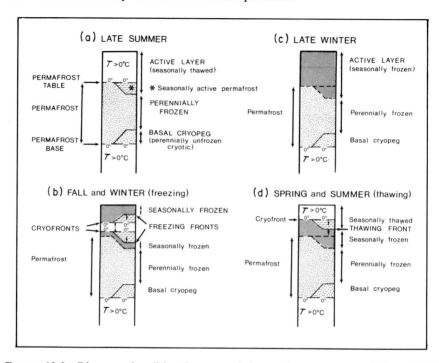

FIGURE 10.3   Diagrams describing the seasonal changes in temperature relative to 0°C, and in the state of water versus depth in a permafrost environment (modified from van Everdingen, 1985)

necessarily be frozen since the freezing point of included water may be depressed several degrees below 0°C. Moisture, in the form of either water or ice, may or may not be present. To differentiate between the temperature and state (i.e. frozen or unfrozen) conditions of permafrost, therefore, the terms 'cryotic' and 'non-cryotic' have been proposed to refer solely to the temperature of the material independent of its water/ice content. Perennially cryotic ground is, therefore, synonymous with permafrost, and permafrost may be 'unfrozen', 'partially frozen' and 'frozen', depending upon the state of the ice/water content (Fig. 10.2).

The active layer is the zone lying above the permafrost and near the ground surface which is subject to annual freeze and thaw. It is subject to two-sided freezing in the fall and winter, and to thawing from above in the spring and summer. Freezing fronts and thawing fronts delimit the extent of seasonally frozen and unfrozen ground (Fig. 10.3).

10.3   PERMAFROST-RELATED PROBLEMS

The terminological subtleties described above are more than mere semantics. Many of the most important problems posed by man's activities in permafrost regions are related, either directly or indirectly, to the water and/or ice content of permafrost. These may be summarized under three main categories.

First, the freezing of water in the active layer at the beginning of winter each year results in ice lensing and segregation. The volume expansion associated with this phase change from water to ice is approximately 9 per cent. As a result, there is an upward expansion of the ground surface each winter, a process termed 'frost heave' (Polar Research Board, 1984). The magnitude of heave varies according to the amount and availability of moisture present in the active layer. Poorly drained silty soils usually possess some of the highest ice or water contents and are termed 'frost-susceptible' by engineers. Frost creep is intimately associated with frost heaving and is a major component of rapid mass wasting in permafrost environments (e.g. solifluction).

Long-continued heaving may also occur at temperatures below 0°C, as unfrozen water progressively freezes. This is sometimes referred to as secondary heave, as distinct from the primary (i.e. capillary) heave described above (Smith, 1985a). Field experiments have demonstrated that moisture migrates through permafrost in response to a temperature gradient, and this causes an ice-rich zone to form in the upper few metres of permafrost (e.g. Cheng, 1983; Mackay, 1983; Smith 1985b). Frost heave can result in significant damage to structures and foundations (e.g. Beskow, 1935; Ferrians *et al.*, 1969). The annual cost of rectifying seasonal frost damage to roads, utility foundations and buildings in areas of deep seasonal frost such as Canada, Sweden and northern Japan is often considerable.

Second, ground ice is a major component of permafrost, particularly in unconsolidated sediments. Frequently, the amount of ice held within the ground in a frozen state exceeds the natural water content of that sediment in its thawed state. If the permafrost thaws therefore, subsidence of the ground equal in volume to the amount of water released from the soil may result. Thaw consolidation may also occur as the thawed sediments compact and settle under their own weight. In addition, high pore water pressures generated in the process may favour soil instability and mass movement. These various processes associated with permafrost degradation are generally termed thermokarst (e.g. French, 1976, pp. 105–133). A related problem is that the physical properties of frozen ground, in which the soil particles are cemented together by pore ice, may be considerably greater than the same material in its unfrozen state (Tsytovich, 1975). In unconsolidated and soft sediments there is often a significant loss of bearing strength upon thawing.

Third, the hydrologic and groundwater characteristics of permafrost terrain are different from those of non-permafrost terrain (Hopkins *et al.*, 1955; Williams and van Everdingen, 1973). For example, the presence of both perennially and seasonally frozen ground prevents the infiltration of water into the ground or, at best, confines it to the active layer. At the same time, subsurface flow is restricted to unfrozen zones or taliks. A high degree of mineralization in subsurface permafrost waters is often typical, caused by the restricted circulation imposed by the permafrost and the concentration of dissolved solids in the taliks. Thus, frozen ground eliminates many shallow depth aquifers, reduces the volume of unconsolidated deposits or bedrock in which water may be stored, influences the quality of groundwater supply, and necessitates that wells be drilled deeper than in non-permafrost regions.

10.4   GROUND ICE AND THERMOKARST

From the geotechnical and engineering viewpoint, the presence of ground ice is the problem most unique to permafrost regions and central to both thermokarst and frost-heave processes. As a simplification, these problems vary directly with lithology, being most serious in silty unconsolidated sediments and negligible in hard consolidated rock.

Although a variety of ground ice types exist (Mackay, 1972), there are three types which are important in terms of thermokarst, principally because of their ubiquitous nature. The most widespread is pore ice. This is the bonding cement that holds frozen soil grains together. It exists, in varying amounts, in virtually all rock types. Secondly, there is segregated ice which forms lenses ranging from a few centimetres thick to massive ice bodies several metres thick. Thirdly, there is wedge ice which forms from surface water penetrating thermal contraction cracks which develop during winter under the intense cold. Clearly, the total amount of ground ice present varies from locality to locality. In parts of the Western Arctic coastal plain of Alaska and Canada, ice wedges and massive ice bodies assume large dimensions. For example, it has been estimated that ground ice represents 47.5 per cent of the total volume of perennially frozen ground in the upper 10.0 m of earth materials on Richards Island, in the Pleistocene Mackenzie Delta (Pollard and French, 1980), and as much as 55 per cent in the upper 5.0 m on the Yukon coast (Harry *et al.*, 1985).

Quantitative estimates of ground ice can be made by either weight or volume. Low ice-content soils are generally regarded as those having ice contents by weight of less than 40–50 per cent. Soils with high ice contents have values which commonly range from 50 to 150 per cent. From a geomorphic viewpoint, the volume of ice contained by the sediment is important. 'Excess ice' refers to the volume of supernatant water present if the sediment is allowed to thaw. Expressed as a percentage of the total volume of sediment, excess ice values indicate the potential morphological modification or volumetric ground loss consequent upon thawing.

Within this context, thermokarst develops as the result of the disruption of the thermal equilibrium of the permafrost and an increase in the depth of the active layer. This can be illustrated with a simple example (Fig. 10.4). Consider a well-vegetated tundra soil with an active layer of 45 cm underlain by supersaturated permafrost which yields upon thawing, on a volumetric basis,

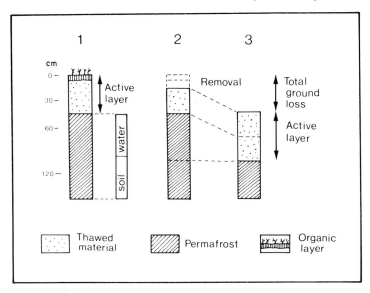

FIGURE 10.4   Diagram illustrating how terrain disturbance of an ice-rich tundra soil can lead to thermokarst subsidence (source: Mackay, 1970)

50 per cent excess ice (water) and 50 per cent saturated soil. If the top 15 cm of soil is removed, the thermal insulating role played by the organic mat disappears. Under the bare-ground conditions that result, the depth of seasonal thaw might increase to 60 cm. Since only 30 cm of the original active layer remains, a further 60 cm of the permafrost must thaw in order to increase the active layer thickness to 60 cm since 30 cm of supernatant water will be released. Thus, in addition to the original 15 cm of material loss from the surface, the terrain subsides a further 30 cm before a new thermal equilibrium is reached.

If we ignore large-scale climatic fluctuations, an infinite number of local geomorphic and vegetation changes may affect thermokarst (Fig. 10.5). These may be either natural or man-induced. For example, natural causes of thermokarst in areas north of the treeline include localized slumping and slope failure, river-bank undercutting and coastal retreat. In forested permafrost regions, fire is an additional factor since by causing rapid changes in vegetation cover the ground thermal regime can be altered substantially. For example, in a part of the Siberian taiga, the active layer increased from 40 to 85 cm in twelve years following a fire in 1953 (Czudek and Demek, 1970), and at Inuvik, in the Northwest Territories of Canada, where a forest fire occurred in 1968, a 40 per cent increase in the active layer depth in the burned over area was recorded in a four-year period (Heginbottom, 1973). Moreover, in an attempt to contain the fire at Inuvik, a number of fire-breaks were bulldozed. In these areas not only were the trees removed but the tops of the underlying frost mounds

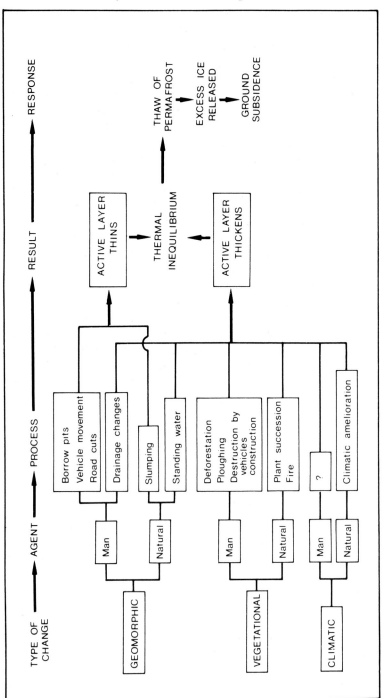

FIGURE 10.5   Diagram illustrating how geomorphic, vegetational and climatic changes may lead to permafrost degradation (modified from Mackay, 1970, and French, 1976, p. 107)

truncated and the surface organic layer of moss and lichen destroyed. In these areas, thaw depths increased by more than 100 per cent.

Once initiated, thermokarst processes are difficult to arrest. Erosion may become concentrated along ice wedges and continued activity may lead to striking badland topography. In areas where massive icy sediments exist, shallow semicircular slumps may form, the headwalls of which may retreat as much as 8–10 m during a single summer. The cause of these slumps is usually some trigger mechanism associated with the exposure of the permafrost, such as stream undercutting or local slope instability.

One type of mass wasting process favoured in certain permafrost regions is the occurrence of various low-angled failures which are termed 'skin flows' (McRoberts and Morgenstern, 1974). These are confined to the active layer and are induced by high pore water pressures which develop following heavy summer rain. The permafrost table acts as a lubricated slip plane. In instances where the natural water content is close to the liquid limits of the material, multiple mudflows may occur which extend downslope for several hundreds of metres. Particularly large and frequently occurring skin flows are characteristic of ice-rich shale known as the Christopher Shale Formation which outcrops widely on the Sabine Peninsula of Eastern Melville in the Canadian Arctic (Barnett *et al.*, 1977). This is an area where sizeable gas deposits have been discovered. Often failure takes place on very low-angled slopes, sometimes less than three degrees. The apparent stability of the tundra surface prior to failure makes the prediction of such events virtually impossible. Since such failures could easily rupture a pipeline traversing such terrain, their occurrence, together with the potential for thermokarst, makes this terrain exceedingly difficult for construction and other activities.

For reasons such as these, permafrost terrain is often regarded as being highly sensitive to disturbance by human activity, and as presenting unusual engineering and geotechnical problems.

10.5 TERRAIN DISTURBANCES

Our understanding of the consequences of terrain disturbance in permafrost regions has increased significantly in recent years. In nearly all instances, the necessity of maintaining the thermal regime of the permafrost has been recognized and measures to prevent thermal change widely adopted. At the same time, the documentation of case histories of disturbance enables the nature and speed of man-induced thermokarst to be determined and the time period necessary for stabilization to be assessed.

The Soviet Union has by far the greatest experience in this respect. As early as 1925, for example, controlled experiments were being undertaken to determine the effects of vegetation changes on the underlying permafrost, brought about either by deforestation or by ploughing. For example, P. I. Koloskov reported how, in the Yenesei region of Siberia, July soil temperatures at a depth of 40 cm increased by 14°C in a semi-bog soil following deforestation and ploughing (Tyrtikov, 1964). In Alaska, similar experimental studies have been undertaken. One of the earliest was in the Fairbanks region and involved the cutting and/or stripping of the surface vegetation by the US Army Corps of Engineers in 1946. In the stripped area, the active layer increased from 1 m to more than 3 m in thickness over a ten-year period. Subsequently, numerous other studies have emphasized the thermal role played by the surface organic layer and/or forest cover (e.g. Babb and Bliss, 1974; Brown *et al.*, 1969; Haugen and Brown, 1970; Kallio and Reiger, 1969). It must be emphasized that even very small disturbances to the surface may be sufficient to induce thermokarst activity. Mackay (1970) for example, describes how an Eskimo dog was tied to a stake with a 1.5 m long chain. In the ten days of tether, the animal trampled and destroyed the tundra vegetation of that area. Within two years, the site had subsided like a pie dish by a depth of 18–23 cm and the active layer thickness had increased by more than 10 cm within the depression.

Without doubt, the most common cause of man-induced thermokarst on a large scale is the clearance of the surface vegetation for agricultural or construction purposes. If this occurs in an area underlain by a polygonal network of ice wedges a distinctive hummocky microrelief forms. This results from the general subsidence of the ground combined with preferential melt along the wedges. A classic example has been described from the Fairbanks region of Alaska where extensive areas were cleared for agriculture in the 1920s. The following thirty years saw the formation of mounds and depressions in the fields, the mounds varying from 3 to 15 m in diameter and up to 2.4 m in height (Rockie, 1942). Subsequently, these features were termed thermokarst mounds (Péwé, 1954). In the Soviet Union, similar undulating surfaces are common in clearings adjacent to small lumber camps in the taiga. In places, shallow troughs, 0.5–1.0 m deep, form hummocky terrain termed 'baydjarakhs' or 'graveyard mounds' (Soloviev, 1973). In other areas underlain by extremely ice-rich sediments, widespread subsidence may lead to the formation of alas depressions. Adjacent to the village of Maya in central Yakutia for example, a depression 5–8 m deep and between 200 and 300 m in diameter had formed in historic times following deforestation and the beginning of the agricultural settlement (French, 1976, p. 114).

These examples illustrate the extreme sensitivity of permafrost terrain to man-induced surface modifications. We can identify at least three controls over such permafrost degradation. These are: first, the ice content of the underlying permafrost and in particular, the presence or absence of excess ice; secondly, the thickness and insulating qualities of the surface vegetation; and thirdly, the duration and warmth of the summer thaw period.

Several case histories illustrate the nature and rapidity of man-induced thermokarst. In all cases, the initial disturbance is associated with borrow pits, where material has been removed for road, airstrip, or other construction purposes. Terrain disturbance at the Deminex Orksut 1-44 wellsite in central Banks Island, in the Canadian High Arctic, has been monitored since initiation in 1973 (French, 1978a). In August of that year, following completion of the well, the site was abandoned and restoration undertaken. In an attempt to infill the kitchen sump, material was removed from an adjacent gravelly ridge and pushed into the depression. In all, an area of approximately 3000 m² was disturbed (Fig. 10.6(a)). Two years later, a crude polygonal system of gullies had developed at the site, reflecting the preferential melt along underlying ice wedges (Fig. 10.6(b)). After two further years, in August 1977, typical thermokarst topography had formed, consisting of unstable hummocks and mounds interspersed with standing water bodies (Fig. 10.6(c)).

The experience at the Orksut 1-44 wellsite can be compared to the man-induced thermokarst which formed adjacent to the airstrip at Sachs Harbour, on southern Banks Island (French, 1975). This airstrip was constructed during summer in the years between 1959 and 1962. In order to grade the proposed strip, thawed material was removed from adjacent terrain and transported, via access ramps, to the strip. In all, a total of 50 000 m² was disturbed and as much as 2.0 m of material removed in places. Subsequent surficial drilling revealed the underlying sediments to be ice-rich sand and gravel with excess ice values of 25-30 per cent and natural water (ice) contents of between 50 and 150 per cent. When examined in 1972 the borrow pits portrayed actively subsiding thermokarst mound topography. When examined recently in 1982, the mounds were not so sharp and plants were beginning to invade the disturbed terrain.

These studies provide some insight into the speed at which man-induced thermokarst develops and the time period over which such terrain remains unstable. They suggest that thermokarst processes are rapid and that the typical hummocky relief forms through preferential subsidence along ice wedges. Stabilization only begins 10-15 years after the initial disturbance and probably is not complete until 30 years or more have passed.

In environments of greater summer thaw, an essentially similar sequence of thermokarst activity takes place, except that the amplitude of the thermokarst mounds is greater, reflecting the greater thaw depths. For example, an area of disturbed terrain adjacent to the Maya-Abalakh road in Eastern Siberia was examined by the writer in July 1973 (French, 1975, pp. 141-143). Material had been removed to provide aggregate for the road. At one locality where construction had taken place three years previously, mounds similar in size to those at Sachs Harbour had formed. In a second area where disturbance had occurred in 1966, the mounds were much larger, with relative relief exceeding 3 m, and were interspersed with deep standing water bodies reflecting the active melt of the ice-rich permafrost beneath. Similar processes operate in the boreal forest zone of northern Canada (Fig. 10.6(d)) although the application of the

FIGURE 10.6   Terrain disturbances adjacent to exploratory wellsites, Western Arctic. (a) (top left) air view of the Deminex Orksut 1–44 wellsite in central Banks Island, (72°23′N; 122°42′W), Western Arctic. In an attempt to infill the sump at the living quarters, material was removed from an adjacent gravelly ridge. (Photograph taken August 1973.) (b) (bottom left) Oblique air view of the Orksut 1–44 wellsite in July 1975. A system of gullies had developed reflecting preferential melt-out of underlying ice wedges. (c) (top right) Ground view of disturbed terrain at Orksut 1–44 wellsite in August 1977. An unstable topography

consisting of irregular hummocks, depressions and standing water bodies had formed. (d) (bottom right) Thermokarst mounds and standing water bodies developed in disturbed terrain adjacent to the SOBC Blackstone D-77 wellsite, interior Yukon (65°40′N; 137°15′W). The well was drilled in the summer of 1962 and the black spruce (*Picea mariana*) and tamarack (*Larix* sp.) vegetation were removed. Thaw depths in disturbed terrain commonly exceed 90 cm; those in undisturbed terrain are 40–45 cm. (Photograph taken, July 1979.)

Territorial Land Use Act and Regulations since 1971 has significantly reduced such disturbances (French, 1984).

A second major cause of man-induced thermokarst relates to the movement of vehicles over permafrost terrain. If this occurs in summer when the surface has thawed and is soft and wet, surface vegetation can be destroyed and deep trenching and rutting can occur. Probably the worst examples of this sort of activity exist on the Alaskan North Slope in the old US Naval Petroleum Reserve No. 4. There, in the late 1940s and early 1950s, the uncontrolled movement of tracked vehicles in summer associated with early well drilling activities led to considerable trenching and thermokarst on account of the ice-rich subsoils. In places subsidence along vehicle tracks has formed trenches as much as 1 m deep and between 3 and 5 m wide (Fig. 10.7(a)). Large areas of the North Slope are permanently scarred by these tracks. Furthermore, the tracks favour continued thermokarst by collecting water and, if located upon a slope, promote gullying by channelling snowmelt and surface runoff.

In Canada, an unfortunate error was made in 1965 when a summer seismic line programme was undertaken in the Mackenzie Delta. Approximately 300 km of seismic lines were bulldozed and long strips of vegetation and soil, approximately 4.2 m wide and 0.25 m thick were removed. Thermokarst subsidence and erosion by running water subsequently transformed many of these lines into prominent trenches and canals over much of their length (Kerfoot, 1974). A more recent example of extensive vehicle track disturbance in Canada occurred during the summer of 1970 on the Sabine Peninsula of Eastern Melville Island. At that time a blow-out occurred at a wildcat well being drilled in the Drake Point area and vehicles were moved, of necessity, across tundra. Where sensitive tundra lowland underlain by soft ice-rich shale of the Christopher Formation was traversed, substantial and dramatic trenching occurred (Fig. 10.7(b)).

In general, however, severe erosion associated with vehicle tracks is rare and one must conclude that, for the most part, vehicle tracks present primarily aesthetic rather than terrain problems (French, 1978b, p. 48). Usually, vehicle operators prefer to traverse gently sloping terrain which does not provide the necessary gradient for subsequent deep gullying. Also, the differential settlement of the ground by thermokarst subsidence creates an irregular surface which precludes the development of an integrated drainage system. Furthermore, the introduction of vehicles equipped with special low-pressure tyres, the restriction of movement of heavy equipment to the winter months by both Canadian and Alaskan authorities, and the initiation of various terrain sensitivity and biophysical mapping programmes in areas of potential economic activity, particularly in Canada (e.g. Monroe, 1972; Kurfurst, 1973; Barnett *et al.*, 1977), are minimizing this sort of damage.

FIGURE 10.7   Terrain disturbance associated with vehicle movement over ice-rich tundra. (a) Old vehicle track, probably created in late 1940s or early 1950s in area of previous United States Navy Petroleum Reserve-4 (NPR-4), Alaska North Slope. (Photo taken, August 1977.) (b) Gully erosion along an old vehicle track made in the summer of 1970 near the site of the Drake Point blow-out (76°26′N; 108°55′W), Sabine Peninsula, Melville Island, N.W.T. The terrain is underlain by ice-rich shale of the Christopher Formation. (Photograph taken, August 1976.)

FIGURE 10.8 Constructions on permafrost. (a) (top left) Old and new buildings exist side by side in the city of Yakutsk in central Yakutia, Siberia. The old buildings, placed directly upon permafrost have experienced subsidence. Note also the frost heaving and tilting of the telegraph poles. (Photograph taken July 1973.) (b) (bottom left) Oblique air view of a recently drilled wellsite in the US National Petroleum Reserve (NPR-4), on the Alaskan North Slope. The operation was carried out on a large gravel pad and no damage to the

surrounding ice-rich tundra occurred. Husky Oils (NPR) Operations Ltd, South Simpson wellsite. (Photograph taken June 1977.) (c) (top right) Modern construction techniques, Yakutsk, Siberia. Piles are inserted into the permafrost and construction takes place above the ground surface. (Photograph taken July 1973.) (d) (bottom right) At Inuvik, Canada, buildings are placed upon wooden piles and services such as water and sewage are effected by a utilidor system which links each building to a central plant. (Photograph taken July 1969.)

10.6   ENGINEERING AND CONSTRUCTION PROBLEMS

The thawing of permafrost and the heaving and subsidence caused by frost action can cause serious damage to roads, bridges and other structures (Fig. 10.8(a)). In Alaska following the realization of these problems in the 1940s there was a determined effort made by federal and state agencies to improve construction practices and to document permafrost problems (e.g. Ferrians *et al.*, 1969; Péwé, 1966; Péwé and Paige, 1963). In Canada, where large-scale development projects in permafrost regions occurred slightly later, it was possible to benefit from Alaskan experience.

With respect to construction in permafrost, a number of approaches are available, depending on site conditions and fiscal limitations. For example, if the site is underlain by hard consolidated bedrock, as is the case for some regions of the Canadian Shield, ground ice is non-existent and permafrost problems can be largely ignored. In most areas, however, this simple approach is not feasible since an overburden of unconsolidated silty or organic sediments is rarely absent. In the majority of cases therefore, construction techniques are employed which aim to maintain the thermal equilibrium of the permafrost.

The most common technique is the use of a pad or some sort of fill which is placed on the surface (Fig. 10.8(b)). This compensates for the increase in thaw which results from either the warmth of the structure or the destruction of the vegetation that might have occurred during construction, or both. By utilizing a pad of appropriate thickness the thermal regime of the underlying permafrost is unaltered. It is possible, given the thermal conductivity of the materials involved and the mean air and ground temperatures at the site, to calculate the thickness of fill required. Too little fill plus the increased conductivity of the compacted active layer beneath the fill will result in thawing of the permafrost (Fig. 10.9(a)). On the other hand, too much fill will provide too much insulation and the permafrost surface will aggrade on account of the reduced amplitude of the seasonal temperature fluctuation (Fig. 10.9(b)). In northern Canada and Alaska, gravel is the most common aggregate used since it is reasonably widely available and is not as susceptible to frost heave as more finely grained sediments.

In instances where the structure concerned is capable of supplying significant quantities of heat to the underlying permafrost, as is the case of a heated building or a warm oil pipeline, additional measures are adopted. Usually the structure is mounted on piles which are inserted into the permafrost (Fig. 10.8(c)). An air space left between the ground surface and the structure enables the free circulation of cold air which dissipates the heat emanating from the structure. Other techniques used include the insertion of open-ended culverts into the pad, the placing of insulating matting immediately beneath the pad and, if the nature of the structure justifies it, the insertion of refrigeration units or 'Cryo-Anchors' (e.g. Hayley, 1982) around the pad or through the pilings.

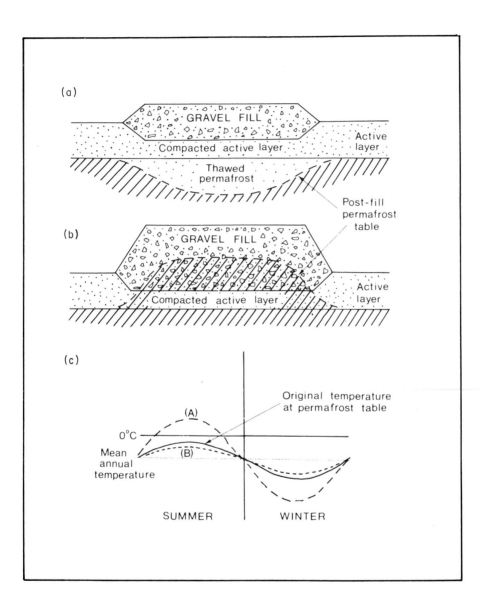

FIGURE 10.9 Diagram illustrating the effects of a gravel fill upon the thermal regime and thickness of the active layer. (a) Too little fill; (b) too much fill; (c) effects of cases (a) and (b) upon the thermal regime—too little fill increases the amplitude of seasonal temperature fluctuation at the permafrost table. (Source: Ferrians *et al.*, 1969)

In Canada, the construction of the town of Inuvik in the Mackenzie Delta in the early 1960s is another example of the careful manner in which large-scale construction projects have to be undertaken in permafrost regions. A major factor governing the location of the town was the presence of a large body of fluvioglacial gravel a few kilometres to the south (Brown, 1970). Rear-end dumping of these gravels was an essential prerequisite to any heavy construction activity. The result was that the entire townsite was developed upon a gravel pad. Today, the gravel deposit has been exhausted and the future growth of the community is dependent upon the exploitation of more distant aggregate sources with their associated higher costs of haulage. The provision of municipal services such as water supply and sewage disposal are particularly difficult in permafrost regions. Pipes to carry these services cannot be laid below ground beneath the depth of seasonal frost as is the case in non-permafrost regions, since the heat from the pipes will promote thawing of the surrounding permafrost and subsequent subsidence and fracture of the pipe. At Inuvik, the provision of these utilities has been achieved through the use of utilidors — continuously insulated aluminium boxes which run above ground on supports and which link each building to a central system (Fig. 10.8(d)). The cost of such utilidor systems is high, involving a high degree of town planning and constant maintenance, and can only be justified in large settlements.

In the Soviet Union, where modern cities with populations greater than 100 000 have developed in recent years in permafrost areas, utilidor systems constructed mainly of wood or cement are widely employed. In Yakutsk for example, many of the recently completed high-rise buildings are connected to a central large utilidor system for sewage which runs beneath the main street and eventually empties into the large Lena River. As in North America, modern construction takes place on pilings, and airspaces are left between the buildings and the ground surface.

Frost heaving of the seasonally frozen zone is the second major engineering problem to be encountered in permafrost regions. Differential heave can cause structural damage to buildings. Equally important is the fact that frost heaving affects the use of piles for the support of structures. While in warmer climates the chief problem of piles is to obtain sufficient bearing strength, in permafrost regions the problem is to keep the pilings in the ground since the frost action heaves them upwards. Since heaving becomes progressively greater as the active layer freezes, it follows that the thicker the active layer the greater is the upward heaving force. In the discontinuous permafrost zone, where the active layer may exceed 2 m in thickness, frost heaving of piles may assume critical importance. In parts of Alaska, for example, old bridge structures illustrate very dramatically the effects of differential frost heave (Fig. 10.10). In these regions, it is not uncommon for a thawed zone to exist beneath the river channel. Thus, piles inserted in the stream bed itself experience little or no frost heave. Likewise, piles inserted within permafrost on either side of the river are unaffected since

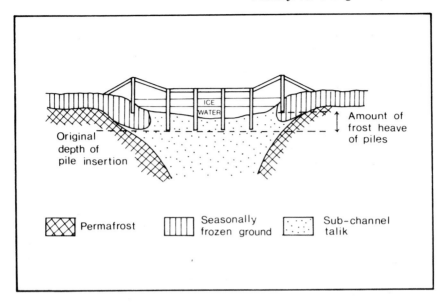

FIGURE 10.10 Diagrammatic illustration of how frost heaving of piles inserted in the layer of seasonal frost can result in bridge deformation. (source: Péwé and Paige, 1963, after an Alaskan example)

the adfreezing of the permafrost to the piles provides a resistance to the upward heaving of the seasonally frozen zone. By contrast, the piles adjacent to the river bank experience repeated heave since they are located in the zone of seasonal freezing. As a result, uparching of both ends of the bridge may occur.

In order to prevent heave, piles are usually inserted to a depth of at least 5 m in the permafrost. In the case of bridges, alternative structures involving minimal pile support are often considered. A case in point is the recent construction of the Eagle River bridge on the Dempster Highway, northern Yukon (Fig. 10.11). The bridge consists of a single 100 m long steel span with the footings on the north side placed in ice-rich permafrost. Drilling prior to construction indicated that permafrost was present on the north bank to a depth of ≈ 90 m and with a temperature of − 3°C. However, a deep near-isothermal talik existed beneath the main river channel while near the proposed south bridge abutment, permafrost had aggraded to depths of 8–9 m and was marginal in temperature (− 0.4°C). The shape of the thaw basin undoubtedly reflected a history of lateral river migration.

In order to maintain the natural permafrost conditions and to provide structural integrity, 15 steel piles were inserted at each abutment. Conventional engineering adfreeze analysis indicated the optimum depth of emplacement of each pile was ≈ 5 m. However, because of the marginal permafrost conditions at the south abutment, no support could be derived from the perennially frozen

FIGURE 10.11 Frost heave and related engineering problems. (a) (top left) The Eagle River bridge, Dempster Highway, Canada, is a single-span structure with minimal pile support in the river. Note also the erosion control measures in force (see text). (Photograph taken July 1979.) (b) (bottom left) The Trans-Alaska Pipeline near Fairbanks. Note the elevation of the pipe, the cooling devices in the VSMs and the gravel access pad. (Photograph taken May 1980.) (c) (top right) General view of the Sodegaura

LNG storage facilities of Tokyo Gas Company, Tokyo Bay, Japan. Insulation of the tanks is essential if frost heave adjacent to the structures is to be avoided. (Photograph taken August 1980.) (d) (bottom right) View of the frost heave test facility owned and operated by Northwest Alaskan Pipeline Company, at Fairbanks, Alaska. A series of cooling devices (Cryo-Anchors) are in the foreground. (Photograph taken May 1980.)

ground and the piles were driven to refusal at a depth of about 30 m. On the north abutment, where the permafrost was colder, the piles were installed in holes augered to a depth of 12 m below the ground surface. All holes were then backfilled with a sand slurry to promote adfreeze. The construction was carried out during the winter of 1976–77 in order to minimize surface terrain damage. Subsequent monitoring has indicated that the piles experienced minimal heave, the thermal regime of the permafrost was not altered, and the bridge structure has performed satisfactorily (Johnston, 1980).

The construction of oil and gas pipelines through permafrost terrain further illustrates the complexity of frost heave and related permafrost problems. For example, the construction of the Trans-Alaska Pipeline System (TAPS) from Prudhoe Bay on the North Slope to Valdez on the Pacific Coast between 1974 and 1977 utilized many procedures designed to minimize permafrost problems (Fig. 10.11(b)). First, an access road was constructed along the proposed route. This consisted of a layer of gravel nearly 2 m thick placed directly upon the tundra. In places, it was insulated to further prevent thaw of the underlying ice-rich materials. Working from this pad, heavy equipment enabled the pipeline to be constructed immediately adjacent to the road (Metz *et al.*, 1982). Since the pipe carries crude oil at temperatures which may exceed 30°C, elaborate measures were taken to ensure that permafrost degradation did not occur. Wherever possible, the pipe was mounted on piles, termed vertical support members or VSMs, inserted into the permafrost. Individual cooling units, using an ammonia solution and capable of airborne monitoring, were mounted on each VSM to conduct heat from the piling to the surrounding cold air. Where the pipeline was buried, the pipe was encased within a specially constructed thermal box surrounded by 60 cm of insulation materials (Heuer *et al.*, 1982).

The proposed construction of buried chilled gas pipelines and LNG storage tank (e.g. Fig. 10.11(c)) presents even more complex problems that are, as yet, not completely resolved. Here, the problem is not that of permafrost degradation but instead one of permafrost initiation and prolonged frost heave adjacent to the pipe or storage facility with the possibility of eventual rupture (Fig. 10.12). This would occur in the discontinuous permafrost zone wherever the pipe crosses unfrozen ground and where there would be relatively unlimited moisture migration towards the cold pipe. Several natural scale experiments are currently in progress, at Calgary, Caen and Fairbanks, Alaska, which aim to study the behaviour of soil around a refrigerated pipeline (Fig. 10.11(d)). The Calgary frost heave test facility has been in operation since 1974 and circulates air at −10°C in a 1.2 m diameter pipe buried to represent a number of possible gas pipelines modes (Carlson *et al.*, 1982). A 'frost bulb' has formed around the pipe and, in the deep burial mode, the pipe has heaved more than 60 cm while frost depths have penetrated to 3 m below the pipe (Fig. 10.13). At Caen, a non-insulated pipe 2.7 m in diameter is buried in initially unfrozen soil with a lateral transition from a frost-susceptible silt to a non-frost-susceptible sand,

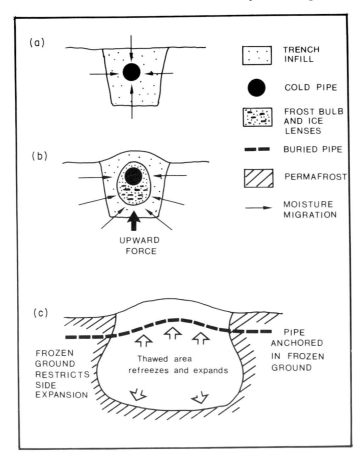

FIGURE 10.12 Frost heave and the problem of chilled gas pipelines. (a) Moisture migration towards cold pipe. (b) Subsurface water migration causes ice lensing (vapour and liquid migrates to the cold pipe) and 'frost bulb'. (c) Pipe stress caused by frost heaving in discontinuous permafrost terrain

thereby simulating a major boundary of soil types common to permafrost terrain. During a first freezing experiment, run in 1982–83 with a pipe temperature of $-2°C$ and a chamber temperature of $-7°C$, the pipe heaved 11 cm on the 16 m long section, and frost penetrated 45 cm beneath the pipe in the sand and 30 cm beneath the pipe in the silt (Burgess, 1985). Results from the Fairbanks test facility are not yet available, but because of its ability to simulate field conditions, they will be of great significance. Preliminary results indicate that in the first 166 days of operation, the pipe heaved at least 10 cm at the critical permafrost/non-permafrost boundary. If these results are representative of what might happen to a chilled buried gas pipeline passing through the discontinuous

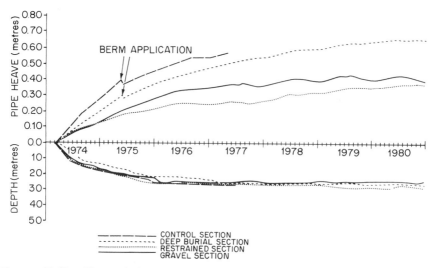

FIGURE 10.13   Observed pipe heave and depth of frost below bottom of pipe at the Calgary frost heave test facility (source: Carlson *et al.*, 1982)

permafrost, then eventual pipeline failure due to frost heave is a problem yet to be resolved.

10.7   HYDROLOGIC PROBLEMS

The groundwater hydrology of permafrost regions is unlike that of non-permafrost regions since permafrost acts as an impermeable layer. Under these conditions the movement of groundwater is restricted to various thawed zones or taliks (Fig. 10.14). These may be of three types. First, a supra-permafrost talik may exist immediately above the permafrost table but below the depth of seasonal frost. In the continuous permafrost zone, supra-permafrost taliks are rare. In the discontinuous permafrost zone however, the depth of seasonal frost frequently fails to reach the top of the permafrost since the latter is often relic and unrelated to present climatic conditions. In these areas, supra-permafrost taliks are widespread and may be several metres or more thick. Second, intra-permafrost taliks are thawed zones confined within the permafrost and, third, sub-permafrost taliks refer to the thawed zone beneath the permafrost.

Given these hydrologic characteristics, a difficult problem in many permafrost regions is the provision of water to settlements. Since supra-permafrost water is subject to contamination and usually small in amount, and intra-permafrost water is often highly mineralized and difficult to locate, the tapping of sub-permafrost water is vital. In the discontinuous permafrost zone, opportunities exist for groundwater recharge. In parts of Alaska and the Mackenzie Valley

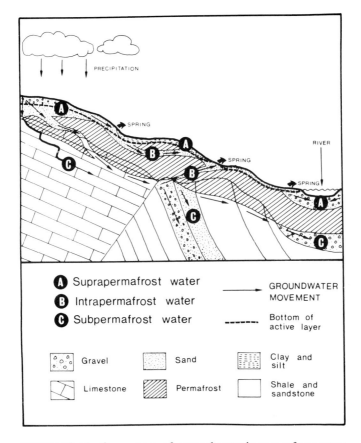

PRECIPITATION

SPRING

RIVER

SPRING

SPRING

A Suprapermafrost water
B Intrapermafrost water
C Subpermafrost water

→ GROUNDWATER MOVEMENT

------ Bottom of active layer

Gravel

Sand

Clay and silt

Limestone

Permafrost

Shale and sandstone

FIGURE 10.14   Occurrence of groundwater in permafrost areas

extensive alluvial deposits provide an abundant source of groundwater. In Fairbanks, houses rely on numerous small diameter private wells (Péwé, 1966, p. 28). In parts of Siberia, the occurrence of perennial springs fed by sub-permafrost water assumes special importance to man since these may be the sole source of water available over large areas. In areas of continuous permafrost which may exceed several hundreds of metres in thickness, drilling is either not possible since the hole would freeze or be too costly. In these areas, surface water bodies, particularly those which do not freeze to their bottoms in winter, must be utilized and great care taken to prevent contamination. It follows that the supply of water is a severe limitation to any large-scale permanent settlement in the continuous permafrost zone.

A different group of hydrologic problems relate to the formation of icings. These are sheet-like masses of ice which form at the surface in winter where water issues from the ground. Icings are of great practical concern as regards highway and railway construction and in fact, are a distinct hazard to any construction activity. These problems are most common in the discontinuous permafrost zone. Although sub- and intra-permafrost waters may be involved, the most frequently occurring icings are those associated with supra-permafrost water. A common occurrence is where a roadcut or other man-made excavation intersects with the supra-permafrost groundwater table (Fig. 10.15(a)). Seepage occurs and a sheet of ice forms, often several tens of square metres in extent. In North American, icings were first encountered on a large scale during the building of the Alaskan Highway (Thomson, 1966). Unless precautions are taken, icings can occur on most northern highways which traverse sloping terrain (Brown, 1970, pp. 109–111). Counter-measures to reduce icing problems include the avoidance of roadcuts wherever possible, the installation of high-arch culverts to divert water from the source of the icing, and the provision of large drainage ditches adjacent to the road. Icings may also block culverts placed beneath road embankments and, by diverting meltwater, initiate washouts in the spring thaw period. The costs of icing control and/or remedial measures can be considerable; for example, van Everdingen (1982) provides a conservative estimate of $20 000.00 for icing control at a locality studied on the Alaska Highway, Yukon, in the 1979–80 winter.

A major consideration in recent road building in northern Canada has been the design of river-crossing facilities. To avoid the expense associated with bridges like the one which crosses the Eagle river (see above), culverts or ferries are usually employed depending on the size of the river crossing. To minimize culvert utilization, modern roads such as the Dempster Highway, follow upland interfluve locations wherever possible (Fig. 10.15(b)). Culvert diameters are chosen to accommodate expected peak flood discharges with recurrence intervals of 50 years. They are usually installed with their bases at or below the natural stream bed elevation to discourage the formation of upstream ponds or downstream plunge pools. Since the higher and sustained water velocities that occur within culverts may block fish migration in the upstream direction, culverts are installed such that the mean water velocity does not usually exceed 0.9 m/s except during the mean annual flood and during the 50-year design flood.

In certain instances, man-made changes to the hydrologic and thermal regime of the supra-permafrost zone can lead to the growth of seasonal icing mounds or frost 'blisters' at locations where they would not occur under natural conditions. For example, van Everdingen (1978, p. 275) describes a situation which occurred during the construction of the Dempster Highway in the winter of 1973–74 in the vicinity of Fort McPherson in the Northwest Territories of Canada. In order to cross a shallow depression which effected subsurface drainage from a nearby lake, an embankment was constructed. The weight of the fill reduced the transmissivity of the water-bearing material of the underlying supra-permafrost talik. In addition, and as a consequence of the fill, permafrost began to aggrade beneath the fill. Thus, subsurface drainage beneath the

FIGURE 10.15   Road construction in permafrost terrain and associated hydrologic problems. (a) Groundwater icing and icing mounds developed in borrow pit used for road aggregate near Churchill, Manitoba. (Photograph taken June 1979.) (b) General view of the Dempster Highway as it traverses the Eagle Plain. Note the interfluve location, the gravel pad, and the culvert installations. (Photograph taken July 1979.)

embankment became restricted. During the winter following construction, therefore, and as a result of the build-up of hydraulic potential in the water-bearing layer, a triangular area upslope of the embankment was uplifted. Repeated ruptures of the frozen soil released flows of water which contributed to a large icing which eventually extended across the highway.

10.8 CONCLUSIONS

Increasingly, man's activities are being directed towards the more remote northern regions of the world underlain by permafrost. It can be demonstrated that permafrost, with its associated terrain, ground ice, and hydrologic characteristics, exerts a dominant influence over man's activities, and poses unique geotechnical problems. At the same time, man is able to seriously disrupt the sensitive equilibrium of permafrost environments, probably with more long-lasting, costly, and devastating results than in any other environment. A major challenge for the future will be to minimize the deleterious effects of man's activities upon the geomorphic processes of these regions.

REFERENCES

Babb, T. A., and Bliss, L. C., 1974, 'Effects of physical disturbance on Arctic vegetation in the Queen Elizabeth Islands', *Journal of Applied Ecology*, **II**, 549–562.

Barnett, D. M., Edlund, S. A., and Dredge, L. A., 1977, 'Terrain characterization and evaluation from Eastern Melville Island', *Geological Survey of Canada, Paper 76–23*, 18 pp.

Beskow, G., 1935, 'Soil freezing and frost heaving with special attention to roads and railroads', *Swedish Geological Society Yearbook*, Series C, 26, 375 pp.

Brown, J., Rickard, W., and Vietor, D., 1969, 'The effect of disturbance on permafrost terrain', *US Army CRREL, Special Report 138*, 13 pp.

Brown, R. J. E., 1970, *Permafrost in Canada; its influence on Northern Development*, University of Toronto Press, Toronto, 234 pp.

Brown, R. J. E., and Kupsch, W. O., 1974, 'Permafrost Terminology', *Technical Memorandum III*, Associate Committee on Geotechnical Research, National Research Council of Canada, Ottawa, 62 pp.

Burgess, M., 1985, 'Permafrost: Large scale research at Calgary and Caen', *Geos*, **1985** (2), 19–22.

Carlson, L. E., Ellwood, J. R., Nixon, J. F., and Slusarchuk, W. A., 1982, 'Field test results of operating a chilled buried pipeline in unfrozen ground', in French, H. M. (Ed.), *Proceedings, Fourth Canadian Permafrost Conference, Calgary, Alberta, March 2–6, 1981*, National Research Council of Canada, Ottawa, pp. 475–480.

Cheng, Guodong, 1983, 'The mechanism of repeated segregation for the formation of thick layered ground ice', *Cold Regions Science and Technology*, 8, 57–66.

Czudek, T., and Demek, J., 1970, 'Thermokarst in Siberia and its influence on the development of lowland relief', *Quaternary Research*, **I**, 103–120.

Everdingen, R. O. van, 1976, 'Geocryological terminology', *Canadian Journal of Earth Sciences*, 13, 862–867.

Everdingen, R. O. van, 1978, 'Frost mounds at Bear Rock, near Fort Norman, Northwest Territories, 1975–6', *Canadian Journal of Earth Sciences*, 15, 263–276.

Everdingen, R. O. van, 1982, 'Management of groundwater discharge for the solution of icing problems in the Yukon', in French, H. M. (Ed.), *Proceedings, Fourth Canadian Permafrost Conference, Calgary, Alberta, March 2-6, 1981*, National Research Council of Canada, Ottawa, pp. 212-228.

Everdingen, R. O. van, 1985, 'Unfrozen permafrost and other taliks', in Brown, J., Metz, M. C., and Hoekstra, P. (Eds), *Proceedings of a Workshop on Permafrost Geophysics, Golden, Colorado, October 1984. US Army CRREL, Special Report 85-5*, pp. 101-105.

Ferrians, O., Kachadoorian, R., and Green, G. W., 1969, 'Permafrost and related engineering problems in Alaska', *United States Geological Survey Professional Paper, 678*, 37 pp.

French, H. M., 1975, 'Man-induced thermokarst, Sachs Harbour airstrip, Banks Island, NWT', *Canadian Journal of Earth Sciences*, **12**, 132-144.

French, H. M., 1976, *The Periglacial Environment*, Longman, London, 308 pp.

French, H. M., 1978a, 'Terrain and environmental problems of Canadian Arctic oil and gas exploration', *Muskox*, **21**, 11-17.

French, H. M., 1978b, 'Why Arctic oil is harder to get than Alaska's', *Canadian Geographical Journal*, **94**, 46-51.

French, H. M., 1984, 'Some terrain and land use problems associated with exploratory wellsites, Northern Yukon Territory', in Olson, R., Hastings, R., and Geddes, F. (Eds), *Northern Ecology and Resource Management*, University of Alberta Press, Edmonton, 365-385.

Harry, D. G., French, H. M., and Pollard, W. H., 1985, 'Ice wedges and permafrost conditions near King Point, Beaufort Sea coast, Yukon Territory', *Geological Survey of Canada, Paper 85-1A*, pp. 111-116.

Haugen, R. K., and Brown, J., 1970, 'Natural and man-induced disturbances of permafrost terrain' in Coates, D. R. (Ed.), *Environmental Geomorphology*, State University of New York, Binghamton, pp. 139-149.

Hayley, D. W., 1982, 'Application of heat pipes to design of shallow foundations on permafrost', in French, H. M. (Ed.), *Proceedings, Fourth Canadian Permafrost Conference, Calgary, Alberta, March 2-6, 1981*, National Research Council of Canada, Ottawa, 535-544.

Heginbottom, J. A., 1973, *Effects of Surface Disturbance upon Permafrost*, Report 73-16, Environmental-Social Committee Northern Pipelines, Task Force on Northern Oil Development, Information Canada, Ottawa, 29 pp.

Heuer, C. E., Krzewinski, T. G., and Metz, M. C., 1982, 'Special thermal design to prevent thaw settlement and liquefaction', in French, H. M. (Ed.), *Proceedings, Fourth Canadian Permafrost Conference, Calgary, Alberta, March 2-6, 1981*, National Research Council of Canada, Ottawa, pp. 507-522.

Hopkins, D. M., Karlstrom, T. N., *et al.*, 1955, 'Permafrost and groundwater in Alaska', *United States Geological Survey Professional Paper 264-F*, pp. 113-146.

Johnston, G. H., 1980, 'Permafrost and the Eagle River bridge, Yukon Territory, Canada', in *Proceedings Permafrost Engineering Workshop, September 27-28, 1979*, National Research Council of Canada (Associate Committee on Geotechnical Research), *Technical Memorandum 130*, pp. 12-28.

Kallio, A., and Reiger, S., 1969, 'Recession of permafrost in a cultivated soil of interior Alaska', *Proceedings, Soil Science Society of America*, **33**, 430-432.

Kerfoot, D. E., 1974, 'Thermokarst features produced by man-made disturbances to the tundra terrain', in Fahey, B. D. and Thompson, R. O. (Eds), *Research in Polar and Alpine Geomorphology*, Proceedings, Third Guelph Symposium on Geomorphology, Geo Abstracts Ltd, Norwich, pp. 60-72.

Kurfurst, P. J., 1973, *Norman Wells, 96E/7, map 22; Terrain Disturbance and Susceptibility Maps*, Environmental–Social Program, Task Force on Northern Oil Development, Information Canada, Ottawa.

Mackay, J. R., 1970, 'Disturbances to the tundra and forest tundra environment of the Western Arctic', *Canadian Geotechnical Journal*, **7**, 430–432.

Mackay, J. R., 1972, 'The world of underground ice', *Annals, Association of American Geographers*, **62**, 1–22.

Mackay, J. R., 1983, 'Downward water movement into frozen ground, western arctic coast, Canada', *Canadian Journal of Earth Sciences*, **20**, 120–134.

McRoberts, E. C., and Morgenstern, N. R., 1974, 'The stability of thawing slopes', *Canadian Geotechnical Journal*, **II**, 447–469.

Metz, M. C., Krzewinski, T. G., and Clarke, E. S., 1982, 'The Trans-Alaska Pipeline workpad—an evaluation of present conditions', in French, H. M. (Ed.), *Proceedings Fourth Canadian Permafrost Conference, Calgary, Alberta, March 2–6, 1981*, National Research Council of Canada, Ottawa, pp. 523–534.

Monroe, R. L., 1972, 'Terrain maps—Mackenzie Valley', *Geological Survey of Canada, Open File Report 125* (Maps, scale 1:250 000 of Blackwater Lake, 96B; Norman Wells, 96E; Mahoney Lake, 96F; and Fort Franklin, 96G, map-areas).

Muller, S. W., 1945, 'Permafrost or perennially frozen ground and related engineering problems', *United States Geological Survey Special Report*, Strategic Engineering Study 62 (2nd edn).

Péwé, T. L., 1954, 'Effect of permafrost upon cultivated fields', *United States Geological Survey Bulletin*, **989F**, 315–351.

Péwé, T. L., 1966, 'Permafrost and its effect on life in the north', in Hansen, H. P. (Ed.), *Arctic Biology*, 2nd edn, Oregon State University Press, Corvallis, pp. 27–66.

Péwé, T. L., 1983, 'The periglacial environment in North America during Wisconsin times', in Porter, S. C. (Ed.), *The Late Pleistocene*, University of Minnesota Press, Minneapolis, pp. 157–189.

Péwé, T. L., and Paige, R. A., 1963, 'Frost heaving of piles with an example from the Fairbanks area, Alaska', *United States Geological Survey Bulletin*, **1111-1**, 333–407.

Polar Research Board, 1984, *Ice Segregation and Frost Heaving*. Ad Hoc Study Group on Ice Segregation and Frost Heaving, Committee on Permafrost, National Research Council, Washington, D.C., National Academy Press, 72 pp.

Pollard, W. H., and French, H. M., 1980, 'A first approximation of the volume of ground ice, Richards Island, Pleistocene Mackenzie Delta, Northwest Territories', *Canadian Geotechnical Journal*, **17**, 509–516.

Rockie, W. A., 1942, 'Pitting on Alaskan farms; a new erosion problem', *Geographical Review*, **32**, 128–134.

Smith, M. W., 1985a, 'Models of soil freezing', in Church, M., and Slaymaker, O. (Eds), *Field and Theory: Lectures in Geocryology*, University of British Columbia Press, Vancouver, B. C., pp. 96–120.

Smith, M. W., 1985b, 'Observations of soil freezing and frost heave at Inuvik, Northwest Territories, Canada', *Canadian Journal of Earth Sciences*, **22**, 283–290.

Soloviev, P. A., 1973, *Alas Thermokarst Relief of Central Yakutia*, Guidebook, Second International Permafrost Conference, Yakutsk, USSR, 48 pp.

Thomson, S., 1966, 'Icings on the Alaskan Highway', *Proceedings, First International Permafrost Conference*, National Academy of Sciences, National Research Council of Canada, 1287, pp. 526–529.

Tsytovich, N. A., 1975, *The Mechanics of Frozen Ground*, McGraw-Hill, New York, 426 pp.

Tyrtikov, A. P., 1964, 'The effect of vegetation on perennially frozen soil', *National Research Council of Canada Technical Translation,* **1088,** Ottawa, pp. 69–90.
Williams, J. R., and Everdingen, R. O. van, 1973, 'Groundwater investigations in permafrost regions of North America: a review', *Permafrost: North American Contribution, Second International Conference on Permafrost*, Yakutsk, USSR, National Academy of Sciences, Washington, D.C., 2115, pp. 435–446.

Human Activity and Environmental Processes
Edited by K. J. Gregory and D. E. Walling
©1987 John Wiley & Sons Ltd.

11

Subsurface Impacts

D. R. COATES

11.1 INTRODUCTION

Mankind's modification of subsurface materials and fluids produce manifold terrain changes and involve a wide spectrum of subdisciplines within the earth sciences. These displacements may be large or small, fast or slow, and surface disruptions may be up, down or lateral. These landform alterations involve materials that may expand, contract and fracture. Such changes may range from mere nuisances to those that produce disasters with accompanying loss of life and property. Total damage inflicted by these human activities amounts to billions of dollars yearly throughout the world and constitutes the greatest single type of man-induced damages.

Problems associated with man-made subsurface impacts received only minor attention by scientists and engineers until the 1950s. Unfortunately changes within the substrate are not visible, so unlike surface soil erosion, impending ground readjustments may be delayed. Thus the insidious gradual build-up of interior stresses has already reached the danger point before the problem is recognized. Below-ground changes are difficult to predict, and their prevention and control are even more demanding. The only solution in many cases may require abandonment of the site. Awareness, perception and monitoring of potential problems provide the safest guardians for man's protection, and require the close cooperation of scientists, engineers and land use managers.

Man lives on the earth's surface and many of his activities affect in some manner the stability of subsurface materials. His buildings may cause settlement, and mining of resources may produce land surface subsidence, fracturing and modification. Changes of the substrate also result in hydrocompaction from irrigation, desiccation during reclamation and even earthquakes from large reservoirs.

The diverse character of subsurface impacts prevents a precise classification of such changes. The character and magnitude of resulting deformation is a function of many variables and their origin can be polygenetic. Controlling factors that influence surface deformation include the type, size and distribution of the human activities. In addition the properties of the earth materials and their environmental setting determine the style and the magnitude of terrain dislocation. These properties are:

- Rock or sediment type, composition, thickness.
- Shape, orientation and arrangement of materials.
- Fabric, structure and occurrence of pre-existing fractures and *in situ* stress conditions.
- Physical conditions of the site such as geologic history, groundwater and other moisture conditions, constraints imposed by boundary conditions of dissimilar environments.

11.2   LOADING EFFECTS

Man introduces a variety of artificial loads on earth materials which would otherwise be in equilibrium within the geological setting. These extra burdens have the common denominator of overloading the system by subjecting the substrate to a surcharge of forces. Invariably the pore pressure of materials is modified, and when the initiating forces exceed the resisting forces, the resulting changes may include compaction of materials, migration of fluids and rupture of confining and surface rocks and sediment. Loading effects are discussed by reference to the consequences of reservoirs, water emplacement and injection, and man-made structures and buildings.

(a)   Dams and reservoirs

The first correlation between reservoir filling and earthquakes was made in Greece in 1931, and since that time more than 40 substantiated cases have been recorded where the man-made lakes have triggered earthquakes. The shocks range from microseisms to earthquakes with 6.4 magnitude on the Richter scale. The Marathon Dam, Greece, started to impound water in 1929 and earthquakes were first noted when the reservoir reached its highest level in 1931. Two damaging earthquakes > 5.0 magnitude occurred in 1938. The earthquake history showed that strongest seismicity was always associated with periods of rapid rise in water level.

Soon after the major filling of the Koyna Reservoir, India, in 1962, tremors were felt in areas previously mapped as aseismic. Although five important earthquakes had been felt prior to 10 December 1967, on that date a 6.3 magnitude event occurred that killed about 200 people, injured more than 1500 and left thousands homeless. The city of Bombay, 230 km from the epicentre, was shaken severely and the shutdown of the hydroelectric plant paralysed industry (Gupta, 1976). Other shocks in 1962–73 showed a correlation with water levels in the reservoir. Whenever the water level reached 652 m and remained at that level for some time, then earthquake activity, after some time lag, was increased. Other reservoirs that have produced significant earthquakes included: Monteynard and Grandvale in France, Mangla in Pakistan, Contra in Switzerland, Kariba in Zambia, Kremasta in Greece, Manic in Canada, Hendrick Verwoerd in South Africa, Nourek in USSR, Kurobe and Kamafusa in Japan, Hsinfengkiang in China, Camarillas in Spain, and Lake Mead in the United States. Although there is some dispute about how the trigger mechanism works, the consensus by seismologists is that the incidence of earthquakes is related to shear fracturing in rocks. When rocks are already under initial shear stress along an existing fracture plane, an increase in pore pressure caused by fluid migration can be sufficient to overcome the frictional resisting force and to induce shear failure and slippage. Not all reservoirs produce earthquakes because the presence of incompetent strata in basement rocks of some areas prevents stress build-up, and large ambient stress differences may be absent in regions that are insufficiently deformed.

A water surcharge in reservoirs in addition to causing earthquakes can also lead to landsliding. This reason is cited as a major contributing cause to the disastrous landslide to the Vaiont, Italy, dam on 9 October 1963 that led to the loss of more than 2000 lives. Jones *et al.*, (1961) investigated 321 landslides that occurred adjacent to Franklin D. Roosevelt Lake, the reservoir of Grand Coulee Dam, and determined that the majority had been triggered by groundwater build-up caused by increased water heights in the reservoir.

(b) Irrigation

Subsidence caused by application of irrigation waters on loose, dry, low-density soils has come to the attention of researchers only during the past 25 years. Large areas in North America, Europe and Asia contain these materials and especially in the western United States this problem has become acute with land lowering of 1–2 m being common, locally reaching 5 m (Lofgren, 1969). This process of hydrocompaction is especially prevalent in areas where there are: either loose, moisture-deficient alluvial deposits ranging from clayey water-laid sands and silts to mudflow materials; or loess and related aeolian sediments; or materials which are reasonably fine-grained with moisture deficiency where seasonal rainfall rarely penetrates below the root zone. Such deposits under natural conditions have sufficient dry strength due to clay bonding, cohesion and stacking to support an overburden of a few hundred metres. However, when the dry strength is disrupted by wetting, the materials are forced to adjust to the new pressure system and a different packing arrangement is produced with the resultant subsidence.

Damages by subsidence from the irrigation water surcharge amounts to many millions of dollars annually. Much of the damage is to irrigation ditches and

FIGURE 11.1   Severe fissures caused by hydrocompaction in irrigated fields. The site is in the west-central section of the San Joaquin Valley, southwest of Mendota, California. Photograph by Nikola P. Prokopovich in 1962 and used by his courtesy

canals, well casings, roads, pipelines and houses (Fig. 11.1). The largest affected area in the United States is in the San Joaquin Valley where more than 500 km² is undergoing hydrocompaction subsidence. Other areas are in the Heart Mountain and Riverton areas of Wyoming, near Billings, Montana, on many alluvial fans near Phoenix, Arizona, and areas within the Missouri Basin and the State of Washington.

(c)   Water injection

When an abnormal surcharge of water is pumped into subsurface materials one result may be the creation of earthquakes. Starting in March 1962 toxic wastes from the Rocky Mountain Arsenal of the US Army were injected into a 3671 m deep well sunk into Precambrian bedrock of weathered schist and fractured hornblende granite and gneiss. In April 1962, Denver, Colorado, felt the first earthquakes, in what had previously been a quiescent area (Evans, 1966). Although the injection programme was halted in September 1965, when proof had established the linkage between pumping and seismic activity, the shocks continued for several years and as late as 1969 there were two events of 3.5 magnitude and 14 events of more than 2.5. During the 1962–67 period, more than 1500 tremors occurred with magnitudes of 0.7 to 4.3 with most epicentres within 8 km of the well.

Other water injection schemes are those associated with pumping into aquifers to reverse salt water intrusion of coastal areas, and into petroleum reservoir strata to enhance oil recovery. Only the latter have produced notable subsurface and surface effects. The Baldwin Hills dam and reservoir in California were commissioned in 1951 at a site surrounded by oil wells of the Inglewood field. In 1954, a pilot water-injection project was so successful in recovering additional oil that an extensive programme began in 1957, and by 1963 there were 22 injector wells (Hamilton and Meehan, 1972). The first fault arising from this operation was noticed in May 1957, and by 1963 eight additional ones had been activated. On 14 December 1963, water burst through the foundation of the earthen dam and in a few hours the 946 000 m³ reservoir emptied onto the communities below. Investigation of the dam failure showed the faults under the reservoir had been activated, had cracked the protective clay liner, had caused 15 cm displacements, and had created sinkholes in the floor. The dam failure released waters that damaged and destroyed 277 homes, killed 5 persons, and caused $15 million of destruction. The Los Angeles Department of Water and Power brought a $25 million lawsuit against the oil company that was settled out of court for $3 875 000.

(d) Buildings and structures

The weight of man-made structures of all types may produce settlement of the substrate when improperly engineered. Thus buildings, streets, dams, landfills, canals etc. may all produce distortion in underlying materials (Fig. 11.2). The classic case is the Leaning Tower of Pisa, Italy, whose tower is tilted 5 m from the vertical. The tower rests on 4 m of clayey sand that is underlain by 6.4 m of sand on top of brackish clay (Legget, 1973). The Washington Monument, Washington D.C., has settled 14.6 cm since construction began. A concrete grain elevator near Winnipeg, Canada, was built on a 0.6 m thick cement slab foundation. The structure was 31 m high, 59 m long, and 23 m wide and weighed more than 20 000 tons. After being filled for the first time, 0.3 m of settlement occurred within an hour after being first noticed. Within 24 hours, the structure tilted 26°53′ from the vertical with settlement of 7.2 m on the west whereas the east side rose 1.5 m (Legget, 1973).

One of the largest artificial fills occurs in the San Francisco Bay area of California where one-third of the original bay has now been in-filled. Filling was initiated in 1849, and measurements in the Market Street area from 1864 to 1964 showed settlement of 3 m. Typical problems that have resulted include

FIGURE 11.2   Settlement of buildings into the artificial fill of the skid row area of San Francisco, California. These former wetlands were filled with muds and silts from San Francisco Bay to provide increased space for urbanization. This house has settled about 2 m as shown by windows which are now below sidewalk level but which originally were located on the second storey. (Courtesy of Donald Doehring.)

the tilting and settling of buildings into the fill (Fig. 11.2) below street level, cracking of walls, vertical separation of building, and ground sinking around piling foundations. Ramps have had to be constructed for gaining entrance to many buildings and 6 m of asphalt has been needed to counteract localized settlement at bridges (Griggs and Gilchrist, 1977).

11.3   WITHDRAWAL EFFECTS

Underground extraction of materials and fluids can produce fissures and faults, surface subsidence and collapse, and land uplift, as well as other dislocations in materials and in groundwater flow regimes. These impacts now occur on a worldwide basis as a result of mining of natural resources, both solid and fluid, such as coal, groundwater, steam, gas, oil and salt. Other types of extraction can occur for engineering purposes as in tunnel construction and the creation of caverns for underground space.

(a)   Groundwater mining

Excessive pumping rates that exceed groundwater recharge produce a lowering of the water table and lead to a condition referred to as 'groundwater mining'.

FIGURE 11.3   Dr Joseph Poland stands at the approximate point of maximum subsidence in the San Joaquin Valley, California. Subsidence of about 9.0 m occurred because of aquifer compaction caused by groundwater overdrafts. The telephone pole is only used to illustrate the position of the ground in 1925, 1955, and 1977 respectively. (Courtesy of Joseph Poland.)

The United States has the world's largest subsidence areas and the San Joaquin Valley, California, is the leader where more than 13 500 km² have been affected. Here the average land lowering exceeds 1 m with one 112 km long area having subsided more than 3 m with a maximum of nearly 10 m (Fig. 11.3). The total volume of subsided material amounts to 186 km³. Subsidence began in the 1920s when extensive use of groundwater for irrigation was initiated and increased until the mid-1950s, at which time the annual rate of land lowering was 0.55 m per year (Poland and Davis, 1969). Thereafter the subsidence rate decreased to 0.33 m per year in 1963–66. By 1973, the rate had become negligible, owing to importation of surface water from northern California via the California Aqueduct system. The water table has now recovered by as much as 60 m. The Santa Clara Valley, California, is another heavily subsided area where the 2.4 m land lowering has caused millions of dollars to be spent realigning canals and ditches, building higher roads, and renovating damaged buildings.

The Houston–Galveston region of Texas is the second largest subsidence area where more than 12 200 km² have lowered from 0.15 m to nearly 3 m. Groundwater pumping steadily increased from the early 1940s and by 1973 had reached 1.9 million m³ per day (Spencer, 1977). Extensive damages include coastal and tidal flooding where 2.4 m of freeboard has been lost (Fig. 11.4). Well installations protrude throughout the area and annual property losses now exceed $50 million per year in a 2450 km² area (Fig. 11.5). In addition to subsidence the region has numerous faults and fissures (Fig. 11.6). The 86 largest faults which have been historically active have an aggregate length of 240 km. Although the majority of faults occur along pre-existing faults their activation has been primarily caused by adjustments from the dewatering of more than 100 m in the underlying aquifers (Verbeek and Clanton, 1981).

Since the Second World War there has been rapid groundwater development in many basins of southern Arizona. More than 50 areas have now been identified covering 8000 km² where subsidence, earth fissures and faults occur. Water tables have been lowered more than 100 m in several areas, with maximum land lowering of 3.8 m. Fissures vary greatly in size but frequently are 1 m in width and several metres deep. Surface faults have lengths of 1 km and the largest has a scarp of 1 m and can be traced 16.7 km. Most ground failures in these unconsolidated sediments result from localized differential compaction. The earth fissures are caused by stretching related to overburden stresses, whereas the surface faults generally form where the compaction is discrete over pre-existing faults. Such failures can sometimes be predicted, either by determining the potential areas of differential compaction or by monitoring surface deformation in areas of ongoing water-level decline. Such techniques are currently being used to determine the location of potential ruptures in the Las Vegas, Nevada, area where fissuring occurs in a 125 km² area and there is damage to roads and buildings (Fig. 11.7).

FIGURE 11.4   House in what is now the extension of Galveston Bay, Baytown, Texas.
This area has subsided 2.4 m from groundwater pumping, and marine waters have
inundated residential areas. (Courtesy of Charles Kreitler.)

FIGURE 11.5   Abandoned water well in Baytown, Texas, of the Houston–Galveston subsidence bowl. The top of the concrete platform was originally at ground level, showing land lowering of 1.5 m caused by groundwater withdrawal. (Courtesy of Charles Kreitler.)

Of course subsidence from groundwater mining is international in scope. For example the rapid population growth of Mexico City led to abnormal pumping amounts. Water is obtained from sand and gravel aquifers which are separated by silt and clay deposits from depths that range from 60 to 500 m. Maximum subsidence now exceeds 7 m (Fig. 11.8). In London, England, a decline in artesian head began about 1820. By 1843 it was 7.5 m deeper and by 1936 had lowered as much as 100 m, with maximum subsidence of 0.21 m. Excessive pumping rates in Japan have caused subsidence below sea level in areas where 2 million people live in Tokyo and where 600 000 live in Osaka. In all these cases the land lowering is attributed to a reduction in fluid pressure to sustain water levels or artesian heads. This produces an increased and excessive load on the skeleton fabric of the granular material. When the bearing pressure cannot be sustained, then intergranular adjustment occurs in the fine-grained sediment resulting in compaction of the stratigraphic section.

Bangkok, Thailand, offers the most recent case of an area in severe jeopardy. The city is built on nearly flat terrain that averages only 1 m above sea level. Here 2000 m of deltaic sediments rest on a crystalline basement. The surface is covered with a soft marine clay that is now undergoing compaction from the heavy groundwater pumping caused by rapid industrial, urban and agricultural growth. Before such expansion the water table was 1 m below the land surface. However, by 1959 it had dropped to 9 m and during the next 20 years it had declined at a rate of about 2 cm per year and was 40 m below ground in 1978 (Rau and Nutalaya, 1982). During this period the surface elevation was lowered as much as 80 cm with rates of 10 cm per year in the southeastern part of the city where 1 million people live. In this locality flooding is an especially serious problem, and drainage and sewerage systems malfunction during the heavy monsoon rains. Sidewalks and roads are differentially subsiding in respect to those buildings constructed on pilings. In addition to extensive damage to some buildings, masonry walls, steps and driveways, saltwater has invaded some aquifers resulting in chloride content that exceeds 7000 ppm. Groundwater pumping is now more than 1 million m^3 per day.

Under certain conditions, it is even possible for land uplift to occur instead of subsidence in heavily pumped groundwater areas. In Arizona, water level decline from irrigation pumping in the 1915 to 1972 period amounted to 48 m in the Lower Santa Cruz basin and 42 m in the Salt River Valley basin (Holzer, 1979). Locally, decline of the water table in excess of 100 m was also common in both basins. First-order levelling surveys of the two regions established that during the 1948 to 1967 period, there was 6.3 cm of uplift in the Lower Santa Cruz area and 7.5 cm in the Salt River Valley. This uplift, or rebound, occurred in a 8075 km^2 area where 4.35×10^{13} kg of groundwater had been removed. The major uplift areas correspond to areas where crystalline bedrock is close to the surface or crops out through alluvium that is dewatered. The cause of the uplift is attributed to elastic expansion of the lithosphere when the groundwater load is reduced or depleted. Prior to the groundwater withdrawal, the Arizona region had been one of surface lowering.

(b) Sinkholes formed from groundwater pumping

A special type of surface disruption in the form of sinkholes can result from pumping of aquifers in carbonate terrain. In a study of Alabama sinkholes,

FIGURE 11.6   Active fissuring in the parking lot of Ellington Air Force Base, Houston, Texas. This fracture is representative of hundreds in the area caused by extraction of fluids, including oil, gas and water. (Courtesy of Charles Kreitler.)

Newton (1976) determined that more than 4000 were man-induced because of pumping operations. They occur where bedrock cavities are filled with residual or unconsolidated sediments that overlie the carbonate rocks. When the water table is lowered the loose materials lose their buoyant strength and migrate downward into the karstic openings. Other states that experience this type of sinkhole development include Florida, Georgia, Maryland, Pennsylvania, South Carolina and Tennessee.

FIGURE 11.7   House at Owens Avenue and A Street, Las Vegas, Nevada, damaged by an earth fissure that developed in December 1961. (Courtesy of the US Geological Survey.)

Sinkhole development caused by aquifer depletion is a hazard to structures and property, and in some instances can lead to disasters with destruction of property and loss of lives. In 1960, a major dewatering programme was initiated in the Far West Mining District near Johannesburg, South Africa, so that gold mining could be extended to greater depths (Foose, 1967). This resulted in the formation of some of the world's largest man-induced sinkholes in the carbonates and overburden. Between December 1962 and February 1966, eight sinkholes larger than 50 m in diameter and 30 m in depth were formed. The largest of the catastrophic sinkholes was 125 m in diameter with a 50 m depth. It occurred without warning. Other sinkholes did extensive damage in the Carletonville area (Fig. 11.9).

FIGURE 11.8   The Guadalupe Shrine, Mexico City, has differentially tilted and subsided into the substrate. Less compaction has occurred from the underlying fine-grained lake beds on the left side of the photograph because bedrock is closer to the surface so the underlying sediments are thinner. (Courtesy Donald Doehring.)

In 1962, 29 lives were lost in the collapse of a 30 m deep sinkhole and in 1964 five more were killed in another collapse site. During the pumping operations, groundwater levels declined from 100 m below ground to more than 550 m in the vicinity of the mines. Springs throughout the area also became dry.

Localities near limestone quarries in Pennsylvania have also suffered sinkhole development from dewatering operations to permit deeper surface mining. In the Hershey Valley, the Annville Stone Company pumped an average of 20 m^3 per minute during the 30 August to 4 September 1948 period and lowered the water table 10 m (Foose, 1953). In May 1949, a new pumping operation discharged groundwater at a rate of 24 m^3 per minute producing a water level decline of 50 m. This set in motion a chain reaction so that water became unavailable for irrigation and crops failed. Springs throughout the area became dry, and during the second month of pumping, sinkholes began to form. They ranged in size from 0.3 to 6 m in diameter with depths of 0.6 m to 3 m. The greatest number of the nearly 100 sinkholes formed in the valley and walls of Spring Creek, whose lower course dried up.

FIGURE 11.9 Cracks and fissure in pavement with both vertical and lateral displacements. These deformations resulted from differential subsidence caused by dewatering operations connected with gold mining in the Witwatersrand District, South Africa. (Courtesy of R. J. Kleywegt, Director, South Africa Geological Survey.)

(c) Oil and gas production

Withdrawal of fluid hydrocarbons is distinct from groundwater mining because the reservoir is invariably in rocks rather than sediments, extraction is from considerably deeper horizons and the area of influence is usually smaller. However, when appropriate conditions prevail, hydrocarbon fluid mining may lead to surface fracturing and subsidence. Such features were first observed in the late 1910s above the Goose Creek oil field in Texas. This was followed by reports of subsidence in the Bolivar oil field of Venezuela in the late 1920s, and then the Wilmington–Long Beach, California, oil field in the 1930s. More recently subsidence is also occurring over the Groningen gas field in the Netherlands, and at other California sites such as at Buena Vista Hills, Sante Fe Springs and Huntington Beach. Some gas fields in the USSR are reported to have subsidence phenomena.

The first evidence of subsidence in the Long Beach area was from groundwater withdrawal as early as 1928, but significant land lowering did not start until major oil production began in 1938 (Mayuga and Allen, 1966). The first important subsidence of 0.4 m was measured in 1940 and by 1945 had increased to 1.4 m. In 1951, the annual rate of subsidence had reached 0.6 m per year and was causing enormous damage to buildings, pipelines, railroads, bridges and roads (Fig. 11.10). Extensive diking and filling, and other engineering methods were used to counteract the land lowering that ultimately reached a total of 10 m (Fig. 11.11). Water injection into the strata, after permissive California legislation had been passed, helped to stabilize the area. In addition to the subsidence, earthquakes and faulting have also occurred. Eight separate periods of seismicity have been associated with the oil production and the resulting faults have severely damaged hundreds of producing wells with slippage as much as 22.8 cm during a single seismic event. Total damages from all subsurface impacts is greatly in excess of $100 million (Prokopovich, 1972).

Early studies of the Goose Creek oil field, Texas, showed oil production affected an area 6.5 km long and 3.9 km wide, and caused land lowering > 1 m. In addition, faulting and earthquakes occurred, with some ruptures 700 m long (Yerkes and Castle, 1976). Subsidence in the oil fields at Lake Maracaibo, Venezuela, was first discovered in 1933 and by 1954 had reached 3.3 m in some areas. The clearest example of extraction-induced seismicity by gas outside of North America is from the Po delta, Italy, where production of methane gas in 1951 caused a series of earthquakes. In addition, subsidence occurs in an area 40 km long and 20 km wide and has caused extensive damage from flooding and necessitated construction of higher levees and new drainages to relieve flooded lands. Subsidence of significant proportions has also occurred from the methane gas wells at Niigata, Japan (Poland and Davis, 1969). The general

FIGURE 11.10   Sheared and failed bridge column at the north end of Heim Bridge, Long Beach, California. This is typical of structural damage to many installations in the area caused by subsidence induced by extensive petroleum withdrawal. Photograph was taken by the Long Beach Harbor Department on 16 September, 1955. (Courtesy of Raymond F. Berbower.)

FIGURE 11.11   This is an aerial photograph of the Long Beach region, California, taken prior to 1970. The contour lines that are superimposed are in feet and illustrate the subsidence bowl that resulted from petroleum withdrawal. (Courtesy Raymond F. Berbower.)

model used by investigators to explain hydrocarbon extraction effects is the compaction of strata by loss of the fluid support. In the rigid rock units, low-angle thrust faults form in the central area with normal faulting along the periphery.

(d) Underground coal mining

Subsurface extraction of rocks and minerals produce larger cavities than withdrawal of fluids. Thus their effects are likely to be more localized and ground lowering may occur more suddenly by collapse of overburden instead of gradual subsidence. There are more than 3250 km^2 of underground coal mines in the United States and 25 per cent have been affected by some form of surface deformation (HRB-Singer, Inc., 1977). Much of the damage and threatened areas are in the eastern United States where the US Bureau of Mines estimates that 162 km^2 occur in urban areas with potential damages in the billions of dollars. In 1968 extensive losses occurred near Wilkes-Barre, Pennsylvania, where bridges, water and gas lines, homes and churches were damaged. Collapses in Coaldale, Pennsylvania, damaged or destroyed 23 homes in 1963. In Scranton, Pennsylvania, subsidence had become such a severe problem by 1964 that engineering stabilization costing $413 million was necessary in a 0.33 km^2 area. These Pennsylvania losses caused the State to pass the Bituminous Mine Subsidence and Land Conservation Act of 1966 (Vandale, 1967). This helped to regulate coal mining by mandating sufficient support for all structures in addition to requiring permits and bonding as well as inspection of facilities. In most instances nearly 50 per cent of the coal was deemed unmineable because extensive pillars must be left to support the mine openings.

Two types of subsidence are recognized above abandoned mines — sinkholes and troughs, sometimes called pits and sags. Sinkholes occur by collapse of the overburden into a mine opening with steep bounding walls and with a diameter that may increase with depth (Figs. 11.12 and 11.13). They generally occur where the overburden is less than 10 or 15 times the thickness of the excavated coal seam. Most are in areas where the cover is less than 15 m. Troughs are shallow dish-shaped depressions that develop by sagging of the overburden in response to the crushing or downward punching of destroyed pillars. They usually occur where the overburden exceeds 15 m. The size of the land displacements ranges from a few metres to tens of metres in the horizontal, and depths of a metre to tens of metres (Fig. 11.14). There can be a lag time from the cessation of mining till when the first surface disruption occurs. For example, in the eastern United States a study of 354 subsidence incidents showed that more than half occurred 50 years or more after mining had stopped (Gray and Bruhn, 1984). Mines abandoned in the Denver Metropolitan area decades ago are now causing destruction

FIGURE 11.12 Sinkholes and troughs caused by collapse and subsidence from underground mining of lignite coal near Beulah, North Dakota. Coal was mined here during the 1918–52 period with 15 m of overburden above the rooms and pillars of the coal seams. Photograph taken in October 1976 by C. Richard Dunrud
(Dunrud, 1984)

of highways, utility lines and buildings in or near the towns of Lafayette, Erie, Louisville and Firestone, Colorado. Littlejohn (1979) has reported that subsidence is still occurring in Great Britain above mines that were abandoned centuries ago.

Underground coal mining can also initiate other distressing repercussions. Mining in the North Fork of the Gunnison River, Colorado, started in the 1930s, but coal production stopped in 1953 after methane gas and water were encountered in quantities too costly to control. After the mine had been sealed, methane rose to the surface through subsidence fractures in the overlying materials, killing scrub oak and all other woody plants on the surface. In addition, springs ceased to flow in the nearby canyons because mining had caused diversions in the underground flow. Two coal beds below the mine are also threatened as mineable reserves because of the intrusion of methane and water from above.

(e) Salt mining

Salt mining, both underground and brine pumping, can lead to surface disruption and even disasters. Poor mining practices and high extraction ratios

FIGURE 11.13 Sinkholes caused by underground coal mining, New South Wales, Australia. Underneath the Homeville Seam attains a thickness of 4 m. Photography courtesy of the Department of Mines taken in May 1975

of mined thickness and overburden have caused the collapse of many areas in the salt mines near Cheshire, England (Bell, 1975). Mining started in the eighteenth century but few mines lasted longer than 40 years before a 'rock pit hole' occurred. The last catastrophic collapse was of the Adelaide Mine in 1928. Brine pumping operations have now caused subsidence as far as 8 km from the site of the producing well and create what are called 'flashes'. These are waterfilled linear hollows developed in the overlying sediments. They may be 10 m in depth and 70 m wide, and they form by collapse of the brine runs. The Cheshire Brine Subsidence Compensation Board was instituted because of the many associated hazards and the damage

FIGURE 11.14   Structural damage to house in Pittsburgh, Pennsylvania, from ground
subsidence caused by underground coal mining. (Courtesy of Jesse Craft.)

to buildings, farmland, piped services and road and rail communications.
Although some damage claims have been settled, it is often difficult to
assess responsibility because effects generally occur at some distance from the
producing well which is responsible.

Drilling through aquifers and salt beds in search of oil, gas and water can
result in salt dissolution and subsequent land lowering. In addition construction
of highways, dams and reservoirs over saline or gypsiferous rock has resulted
in surface ruptures, subsidence and water loss, and even dam failure, as in the
case of the St Francis Dam, California. Walters (1977) discussed 13 cases
of subsidence in central Kansas associated with solution salt mining. One
dramatic case occurred at Hutchinson, Kansas, where a sinkhole 90 m in
diameter formed during a three-day period and left railroad tracks suspended
in air. Here the salt had been mined from a 105 m thick Permian salt
stratum whose top was 120 m below the ground. Another case happened on
24 April 1959 in Barton County, Kansas, where a 90 m wide sinkhole formed
during a 12-hour period.

Ege (1984) described sinkholes that formed at Grosse Ile, Michigan, in 1971.
Solution mining of salt had been occurring for 20 years at the site. Here 18 m
of glacial clay overlies 150 m of impure dolomite bedrock. Below this there is
a salt section 220 m thick. In January 1971 a depression 7 × 9 m first formed
and progressively enlarged over a period of several months to a 60 m diameter
sinkhole at which point it stabilized. Thereupon a second sinkhole began to

develop 800 m away and again after several months stabilized with a 135 m diameter and depth of 50 m. Fortunately damages to surface installations were minimal. This, however, was not the case in the area overlying a salt mine at Lake Peigneur, Louisiana (Fig. 11.15). Underground salt mining had formed rooms that were 24 m high and 20 m wide at a depth of 390 m below the lake. While drilling for petroleum at the site, a Texaco drill bit penetrated the roof of the salt mine on 20 November 1980. Within a 2½-hour period, water from the lake churned into the hole and a giant whirlpool had developed. It opened to a diameter of 1 km and swallowed 11 barges, one tugboat and demolished three buildings (Fig. 11.16). In addition 26 ha of lake frontage was destroyed by catastrophic flooding and erosion by streams, and the faulting and earth slumping of the Pleistocene sediments (Fig. 11.17). The final indignity was the fire that raged in an adjacent gas well. Miraculously no lives were lost but 200 miners lost their jobs as an aftermath of the destruction (Martinez *et al.*, 1981).

(f) Other disturbances from mining operations

Underground mining of other earth resources can also lead to surface and subsurface impacts. The character of the mining methods may inadvertently, or in some cases deliberately, produce surface modification such as that at the San Manuel, Arizona, copper mine (Fig. 11.18). Geothermal fluid withdrawal at Wairakei, New Zealand, has produced vertical displacements of 4.5 m and horizontal displacements of 0.5 m in a subsidence bowl that is 1500 m in diameter. To date The Geysers, California, geothermal production is the largest in the world and the extraction process has caused a maximum subsidence of 0.13 m.

Mining activity can also precipitate sudden energy changes that can be grouped into three categories, namely, rockbursts, which are violent failures of explosive character, bumps, which are less violent rock changes, and outbursts, which involve the rapid release of gas that can eject rock with damaging effects (Cook, 1976). Microseismic events can also be associated with these phenomena. An early example of the relation between the incidence of rockbursts and the rate of mining occurred in mines of the Brezove Hoty District, Czechoslovakia. Here the annual frequency of rockbursts during the 1913–38 period closely coincided with the amount of rock extracted. Similar relationships between damage and mining activity have been reported in the nickel mines of Sudbury, Canada, the gold mines in Kolar, India, the gold mines in the Witwatersrand District, South Africa, and the zinc mines in the Coeur d'Alene District, United States. Such forces can produce seismic magnitudes as high as 5.0, and in 1971 there were 1600 tremors produced in the Witwatersrand that ranged from 2.0 to 4.2 magnitude. Gypsum mining at Garbutt, New York, has created both subsidence and sinkholes throughout the mined area, and

FIGURE 11.15

FIGURE 11.16

FIGURE 11.17   Lake Peigneur has completely drained into underground openings. Barges and docks have foundered and been displaced and rotational slump blocks have developed within the Pleistocene strata across the collapsed area. (Courtesy Joseph Martinez.)

sandstone mines in the northeast part of Prague have led to land lowering that endangers buildings in the city.

11.4   SURFACE EXCAVATION EFFECTS

Whenever man cuts into the ground, rearranges landforms or alters surficial water and drainage, changes can be produced in both surface and subsurface materials. These modifications result because the equilibrium that they formerly possessed has been destabilized by the man-made stresses. Depending on the character of the dislocation and the type of earth materials, they may expand, contract or shear.

FIGURE 11.15   *(opposite above)* Jefferson Island, Louisiana, and Lake Peigneur aerial photograph taken prior to the catastrophic event of 20 November 1980. Here a salt dome was being mined about 400 m below the surface. (Courtesy of Joseph Martinez.)

FIGURE 11.16   *(opposite below)* Lake Peigneur has drained catastrophically and newly incised channels have developed with steep gradients into the barren lake bottom. Note scarps that have begun to develop. (Courtesy Joseph Martinez.)

FIGURE 11.18 Surface faulting with slickensides have developed in these surface materials from underground copper mining at San Manuel, Arizona. (Courtesy of Allen Hatheway.)

(a) Construction for roads, buildings, resources and services

Piping is a severe problem when excavations are made in some fine-grained and loosely compacted soils such as loess or lacustrine strata. Aghassy (1973) described piping that evolved prior to 1948 in the southern coastal plain of Israel. These below-surface conduits and passageways ultimately led to a man-created badland topography. The bedouins used camel-drawn ploughs that incised deep furrows in the loess and led to subsurface infiltration producing vertical pipes that connected to underground lateral pipes. This produced 2–3 m of collapse in surface materials during a single cycle. Additional badland topography by piping formed as a result of construction of the Ottoman Railroad prior to the First World War. Along the rail route, incision and side-drainage channels created oversteepened slopes. Percolating waters developed steeper gradients with the creation of pipes and the subsequent collapse of overburden into the gullies and rills. Many roads along the desert basins in western United States have caused piping adjacent to and under the roadbeds. Coates (1977) described piping in fine-grained glacial sediments near Binghamton, New York, and in the USSR construction of the Khodza–Kola Canal has caused piping and subsidence in every kilometre of the 20 km long excavation (Figs. 11.19 and 11.20).

FIGURE 11.19 Landslide in fine-grained glacial sediments near Binghamton, New York. The roadcut oversteepened the hillslope and with increased gradients percolation water and drainage from upper road initiated piping. (Source: Coates, 1977.)

Construction in permafrost terrain can also produce many undesirable effects when improperly engineered. In Fairbanks, Alaska, the cutting of trees and brush has caused a lowering in the permafrost table of 3.8 m in 10 years and consequent subsidence of the ground surface (Cooke and Doornkamp, 1974). It is not unusual for roads to subside as much as 2 m when no allowance is made for permafrost conditions. Several hundred million dollars in extra construction costs were required in building the Trans-Alaska Pipeline because of potential damage that would result from dislocations produced by permafrost processes.

Problems related to expansive sediments (soils) and rocks, were not seriously recognized by soil engineers until the latter part of the 1930s. Previously to this,

FIGURE 11.20 View of piping in the glacial sediments of Fig. 11.19. This site is in the oversteepened exposure immediately upslope from the lower road. Mattock is 45 cm long and piping conduit extends about 6 m into hillside. (Source: Coates, 1977.)

FIGURE 11.21   Rupture and displacement of concrete wall in Waco, Texas. This damage was caused by expansion and swelling forces of the underlying Eagle Ford Shale when wetted excessively. (Courtesy of Robert Font.)

damage to buildings and roads was generally attributed to faulty construction and materials. The US Bureau of Reclamation first investigated the soil swelling problem in 1938, and since then damages have been assigned to settlement or expansion properties of the soil. The increased use of concrete slab-on-ground construction methods after 1940 has further increased structural damage from expansive soils. Annual damages in the United States linked to expansive soils exceeds $2255 million (Chen, 1975), which at the time the survey was made was greater than losses from such hazards as floods and hurricanes (Fig. 11.21). The principal damage is caused by high montmorillonite clay content within the soil. The following solutions can be used to remedy or prevent such losses:

- Replace clay soils with non-swelling types.
- Provide sufficient deadload pressure on all footings and slabs to withstand cracking.
- Flood the area prior to construction.
- Decrease the density by compaction.
- Change clay properties by chemical treatment and injection.
- Isolate the soil so that no moisture change can occur.

When excavation is made into bedrock, allowance should be made for possible rock readjustments because of the relief in pressure on the floors and walls of the excavation. Depending upon the amount of *in situ* stress within the rocks, the lithologic composition and the character of bedding planes and fractures, it is possible for excavation to create creep and microfaulting. For example, rock creep and swell can amount to many centimetres. Pop-ups and rock bursts can also occur during quarry operations.

(b) Reclamation

Drainage of organic-rich sediments for reclamation purposes invariably causes subsidence by the desiccation–dehydration process. The subsidence rate and amount depends on the type of peat or organic matter, the depth of the water table, and the temperature. The consequence in loss of mass and lowered land surface result in: loss of plant rooting depth, increased pumping for continued drainage, instability of roads and structures, increased outflow of nutrients, colder surface temperatures during winter nights and increase of CO^2 for the global atmosphere. To prevent such losses the water table should be kept as high as crop and field conditions permit.

The oldest record of organic subsidence comes from studies of old polders in the western Netherlands. The low peat soils were reclaimed between the ninth and thirteenth centuries when land surface was equal to or above sea level and adjacent rivers. Therefore excess water could be discharged by sluices into nearby water bodies. However, this system of gravity flow failed by the sixteenth century when the surface had subsided below sea level, thereby requiring removal of excess water artificially by windmills. After steam pumping stations had been installed, about 1870, subsidence was accelerated and there was surface lowering at a rate of 6 mm per year during the 1877–1965 period (Schothorst, 1977).

The peaty fens region bordering the Wash on England's east coast has been the site of reclamation drainage for about 400 years (Allen, 1969). At Holme Post, after the first pumps for drainage were installed in 1948, the peat sank 1.5 m during a 12-year period and then sank another 0.9 m in the next eight years. Total subsidence by 1932 was 2.7 m and the original peat thickness was reduced by desiccation from 6.7 m to 3.4 m. Prior to the artificial drainage, where silt sediments cropped out they were 2.1 m lower than the peat exposures. However, after the drainage, the silt areas stood 3 m higher, owing to difference in subsidence rates and the differentials in dewatering. In Norway the cultivated bog soils at the Stend Agricultural School lost 1.5 m in 65 years by drainage, and the peat bog soils of the Russian Minsk Experiment Station lost 1 m in thickness from 1914 to 1961.

In the United States, the Everglades comprise the largest single area of organic soils in the world, covering 7770 km^2. Lowering of the water table by surface drainage began in 1906 and caused volume losses in the peat through the biochemical actions under aerobic conditions. All of the soil shrinkage has occurred by oxidation above the water table. The arable organic soils subsided more than 1 m in a 30-year period, and after 40 years there had been a 40 per cent loss in soil volume. At some control positions concrete monuments with tops originally flush to the ground now protrude 1.8 m after 57 years (Stephens *et al.*, 1984). The delta confluence of the Sacramento and San Joaquin Rivers, California, is the second largest peat area in the United States. Here more than 1000 km^2 of peat lands have been reclaimed by draining and then cultivated. This area was above sea level before the start of reclamation 100 years ago. By 1953, most of the land surface was more than 3 m below sea level, and protective levees had to be constructed to protect the saucer-shaped lands surrounded by a marine environment. New Orleans, Louisiana, is in a wetland area with most land below sea level. An extensive levee system had to be engineered to prevent flooding. The city must continually operate water pumps to discharge rainfall runoff and groundwater seepage from the land. Subsidence has occurred in a 150 km^2 area with a maximum of almost 1 m. The pumping system is the largest in the world, composed of 21 stations with daily capacity to pump 9.5×10^6 m^3 of water (Wagner and Durabb, 1977).

11.5   OTHER MAN-INDUCED CHANGES

Although not central to the theme of this chapter, there are many additional subsurface side effects, as well as resource benefits, that result from man's involvement with substratal materials. In coastal areas, excessive groundwater pumping of fresh water has permitted saline intrusion by marine waters into aquifers. This has caused severe problems in New York, Florida, Texas and California. Many landslides result from subsurface influence by man when excessive cuts are made for highways. Font (1977) and Mathewson and Clary (1977) discuss such landsliding in Texas initiated by highway construction that destabilized overconsolidated shale. In California, watering of lawns and septic tank effluent have percolated subsurface materials, and this surcharge lowered the frictional resistance of the clayey substrate, causing landslides.

Another range of impacts occurs with detonation of explosives within the earth which can cause earthquakes and may produce damage to structures and create subsurface changes. In the United States, numerous claims are made by property owners who contend that their wells lost production or their walls have been cracked from dynamite used in highway construction. Nuclear testing can create earthquakes of large magnitude. For example a 1 megaton explosion may cause seismicity of about 7.0 magnitude. At the Yucca test facility in Nevada, movement along a pre-existing fault occurred for a distance of 600 m after a nuclear device was exploded underground in 1968. At the site there was a maximum vertical displacement of the ground amounting to 4.5 m.

Of course, it is important to remember that the subsurface should be considered as an important resource to man. In summer, groundwater is cooler than surface water so is widely used for air-conditioning purposes. Furthermore the underground is warmer than surface materials in winter so can be used as a heat pump to warm buildings. Kansas City, United States, and Sweden have made extensive use of the underground to house storage facilities and for office and business purposes. Finally, it can be argued that the subsurface provides important benefits for the incarceration of municipal, industrial and atomic waste. Sanitary landfills have become one of the most popular in-ground waste disposal systems for many cities. For 10 years the United States government has been intensively studying the most effective ways to dispose of radioactive wastes, and the decision has been reached to bury them in deep underground vaults which it is hoped will be sufficiently secure for modern man not to become contaminated.

11.6 CONCLUSION

Man-induced subsurface changes mount each year and the damages increase. The expansion of activities into areas with fragile soils, as in semiarid, permafrost and wetland environments, causes changes in the substrate. The construction of buildings, roads and reservoirs impose additional stress on subsurface materials. Knowledge of these changes has lagged because the disciplines of rock and soil mechanics have only developed in the past few decades and earthquake monitoring is only now reaching high levels of sophistication. There is a plethora of ways that man changes the subsurface materials, fluids and processes, but they can be grouped into such principal categories of: first, placing additional loads on or within the earth, secondly, removal of underground rocks and fluids, and thirdly, surface excavation and drainage. These activities produce a wide variety of responses, which include land subsidence or collapse, earthquakes, fissuring and faulting, expansion or contraction of material, land uplift, altered groundwater or surface water flow systems, sudden bursts of material, the release of gas and the creation of new topographies. These changes have killed people, destroyed dams and buildings, ruined wells, altered road alignments, reversed flow in canals and ditches and caused soils to lose productivity.

Multidisciplinary efforts are necessary to prevent, predict and control damage by subsurface changes. Engineers must be continually aware of the various feedback mechanisms that operate to produce chain-reaction effects. This requires perception and investigation by a variety of earth scientists and enlightened planning by land managers.

REFERENCES

Aghassy, J., 1973, 'Man-induced badlands topography', in Coates, D. R. (Ed.), *Environmental Geomorphology and Landscape Conservation, Vol. 3, Non-urban Regions*, Dowden, Hutchinson & Ross, Stroudsburg, Pa., pp. 124–136.

Allen, A. S., 1969, 'Geologic settings of subsidence', Varnes, D. J., and Kiersch, G. (Eds), *Reviews in Engineering Geology*, Vol. 2, Geol. Soc. America, Boulder, Colorado, pp. 305–342.

Bell, F. G., 1975, 'Salt and subsidence in Cheshire, England', *Engng. Geol.*, **9**, 237–247.

Chen, F. H., 1975, *Foundations on Expansive Soils*, Elsevier, Amsterdam.

Coates, D. R., 1977, 'Landslide perspectives', in Coates, D. R. (Ed.), *Landslides: Reviews in Engineering Geology*, Vol. 3, Geol. Soc. America, Boulder, Colorado, pp. 3–28.

Cook, N. G. W., 1976, 'Seismicity associated with mining', *Engng. Geol.*, **9**, 99–122.

Cooke, R. U., and Doornkamp, J. C., 1974, *Geomorphology in Environmental Management*, Oxford University Press, London.

Dunrud, C. R., 1984, 'Coal mine subsidence—western United States', in Holzer, T. L. (Ed.), *Man-induced Land Subsidence: Reviews in Engineering Geology*, Vol. 6, Geol. Soc. America, Boulder, Colorado, pp. 151–194.

Ege, J. R., 1984, 'Mechanisms of surface subsidence resulting from solution extraction of salt', in Holzer, T. L. (Ed.), *Man-induced Land Subsidence: Review in Engineering Geology*, Vol. 6, Geol. Soc. America, Boulder, Colorado, pp. 203–221.

Evans, D. M., 1966, 'Man-made earthquakes in Denver', *Geotimes*, **10**, 11–18.

Font, R. G., 1977, 'Engineering geology of the slope instability of two overconsolidated north-central Texas shales', in Coates, D. R. (Ed.), *Landslides: Reviews in Engineering Geology*, Vol. 3, Geol. Soc. America, Boulder, Colorado, pp. 205–212.

Foose, R. M., 1953, 'Groundwater behavior in the Hershey Valley, Pennsylvania', *Bull. Geol. Soc. America*, **64**, 623–646.

Foose, R. M., 1967, 'Sinkhole formation by groundwater withdrawal: Far West Rand, South Africa', *Science*, **157**, 1045–1048.

Gray, R. E., and Bruhn, R. W., 1984, 'Coal mine subsidence—eastern United States', in Holzer, T. L. (Ed.), *Man-induced Land Subsidence, Reviews in Engineering Geology*, Vol. 6, Geol. Soc. America, Boulder, Colorado, pp. 123–149.

Griggs, G. B., and Gilchrist, J. A., 1977, *The Earth and Land Use Planning*, Duxbury Press, North Scituate, Massachusetts.

Gupta, H. K., 1976, *Dams and Earthquakes*, Elsevier, Amsterdam.

Hamilton, D. H., and Meehan, R. L., 1972, 'Ground rupture in the Baldwin Hills', *Science*, **72**, 333–344.

Holzer, T. L., 1979, 'Elastic expansion of the lithosphere caused by ground-water depletion', *Jour. Geophysical Research*, **84**, n. B9, 4689–4698.

HRB-Singer, Inc., 1977, *Nature and distribution of subsidence problems affecting HUD and urban areas*, US Department Housing and Urban Development, US Govt. Printing Office, Washington D.C., 113 pp.

Jones, F. O., Embody, D. R., and Peterson, W. L., 1961, 'Landslides along the Columbia River Valley, northeastern Washington', *US Geol. Survey Prof. Paper 367*, 98 pp.

Legget, R. F., 1973, *Cities and Geology*, McGraw-Hill, New York.

Littlejohn, G. S., 1979, 'Consolidation of old coal workings', *Ground Engng*, **12**, 4, 15–21.

Lofgren, B. E., 1969, 'Land subsidence due to the application of water', in Varnes, D. J., and Kiersch, G. (Eds), *Reviews in Engineering Geology*, Vol. 2, Geol. Soc. America, Boulder, Colorado, pp. 271–303.

Martinez, J. D. *et al.*, 1981, 'Catastrophic drawdown shapes floor', *Geotimes*, **26**, 3, 14–16.

Mathewson, C. C., and Clary, J. H., 1977, 'Engineering geology of multiple landsliding along I-45 road cut near Centerville, Texas', in Coates, D. R. (Ed.), *Landslides: Reviews in Engineering Geology*, Vol. 3, Geol. Soc. America, Boulder, Colorado, pp. 213–223.

Mayuga, M. N., and Allen, D. R., 1966, 'Long Beach subsidence', in Proctor, R., and Lung, R. (Eds), *Engineering Geology in Southern California*, Association Engng. Geology, California, pp. 281–285.

Newton, J. G., 1976, 'Early detection and correction of sinkhole problems in Alabama, with a preliminary evaluation of remote sensing applications', *Alabama Highway Research Report*, 7, 83 pp.

Poland, J. F., and Davis, G. H., 1969, 'Land subsidence due to withdrawal of fluids', in Varnes, D. J., and Kiersch, G. (Eds), *Reviews in Engineering Geology*, Vol. 2, Geol. Soc. America, Boulder, Colorado, pp. 187–269.

Prokopovich, N. P., 1972, 'Land subsidence and population growth', *Twenty-fourth Int. Geol. Cong. Proc.*, **13**, 44–54.

Rau, J. L., and Nutalaya, P., 1982, 'Geomorphology and land subsidence in Bangkok, Thailand', in Craig, R. G. and Craft, J. L., (Eds), *Applied Geomorphology*, George Allen & Unwin, London, pp. 181–201.

Schothorst, C. J., 1977, 'Subsidence of low moor peat soil in the western Netherlands', *Inst. for Land and Water Management Research*, Tech. Bull 102, Wageningen, Netherlands, pp. 265–291.

Spencer, G. W., 1977, 'The fight to keep Houston from sinking', *Civil Engng.*, **47**, 69–71.

Stephens, J. C., *et al.*, (1984), 'Organic soil subsidence', in Holzer, T. L., (Ed.), *Man-induced Land Subsidence: Reviews in Engineering Geology*, Vol. 6, Geol. Soc. America, Boulder, Colorado, pp. 107–122.

Vandale, A. E., 1967, 'Subsidence—a real or imaginary problem?', *Mining Engng.*, **19**, 86–88.

Verbeek, E. R., and Clanton, U. S., 1981, 'Historically active faults in the Houston metropolitan area, Texas', in *Houston Area Environmental Geology—Surface Faulting, Ground Subsidence, Hazard Liability*, Houston Geol. Soc. Special Publ., Houston, Texas, pp. 28–68.

Wagner, F. W., and Durabb, E. J., 1977, 'New Orleans: the sinking city', *Env. Comment*, **June**, 15–17.

Walters, R. K., 1977, 'Land subsidence in central Kansas related to salt dissolution', *Kansas Geol. Survey Bull. 214*, 82 pp.

Yerkes, R. F., and Castle, R. O., 1976, 'Seismicity and faulting attributable to fluid extraction', *Engng. Geol.*, **10**, 151–167.

PART IV
PEDOSPHERE

INTRODUCTION

In view of the crucial importance of the soil to the growing of crops and therefore to providing food for the world's population, attention has frequently been directed to human impact on the pedosphere. Furthermore, since reduced crop yields provide a sensitive and fundamental indicator of detrimental changes, the significance of human impact has been recognized more readily than in many other compartments of the environment. History points to several examples of declining civilizations which may in turn be related to problems of declining soil fertility and the collapse of agricultural production.

The role of man in 'improving' soil properties has long been appreciated, since such changes frequently result in improved crop yields. Drainage of marshland soils has been widespread and in Finland, for example, more than 90 per cent of the agricultural land has been drained. The application of lime and sea sand to adjust soil pH and of animal manures to increase the organic matter content and nutrient status of the soil has a long history. Use of chemical fertilizers is more recent, dating from the pioneering research of the German agricultural chemist Justus von Liebig in the mid-nineteenth century, but current global usage amounts to in excess of 120 million tonnes or approximately 25 kg per capita. To these essentially positive changes must be added the more detrimental effects of human impact including salinization, structural deterioration, depletion of organic matter and nutrients and build-up of toxic substances.

Where man has disturbed the natural stability of the soil and its vegetation cover, rates of soil erosion by wind or water in excess of those of soil formation may lead to the disappearance of the soil cover and the pedosphere. Recent estimates reported by Brown (1984) suggest that current rates of soil erosion throughout the world are such that soil loss from cropland exceeds the rate of new soil formation by about 23 billion tonnes. In the USSR alone, ½ million hectares of agricultural land are abandoned each year as a result of destruction of the soil cover by erosion and it is salutary to reflect on the report that scientists at the Mauna Loa observatory in Hawaii are able to date the onset of spring ploughing in North China by an increase in the levels of atmospheric dust. In developing and promoting soil conservation measures to counter these problems man is inevitably introducing further changes to the pedosphere.

The importance of human activity in modifying soil properties and pedogenic processes has been highlighted by Bidwell and Hole (1965) in their paper 'Man as a factor of soil formation' and Yaalon and Yaron (1966) have advocated the term *metapedogenesis* to connote the modification of soil properties by human agency. Many of these modifications merit detailed discussion in this section but the scope of the two chapters which it contains is inevitably selective. In the first chapter S. T. Trudgill reviews the basic processes responsible for soil formation and development and emphasizes the importance of man's impact on these processes. In the second chapter A. C. Imeson focuses on the nature, causes and incidence of soil erosion and the various approaches to soil conservation which have been employed.

REFERENCES

Bidwell, O. W., and Hole, F. D., 1965, 'Man as a factor of soil formation', *Soil Science*, **99**, 65–72.

Brown, L. R., 1984, 'Conserving soils', in *State of the World 1984*, Norton/Worldwatch, New York, pp. 53–73.

Yaalon, D. H., and Yaron, B., 1966, 'Framework for man-made soil changes—an outline of metapedogenesis', *Soil Science*, **102**, 272–277.

Human Activity and Environmental Processes
Edited by K. J. Gregory and D. E. Walling
©1987 John Wiley & Sons Ltd.

12

Soil Profile Processes

S. T. TRUDGILL

12.1 INTRODUCTION

Throughout history, man has sought to improve on nature for his own benefit. As part of this endeavour, soils have been claimed from nature and cultivated for food production. But, as is often the case with man's manipulation of his environment, actions intended as improvements and developments have not always had the desired effect: soil management can lead to the degradation of the soil resource as well as to its improvement. In many cases, degradation has been the result of an incomplete understanding of natural processes and there have been unforeseen side effects and repercussions resulting from management actions. Thus, man has become aware that the understanding of natural systems and their processes is a prerequisite of successful environmental management because it is often more effective and cheaper, especially in the long run, to work with nature than against it. In other cases, the knowledge of how to improve soil conditions for agriculture without degrading the soil resource may, in fact, be present but it may not be applied because of conflicting interests. Often, short-term economics may be in conflict with a longer-term husbandry of the soil; wider political and sociological factors can also be over-riding influences on soil management. Moreover, improvement of soil resources for agriculture may be at variance with the interests of conservation and wider environmental issues. Here, the endeavour is to maximize agricultural production while minimizing any wider environmental effects. The modification of soil profile processes is thus a central theme in the broader issue of the modification of the environment by man, involving both beneficial and detrimental effects on soils and on the environment as a whole. The purpose of this chapter is to discuss man's alteration of the soil and to demonstrate the relevance of the knowledge of such alterations for the management of environmental processes in the future.

Soil is difficult to create. It is true that good agricultural land can be claimed from the sea, for example, but man is largely dependent upon a legacy of deposits and weathering residues which have evolved over time in relation to the forces acting upon it (Catt, 1979). An important aspect of this dependency is, however, that some soil properties inherited from the past are more difficult to change than others. Thus, the basic mineral grains—their size and mineralogy—are difficult to change. Other aspects, such as chemistry, structure, organic content and moisture content can be changed more readily. This means that man's impact on soils is most marked in these more readily changeable properties; in the other aspects, man is left to work with the soil material that has been created by natural processes. Thus it is the distribution of chemical elements, structures, humus and water within the soil profile that has been most altered, rather than the basic nature of the soil mineral material. The major exception to this is soil erosion (discussed in Chapter 13), when the actual soil material itself is lost.

The main horizontal layers, or horizons visible in a soil profile are shown in Fig. 12.1. Here, the basic parent material, rock or unconsolidated material is shown as the C horizon; altered mineral material occurs above it in the B horizon. Organic matter is found at the soil surface in the 0 horizon, and mixed with mineral matter in the A horizon. These layers, described further by Courtney and Trudgill (1984), form the basic units for the study of the visible effects of soil profile processes and of man's effect on them—together with the less visible effects, revealed only by physical and chemical analyses.

12.2   SOIL PROFILE PROCESSES

In order to assess the significance of the interrelationships of man with soil profile processes, an important step will be to compare the man-induced processes with

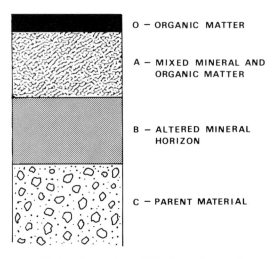

FIGURE 12.1   The major soil horizons in a soil profile

naturally occurring soil profile processes. The latter are traditionally studied in the contexts of soil formation and development but they can also be seen in terms of fundamental processes from which man-induced departures can be assessed. In most soil textbooks, for example Courtney and Trudgill (1984) or Bridges (1970), soil profile development is seen in terms of the extent of movement of material down (and sometimes up) the soil profile: thus the processes of leaching, lessivage (translocation of clay), podsolization, ferralitization (laterization), calcification and salinization are commonly discussed and will be outlined below. Other important processes, not necessarily involving vertical movement but often closely related to it, include the redistribution of organic matter, structural organization and gleying.

Where rainfall exceeds evaporation on an annual basis, the net movement of water in the soil profile is downwards. This downward movement will carry soluble material and very fine particulate material down the profile. In general, this leaves an *eluvial* (E) horizon, near the surface between the A and B horizons, from which material has been removed, and an *illuvial* horizon below, into which material has been washed (Fig. 12.2). Eluvial horizons include the pale brown (Eb) and more leached, ash-like or white (albino or albic) Ea horizons. Illuvial horizons are regarded as a type of B horizon, and include those distinguished

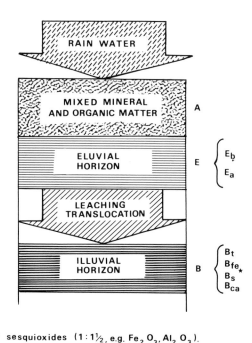

$^{*}B_s$ – sesquioxides $(1 : 1\frac{1}{2}$, e.g. $Fe_2O_3, Al_2O_3)$.

FIGURE 12.2 Movement and redeposition in a soil profile

by their texture (Bt), with inwashed clays, or those with accumulations of calcium carbonate (Bca) or iron (referred to as Bfe horizons for ferric material, Bi for iron or Bs for sesquioxides such as Al_2O_3 or Fe_2O_3, with a ratio of iron or aluminium to oxide of 1:1½; Bfe and Bs are the terms commonly used).

The progression from leaching to lessivage and then to podsolization implies a sequence of degree of movement, though the latter two can also be seen as marked forms of leaching. In simple *leaching* (Fig. 12.3), only the most soluble

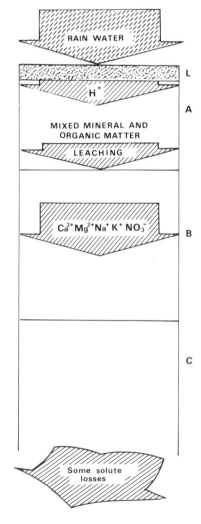

FIGURE 12.3 The process of leaching

material is moved. Here, water soluble elements, especially nitrogen (N) are involved (an important point when nitrogen fertilizers are added to soil, as discussed below). Some weak acidity is derived from atmospheric carbon dioxide dissolved in water (Trudgill, 1983) by the process:

$$CO_2 + H_2O \rightarrow H_2CO_3 \rightarrow H^+ + HCO_3^- \tag{12.1}$$

Some weak acidity is also derived from the dissociation of organic acids. The solutes moved include calcium, magnesium, sodium and potassium.

The transport of clays — *lessivage* or *clay translocation* (Fig. 12.4) — can be regarded as a more intense form of leaching; certainly leaching of solutes can occur without clay movement but clay movement is thought to occur under acid

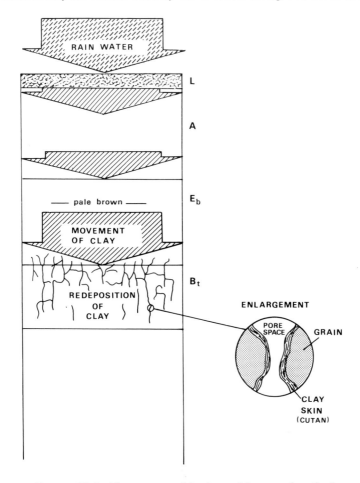

FIGURE 12.4 The process of lessivage (clay translocation)

conditions when solute movement would also be occurring. However, there are two possible processes involved in clay movement, mechanical and chemical. Firstly, a purely mechanical transport, where clay particles are physically washed down the larger pores of well-structured soils. This process is dependent upon rapid, free water movement. Secondly, clays may become solubilized in mobile water under acid conditions and reprecipitated further down the soil profile. There is some debate over these mechanisms and the distinction is probably a blurred one, with transport of a range of sizes of clay colloids in water, which

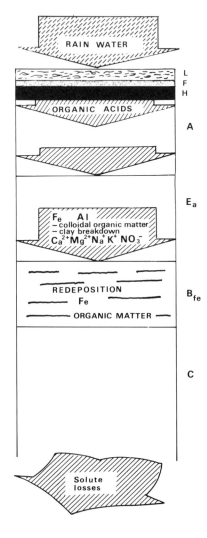

Figure 12.5 The process of podsolization

could be regarded as ranging from very fine particles in suspension to colloids in a solution. It is clear, however, that under acid conditions, clays can break down and become mobilized.

The distinction between *podsolization* and simple leaching is that in the former iron and aluminium are moved in conjunction with organic acids, a process lacking in simple leaching. The important property of the organic acids lies not so much in their dissociation and the provision of acidity, as is the case in simple leaching, but in their chelatory properties: that is, the iron and aluminium are

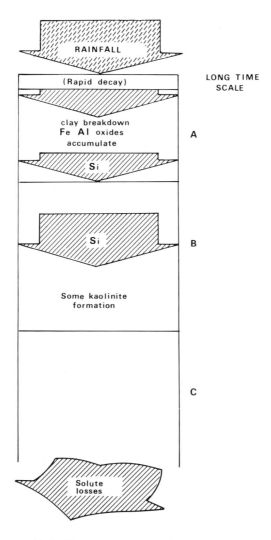

FIGURE 12.6 The process of ferralitization (laterization)

transported in chemical combination with the organic compound. Iron and aluminium are much more mobile under this mechanism than they are in pure water or even in weakly acidified water. Thus, podsolization occurs when organic matter accumulates at the soil surface, yielding organic acids with chelatory properties (Fig. 12.5); in addition, all the solutes mobilized in simple leaching are also mobilized.

Podsolization contrasts with *ferralitization* (Fig. 12.6) where iron and aluminium are less mobile than silica. Clearly, the inference is that chelatory organic acids are not important in the ferralitization process. Indeed, ferralitic soils with accumulations of surface iron are found in the tropics where the decomposition of organic matter is liable to proceed so that deep accumulations of humus are not found on the soil surface. Thus the chelatory acids are not

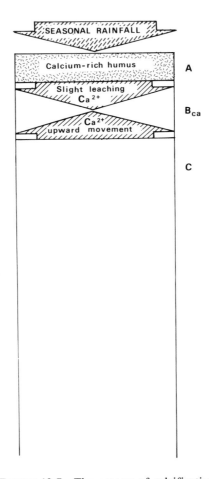

Figure 12.7   The process of calcification

abundant as they are in cool, wet conditions where organic matter decay is limited. In addition, strongly oxidizing conditions are often present at the soil surface and these immobilize iron in the insoluble, oxidized ferric (iron(III)) form. Under wet, cool podsolic conditions, the mobile, reduced, ferrous (iron(II)) form may be present and, indeed, the mobilization of iron by organic acids is often achieved by simultaneous chelation and reduction. Tropical soils are also deeply weathered, having endured soil profile-forming processes over very long time scales, often with little interruption since Tertiary times. They can therefore be seen very much as residual soils, with residual surface iron oxide accumulation, giving a markedly red soil.

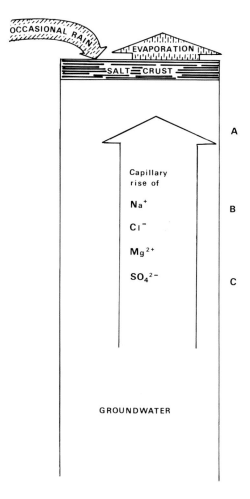

FIGURE 12.8 The process of salinization

Where the soil water balance is such that evaporation exceeds precipitation, either on an annual basis or seasonally, some degree of net upward movement of water and solutes can be expected. Under *calcification* (Fig. 12.7) wet season rainfall is sufficient to leach the most soluble salts, such as sodium salts, to some depth in the profile, but the slightly less soluble substances, such as calcium, remain at shallow depths in the soil. To some degree this is offset by dry season evaporation and upward movement of water in the soil, leading to the formation of a calcium-rich humus soil. Here, humus decay is not limited by low biological activity, as it is in cool, wet conditions, but by the formation of stable calcium-humates.

Finally, in more arid areas, the net upward movement of water results in *salinization* (Fig. 12.8), the accumulation of soluble salts at or near the surface. These salts include sodium chloride and magnesium sulphate.

As mentioned earlier, there are three further aspects of soil profile processes which are not a direct reflection of vertical movements but which should also be considered; these are the redistribution of organic matter, structural organization and gleying.

The depth of *soil organic matter* is, in many ways, an index of the amount of cycling of nutrients and energy which is occurring in the whole ecosystem of which soil is a part (Fig. 12.9). In productive, fertile soil in a warm, moist climate and under natural conditions, little organic matter will be present on the soil surface (Fig. 12.9(a)). This is because soil and climate conditions permit the breakdown of litter rapidly: an active population of soil organisms, notably earthworms, soil arthropods, insect larvae, fungi and bacteria, break down and incorporate the humus in the soil. This incorporation is strongly determined by the activities of surface-feeding earthworms which take leaf litter from the

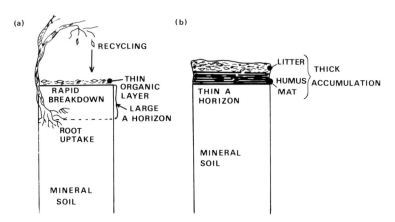

FIGURE 12.9   The relationship between the depth of organic matter in a soil profile and the efficiency of recycling

surface to some depth in the soil. Only one to two years of accumulated litter are present on the soil surface. The energy and nutrients from the litter are rapidly released and the nutrients, especially N, P, K, Ca and Mg, are made available for plant root uptake and thence rapidly returned to the soil again in litter and recycled. If, however, soil organism activity is curtailed, by acidity, wetness or cold, or some combination of these, organic matter accumulation occurs (Fig. 12.9(b)). Under these conditions, fungi often dominate the soil biota and these are efficient in the first stages of decay, but the latter stages are minimal. Thus, a thick accumulation of raw, only partially decomposed humus, called *mor* is present. This can be taken as an indicator of very slow recycling in the ecosystem. The only variation on this is the one already mentioned under calcification, where thicknesses of stable calcium humates form, leading to a build-up of humus, but in this case it is a well-decomposed, crumbly humus, termed *mull*.

Soil structure is the aggregation of primary soil particles into agglomerations termed *peds*. The existence of peds in the soil profile is determined by the cohesiveness of the grains — clay being cohesive, as opposed to uncohesive silt and sand — and by substances acting to hold together or cement the particles, such as iron oxides, organic matter and calcium. These substances also increase the stability of soil structures. Some peds are very stable and massive, such as is the case with heavy clay soils. Other soils have a large number of small peds, especially where there are mixtures of sand, silt and clay, high humus contents and proportions of iron oxides and/or calcium. In general, structures are finer towards the surface of mineral soils where the actions of wetting and drying and of plant roots combine with high organic content to produce a better structure. Mixtures of fine and medium-sized structures up to a few centimetres diameter are beneficial because they permit root penetration, aeration and drainage of excess water, but they also allow water retention for plant use; large, cloddy or platy structures impede root growth, aeration and drainage.

Gleying is a term used for soil processes whereby iron is present in the soil profile in the reduced (ferrous, iron(II)) form. This occurs in soils where oxygen does not penetrate easily, usually because of waterlogging or poor structure or both. The reduced iron imparts a blue-grey colour to the soil, most other soils being red or brown because of the presence of oxidized iron and also humus. In gleyed soils, the colour may be a uniform grey, but more commonly, there may be red mottling where iron has been oxidized locally because of dissolved oxygen present in water or by the penetration of air down cracks and pores; root channels are also common oxidation sites. The fundamental processes of gleying are, then, ones of oxidation and reduction, usually in relation to drainage of the profile.

The outlines of soil profile processes can now be used as a basis for discussing the interrelationships between man and soil processes.

12.3   MAN AND SOIL PROFILE PROCESSES

(a)   The main effects

To recap, the main soil profile processes so far discussed are leaching, lessivage, podsolization, ferralitization, calcification and salinization—all involving vertical water movement; additionally, the processes which influence the disposition of soil organic matter, the nature of soil structure and gleying have been discussed. Man's impact can be summarized as involving the removal or other modification of vegetation cover and also the effects of cultivation and management practices. These, in general, can be seen as either reinforcing or counteracting the natural processes; alternatively they may be seen as processes which influence the impact of man.

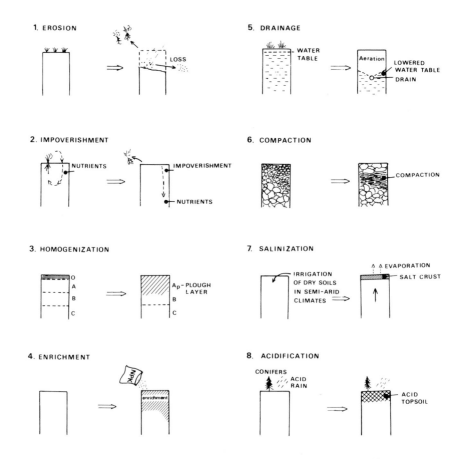

FIGURE 12.10   The impacts of human activity on soil profiles

The main effects of man on soil profiles are shown in Fig. 12.10. The net outcome of these changes is that when a soil profile is studied at the present day, it will often be difficult to interpret it purely in terms of natural processes. In particular, truncation by erosion, impoverishment and homogenization will have the most marked visible effects. In all three cases, it is the distribution of material in the surface organic-rich layers which will have been affected. In the case of soil erosion, it is these very layers which will have been lost (as discussed further in Chapter 13).

(b) Vegetation removal

Some of the most profound impacts of man on soil profile processes are derived from the removal of vegetation cover and from the cultivation of soil for agriculture (Goudie, 1981). Clearing of vegetation may mean that the soil surface is more exposed to the agents of soil erosion such as wind and running water. It also means that the processes of recycling of chemical elements by root uptake, litter fall and the decomposition of organic matter are disrupted. This, in turn, may mean that nutrients which are washed down through the soil by leaching are not returned to the surface by recycling, in many cases leading to the impoverishment of the nutrient status of the upper soil horizon.

Thus, in the case of vegetation removal, organic matter input decreases and levels of soil humus become gradually lower unless replenished by crop residues or animal dung. In some cases, the impoverishment of the soil can make the soil acid enough to decrease soil biological activity (Jackson and Raw, 1966). If this occurs, any organic matter that may arrive at the soil surface will tend to be decomposed only slowly and may be left as a discrete mat of raw humus at the surface. This is especially the case on acid rocks in upland areas with high rainfall and where pasture is not well-maintained and is allowed to revert to a poorer vegetation type. This process may be self-reinforcing since organic accumulations facilitate the production of organic acids, further acting to acidify and impoverish the soils. Thus, vegetation removal does not, by itself, automatically mean that the soil fertility will decline: natural recycling has, however, to be replaced by management actions, and manures and fertilizers can be used to match or improve upon the natural recycling. Such measures are, however, only undertaken at a cost and it is thus on marginal lands, with a poorer return on investment, that reversion and impoverishment are likely. In addition, fertilizers added to soil may be readily leached—again especially if recycling is limited—and this may have an environmental cost in that water quality may be affected. This is especially true of nitrate which is liable to leaching more than other nutrients because it is not held tightly in clay–humus complexes. These complexes have a surface negative charge which holds cations strongly, but nitrate is an anion and is thus unable to be held in this way.

Such arguments have been used as possible guidelines for future environmental management. Pointing to vegetation clearances and the impoverishment of upland soils in high rainfall areas in prehistoric times (Simmons, 1982), the argument is that in comparable present-day areas under natural forest and with a high rainfall, the best policy would be to leave them as forest. If the forest were to be cleared, the loss of recycling would soon mean that leaching under heavy rainfall would impoverish the soil (Lal and Russell, 1981). This could only be offset by the use of fertilizers which are costly and may well be, at least in part, lost by leaching—representing a loss to the farmer as well as a degradation of runoff water quality. If such forest areas are to be exploited, they would perhaps best be used as a source of timber, with replanting as a means of maintaining fertility.

(c)   Cultivation and mangement

In the case of cultivation, any discrete (separate) surface organic horizons will have become mixed with the mineral matter of the soil to form an A_p horizon. This is a plough layer, denoted by the suffix 'p', a variety of A horizon, defined in Fig. 12.1 as a mixed mineral–organic horizon. Cultivation of the topsoil generally thus leads to the homogenization of the upper part of the soil profile. Allied management activities, such as the addition of fertilizers and the irrigation and drainage of soils, alter the soil's chemical and hydrological status. The use of heavy machinery and overstocking can also lead to soil profile compaction. Many of these effects can be seen in the soil profile, but some can only be seen by detailed chemical analysis.

Since many of the soil profile processes discussed in Section 12.2 depend upon the vertical movement of water, some of the major impacts of man on soil profile processes involve those which affect water flow in soil. Principally, the amount of water entering the soil profile in any given climatic zone will vary with the ratio of overland flow to percolation water, as governed by the infiltration capacity of the soil (Hillel, 1982), (Fig. 12.11(a)). The infiltration capacity refers to the amount of rain water which can be accepted by the soil surface. Any water in excess of this stays on the soil surface either to runoff as overland flow or to re-evaporate. The water accepted by the soil surface percolates down the profile to groundwater or flows laterally through the soil, where it is termed throughflow.

There are several soil characteristics which influence soil infiltration capacity and these, in turn, can be substantially modified by human activity. Clay soils have a low infiltration capacity, unless they are dry and cracked; by contrast, open porous sandy soils facilitate infiltration. The nature of the soil humus and litter layers also has a marked influence—open porous humus and coarse litter allow more water into the soil than does a dense, compact humus. Man's impact in altering or removing vegetation, and in cultivation, can alter the infiltration

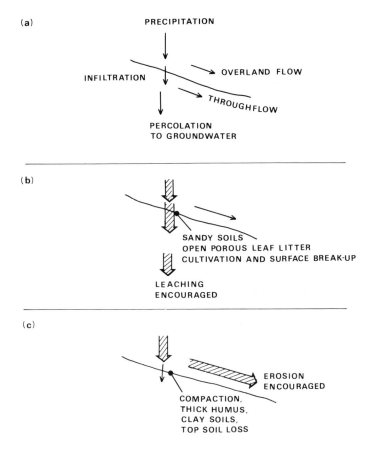

FIGURE 12.11   Soil water movement and infiltration

capacity, in that changing the vegetation type can lead to a change in humus type. Compaction can decrease the capacity, while break up of the surface can increase it. Thus, increases in infiltration capacity can result in more water becoming available for leaching because percolation increases relative to overland flow (Fig. 12.11(b)). Conversely, diverting water to overland flow by reducing the infiltration capacity decreases the amount of water available for leaching, but it can also increase the amount of soil erosion by surface wash (e.g. Kirkby and Morgan, 1980) (Fig. 12.11(c)). In addition, while a dense humus mat may encourage overland flow in many cases, humus is often more absorptive than a bare soil surface and thus, loss of humus may also increase surface wash.

Compaction occurs when wet soils are compressed by machinery or livestock, thus stocking and cultivation should only take place when soils are below field capacity, that is when all the gravitational water has drained off after winter or after heavy rain at other times (Emerson *et al.*, 1978; MAFF, 1970). As will be evident from the discussion of structural stability above, soils with poor structural cementation, i.e. those which are low in ferric iron, calcium and humus, will be the most prone to compaction. Silt soils can also readily be compressed but sandy soils are more resistant to compaction. Clay soils can be smeared when wet. Compaction and smearing give rise to impermeable layers within the soil, with poor aeration and root growth, as well as low surface infiltration capacity.

Alteration of surface humus and structural properties can thus, markedly affect soil hydrological properties, and therefore also soil solute transport. The effects of natural soil water processes on man's management can also be marked, even if man does not directly affect soil hydrological properties. Man adds soluble substances to the soil surface, such as herbicides, pesticides and fertilizer. Leaching can move these from the topsoil to the B horizon, and beyond into groundwater and streams, just as it can move the naturally occurring solutes, as shown in Figs. 12.2 and 12.3. Such movements are especially marked with anionic substances, such as NO_3^- derived from nitrogen fertilizers, since cations (ions with a positive charge) are more liable to be held on the negatively charged surfaces which are present on clays and humus. Anions are not held in such a way and more readily pass through the soil profile. The process of lessivage (Fig. 12.4), while applying to clay under natural conditions, can also apply to added substances, such as sewage sludge and other larger particles, moving them to depths in the soil profile and possibly leading to groundwater contamination. Added substances can also accumulate in the soil, unless broken down, especially the cationic substances, as mentioned above. This is true for the persistent pesticides and herbicides and is also the case with some phosphate compounds. These have a low mobility and phosphate-rich soil horizons can be evident in soils which have a long history of cultivation.

Drainage is also a management activity which can markedly affect soil hydrology (Baver *et al.*, 1972), often increasing the rate of flow of water in soils. In many ways, this can act to increase solute losses from soils. However, drainage does not automatically mean that solutes are more readily lost from the soil profile because improved aeration leads to greater plant root development. This can mean that solute uptake by root systems is more efficient, thus decreasing solute losses on drained soils (Fig. 12.12).

Drainage also interacts with the gleying processes which were discussed above in Section 12.2. Aeration consequent upon drainage can lead to the oxidation of iron compounds. This tends to decrease the mobility of the iron as the oxidized iron(III) (ferric iron) is less soluble than the reduced iron(II) (ferrous iron). This is why there is often a red staining visible in soil drain outflow sites and ditches where anaerobic conditions change to aerobic ones. In addition, precipitation of ferric iron may often stabilize structures which would be prone to smearing when wet if only the soluble ferrous iron was present. However, the oxidation

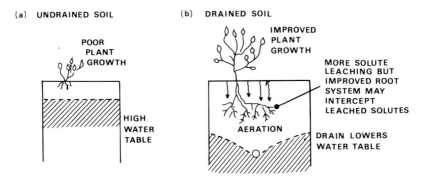

FIGURE 12.12 The effects of soil profile drainage

process can also lead to the acidification of soils and drainage waters. This is because the reduced compounds in anaerobic soils often include sulphide species which can be oxidized to sulphuric acid upon aeration and drainage (Vear and Curtis, 1981; Dent, 1980). These soils are termed acid sulphate soils and the process of oxidation and acidification can be described as:

$$2FeS_2 + 4H_2O + 7\tfrac{1}{2}O_2 \rightarrow Fe_2O_3 + 4H_2SO_4 \qquad (12.2)$$

(after Curtis, 1976), producing ferric oxide and sulphuric acid (actually ionized as $4SO_4^{2-}$ and $8H^+$) in solution. Thus, acid sulphate soils should only be drained with care, if at all, and substantial liming is often needed to counteract acidity. Thus, man's activity, through draining, can substantially interact with gleying processes, fundamentally affecting the chemistry of soils and soil drainage waters.

Oxidation of iron can also lead to hardening, or induration, of lateritic soils in the tropics (Young, 1976). Here, removal of vegetation, cultivation and drainage may facilitate the aeration of soils, leading to the formation of oxidized iron to such an extent that, far from giving a beneficially stable structure, the soil can become so well cemented that it becomes indurated and uncultivable. Thus, man can accelerate the ferralitization process described in Fig. 12.6. Indeed this process can be used for benefit in a non-agricultural situation and that is for house building: iron-rich clay blocks can be left to harden in the sun and thus can be made into building blocks. The word laterite itself comes from the Latin word *later*, meaning a brick.

Drainage is also an important consideration in irrigation in climates where evaporation exceeds precipitation. While irrigation is necessary under these conditions in order to permit crop growth, irrigation with insufficient drainage can render the soil useless for crop production in the long term. This is because if drainage is impeded, irrigation merely adds dilute solutions to the soil and subsequent evaporation acts to concentrate the salts present in the irrigation water in the soil. This thus reinforces the salinization process described in Fig. 12.8. Drainage increases the throughput of water in the soil, facilitating salt removal. Irrigation should therefore be of water of as low a salt concentration as possible and it should be calculated to be of an amount exceeding the evaporation–precipitation deficit in order to incur runoff. This is because:

$$D = I - (E - P) \tag{12.3}$$

where D = drainage, I = irrigation, E = evaporation and P = precipitation, all in comparable units, such as millimetres. Thus, if $I < (E - P)$, no drainage will occur and salts will accumulate in the soil; if $I > (E - P)$, drainage will have a positive value and salt leaching will be facilitated. Careful monitoring of the salt content of the input and output waters is also necessary in order to assess any accumulations of salts in the soil during water throughput (Young, 1976).

Cultivation and drainage thus can have profound effects on soil water relations and solute movement; the processes of leaching, lessivage, podsolization, ferralitization and salinization can all interact with man's manipulative actions, often leading to soil profile modifications or to effects on the mobility of added substances.

(d)   Acidification

It has already been seen that acidification can follow from drainage of water-logged soils. Vegetation removal can clearly also lead to soil impoverishment,

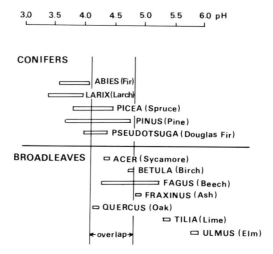

FIGURE 12.13   The acidity of conifers and broad-leafed trees (after data from Handley, 1954)

allowing rainfall to leach solutes not then recycled by vegetation. This is often in conjunction with organic acids derived from organic matter accumulation which can occur once conditions become too acid for decomposing organisms. In addition, the felling of deciduous trees and replanting with coniferous trees can lead to the acidification of the soil since conifer litter is, in general, more acid than deciduous litter (Trudgill, 1979), (Fig. 12.13). While there is an overlap between pH 4.1 and 4.8, the most acid values shown in the ranges are from conifer species. In addition to acidity derived from conifers, atmospheric sources of acidity can be derived from the burning of fossil fuels, notably from sulphuric acid (see Chapter 3). The effects of these sources of acidity depends on the buffer capacity of the soil (Bear, 1964). In this context, the term buffer capacity refers to the reserves of minerals and adsorbed ions in the soil which can offset the effects of external source of acidity. In a soil with a high buffer capacity, considerable quantities of acids can be added to a soil before the actual acidity of the soil is altered because chemical processes in the soil can counteract the effects of the added acids. Thus, soils with a high buffer capacity are not so prone to acidification by conifer needles and acid rainfall when compared with those with a low buffer capacity. Buffer capacity can be measured in the laboratory by adding acid to a soil paste drop by drop (usually dilute hydrochloric acid of, say, 0.01M concentration), stirring the paste and remeasuring pH between each addition. An example of results from such a procedure is shown in Fig. 12.14. Thus, soil 'A' resists pH change relative to soil 'B', which only needs a few millilitres of acid to reduce the pH. Clearly, soil 'B' is more susceptible to acidification than soil 'A'. Soils of the type 'B' include weakly buffered, mesotrophic (moderate nutrient status) brown earths. Other brown earths which are richer in nutrients, especially the calcareous brown earths, are able to offset acidity by the weathering of minerals and by ion exchange. By contrast, podsolic soils are already acid and so there is often little additional effect of acid inputs onto the soil. Clearly, acidification can be seen as a

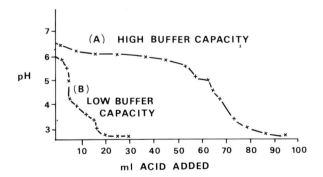

FIGURE 12.14   Two examples of soil buffer capacity: (a) strongly buffered, (b) weakly buffered

reinforced form of podsolization or leaching, as already discussed, the precise effect depending on the nature and strength of the added acid and on the inherent soil properties.

Thus, in these case studies discussed above—vegetation removal, cultivation and other management actions—the impact of man primarily includes the alteration of the nutrient and water status of soils. In many cases the processes involved are quite well understood and the likely effects can therefore be predicted. If this knowledge is not used to advantage, both in terms of advantage to man and of minimizing the impact on the wider environment, it can often be concluded that the knowledge is not necessarily being fully applied or that there may be some conflict between short-term gain and a wider, long-term effect, which can often be deleterious.

12.4   THE CONSERVATION ETHIC

As well as man's striving to improve on nature for his own benefit, and the varied consequences of such striving, there is another facet to man's activities: that of cherishing the naturalness of the environment. Conservation and good husbandry have come together with development in many ways in recent years, not only in terms of trying to preserve those areas of land which are relatively untouched by man but also in trying to maximize the wildlife value of exploited, agricultural areas. The attractive idea is that the needs of conservation and the needs of agriculture will, in the long term, be best served by the same measures. In many cases this is true: for example, shelter belts used for decreasing the erosion of soil profiles also provide welcome habitats for wildlife.

In other cases, it is difficult to see how any compromise can be reached: for example, either a field is drained or it is not; there are no half measures. If it is drained, its value as a wetland wildlife habitat has substantially gone. However, it is generally true that there is a great deal of lip service paid to—and a not inconsiderable amount of implementation of—plans whereby soil management actions are increasingly concerned with their effects on the wider environment and its wildlife. The important point is not to look at soil profile processes from the narrow view of soils, i.e. from the point of view of a soil scientist, nor just from the point of view of raising agricultural production, i.e. from the point of view of an agronomist. In addition, the point of view of the ecologist, environmental scientist or geographer is important to balance out these activities by considering the wider impacts of man's influences on soil profile processes. Maximizing (or optimizing) agricultural production without degrading soil resources and the environment are the new ideals—ideals which, however, must take into account political, social and economic realities.

REFERENCES

Baver, L. D., Gardner, W. H., and Gardner, W. R., 1972, *Soil Physics*, Wiley, New York.

Bear, F. E., 1964, *Chemistry of the Soil*, Van Nostrand, Rheinhold, New York.

Bridges, E. M., 1970, *World Soils*, Cambridge University Press, London.

Catt, J. A., 1979, 'Soils and Quaternary geology in Britain', *Journal of Soil Science*, **30**, 607–642.

Courtney, F. M., and Trudgill, S. T., 1984, *The Soil*. Edward Arnold, London.

Curtis, C. D., 1976, 'Stability of minerals in surface weathering reactions: a general thermochemical approach', *Earth Surface Processes*, **1**, 63–70.

Dent, D., 1980, Acid sulphate soils: morphology and prediction', *Journal of Soil Science*, **31**, 87–99.

Emerson, W. W., Bond, R. D., and Dexter, A. R., (Eds) 1978, *Modification of Soil Structure*, Wiley.

Goudie, A. S., 1981, *The Human Impact*, Blackwell, Oxford.

Handley, W. R., 1954, *Mull and Mor in relation to forest soils*. Forestry Commission Bulletin, 23. HMSO, London.

Hillel, D., 1982, *An Introduction to Soil Physics*. Academic Press, New York.

Jackson, R. M., and Raw, F., 1966, *Life in the Soil*. Edward Arnold, London.

Kirkby, M. J., and Morgan, R. P. C., 1980, *Soil Erosion*. Wiley, Chichester.

Lal, R., and Russell, E. W., (Eds), 1981, *Tropical Agricultural Hydrology*, Wiley, Chichester.

MAFF (Ministry of Agriculture, Fisheries and Food), 1970, *Modern Farming and the Soil*, HMSO, London.

Simmons, I., 1982, *Biogeographical Processes*, George, Allen & Unwin, London.

Trudgill, S. T., 1979, 'Soil profile processes', Ch. 11 in Gregory, K. J., and Walling, D. E., (Eds), *Man and Environmental Processes*, Dawson, Folkestone.

Trudgill, S. T., 1983, *Weathering and Erosion*, Butterworths, London.

Vear, A., and Curtis, C. D., 1981, 'A quantitative evaluation of pyrite weathering', *Earth Surface Processes and Landforms*, **6**, 191–198.

Young, A., 1976, *Tropical Soils and Soil Survey*. Cambridge University Press, London.

Human Activity and Environmental Processes
Edited by K. J. Gregory and D. E. Walling
©1987 John Wiley & Sons Ltd.

13

Soil Erosion and Conservation

A. C. IMESON

13.1 INTRODUCTION

Almost everywhere, natural geomorphological processes of weathering and erosion have been accelerated by man. Sometimes this acceleration has resulted in serious and obvious soil erosion which permanently damages land and causes problems of sedimentation and flooding. At other times the acceleration is slight and difficult to ascertain, involving alterations of sediment and water pathways so subtle that the consequences go unnoticed. The effects of soil erosion are most devastating in developing countries where social, economic and political conditions hinder the implementation of technical measures. Non-technical constraints to soil conservation are also present in developed countries, but the economic base and political organization can support and sustain a wide range of ameliorating measures beyond the reach of peasant communities.

The causes of soil erosion are complex. The rates of erosion by the agents of water and wind are generally slow in natural ecosystems. The degree to which erosion rates are accelerated by human activity is dependent on the type of activity, the general soil environment, which determines the inherent susceptibility of the soil to erosion, and on the general climatic environment, which determines the meteorological inputs that the altered environment has to accommodate. The impact of human activity will vary considerably from one region to the next and is difficult to predict without empirical data.

To understand how man causes soil erosion, it is useful to adopt a holistic or ecological approach, whereby the landscape in which erosion occurs is viewed as an erosional system. This system with its biotic and abiotic components is immensely complex and its study involves many disciplines. Developing mathematical models to simulate erosional systems and to study their structure, remains a future challenge since surprisingly little is known about the pathways that sediments follow through terrestrial ecosystems (Walling, 1983). Even without mathematical simulation, the application of an ecological or systems approach to the erosion problem has recently led to a better understanding of many processes and to improved methods of soil conservation (Gray and Leiser, 1982).

It is a paradox that our lack of knowledge of soil erosion processes has not prevented erosion from being controlled by a wide range of measures which have been available for several decades and in many cases for several millennia (Hudson, 1971). Today, empirical formulae can be used to predict the rates of erosion, or potential erosion hazards for different crops and methods of cultivation, and to design control works such as terraces and check dams (FAO, 1977).

Writing on soil erosion and conservation six years ago, Bryan (1980) pointed out that in spite of this technical knowledge, the soil erosion problem had not been solved and that soil erosion was more widespread and serious than ever before. This is even more so today. Soil conservation measures will only be adopted if they are profitable and acceptable to the users of the land. It still remains necessary to develop inexpensive and practical soil conservation techniques that peasants with a negligible cash income can afford and which utilize our understanding of the effects of human activity on environmental processes. In many underdeveloped agricultural communities, traditional agricultural systems are well adjusted to the erosional environment, but the need for more food, or firewood, as well as external market factors, result in human activities which inevitably lead to erosion (Fig. 13.1).

In the first part of this chapter the processes of erosion by water and wind will be briefly described before considering how erosional systems respond to human impact and how the soil erosion hazard can be predicted. In the second part soil conservation measures will be reviewed which provide an illustration of how man can manipulate the erosional system to reduce soil erosion.

13.2   WATER EROSION

For convenience, water erosion is subdivided into rainwash and gully erosion. Rainwash occurs when soil particles on the land surface are detached and transported by rainsplash or overland flow. The process of overland flow was investigated by Horton (1945), who distinguished several categories of overland flow according to a number of hydraulic properties. More recently, overland flow has been treated by Emmett (1978), Savat (1977) and Hodges (1982).

Horton envisaged overland flow as occurring when rainfall intensity exceeds the infiltation capacity of the soil and the ponded water, in excess of the amount that can be stored in surface depressions, flows downslope. 'Saturated overland flow' occurs when perched water tables rise to the surface (Kirkby and Chorley, 1967). Emerging groundwater has also been observed to produce overland flow (Tanaka, 1982).

On agricultural land overland flow frequently becomes concentrated in shallow channels or rills. A distinction is therefore made between erosion in rills and

FIGURE 13.1   In the northern Rif Mountains of Morocco almost all possible sites are
cultivated in spite of the obvious erosion risk

erosion on the intervening inter-rill areas (Foster and Meyer, 1975). Detachment
of soil particles on the areas of shallow overland flow, which occur on inter-rill
or non-rilled soils, is predominantly by rainsplash. In the rills detachment occurs
by forces exerted by the concentrated flow. The amount of entrainment of
sediment by rainsplash is dependent upon the impact force of the raindrops,
the depth of any ponded waterfilm and the roughness of the soil surface. The
impact force (and kinetic energy) of rainfall is dependent upon raindrop size,
form and velocity. Raindrop velocity can be increased considerably by the wind.
Thick water films dampen impact forces but more splash occurs in the presence
of a thin waterfilm than on a dry soil. Splash is at a maximum when a waterfilm
approximately equal to the raindrop diameter is present (Palmer, 1963; Mutchler
and Young, 1975). Most rainsplash entrainment and transport is the result of
a relatively small number of large raindrops. Since the number of large raindrops
increases with rainfall intensity, rainfall momentum, impact force and kinetic
energy can be estimated from rainfall intensity data.

The detachment of material by overland flow is related to a number of
hydraulic properties (roughness, slope, depth of flow, kinematic viscosity and
Reynolds number). These flow properties are very difficult to measure directly.
Runoff hydrographs of overland flow can be modelled using the continuity and
momentum equations. The momentum equation is simplified in many procedures

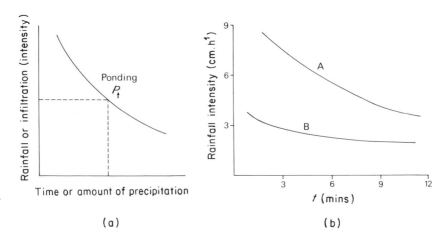

FIGURE 13.2   Infiltration envelopes. (a) The infiltration envelope describes the relationship between the time or amount of rain required to pond the soil and rainfall intensity. (b) Infiltration envelopes for a soil from northern Morocco. At site A the surface is rough and aggregated, at site B the surface is crusted

by a kinematic wave approximation (Woolhiser and Liggett, 1967). Solving the flow equations either analytically (Parlange *et al.*, 1981) or numerically enables runoff equations to be obtained which can be fitted to field data to estimate parameters which cannot be measured in the field. Croley (1982, 1983) and Croley and Foster (1984) have extended this approach to include sedimentation and deposition on rilled slopes.

Before overland flow can take place, water has to pond on the soil surface. The amount of rain required to pond a soil, or the time to ponding, varies with rainfall intensity in a way which is described by the infiltration envelope (Fig. 13.2). Infiltration envelopes have been described by Rubin (1968) and Smith (1972) and can be calculated, for example, from the Phillips infiltration equation (Smith and Parlange, 1978) or from the Green & Ampt equation (for a review see Mein and Larson (1971) and Moore *et al.*, (1980)). Infiltration envelopes can be compared with rainfall intensity data to predict the frequency of ponding. Allowances can be made for variable rainfall intensities (Chu, 1978) and, when runoff is being predicted, for surface microtopography (Moore *et al.*, 1980). Ponding times measured in the field and applied to soil erosion problems have been described by Scoging (1982), Imeson and Kwaad (1982) and Hamilton *et al.* (1983). Fig. 13.3 shows a ponded soil from a location near Roma, Lesotho, following heavy rainfall.

If certain thresholds are reached and provided that the flow can entrain all size fractions (Boon and Savat, 1981), overland flow becomes channelled into rills. De Ploey and associates at the University of Louvain have recently reported

FIGURE 13.3 A ponded cultivated soil in an area of severe gully erosion near Roma, Lesotho (photo: Th. Faber)

experimental laboratory and field results on threshold conditions related to rill erosion on loamy soils. For conditions pertaining in Belgium, Boon and Savat (1981) developed a nomogram for predicting the rill erosion hazard from easily obtained parameters.

Although soil erosion is usually considered to result from mechanical erosion processes, it is being increasingly realized that chemical processes also need to be considered. Of particular importance are chemical conditions which encourage spontaneous dispersion of soil material and which lead to very low infiltration rates (Shainberg and Letey, 1984) and to hyperconcentrations of suspended sediment (Imeson and Verstraten, 1981). Chemical processes also account for certain soils being susceptible to gully erosion and badland development (Imeson *et al.*, 1982; Bryan *et al.*, 1984) and explain why sediment concentrations and hydrologic variables are sometimes seemingly unrelated.

13.3 GULLY EROSION

Gully erosion is a widespread and frequently indirect effect of human activity. Definitions of gully erosion are often imprecise. The US Soil Conservation Service (1966) defines gully erosion as 'the removal of upland soil and parent materials and formation of channels by concentrated flow of water. A channel formed in this manner is classified as a gully when it cannot be obliterated by normal tillage operations'. However, this definition excludes a number of essential characteristics of gullies distinguishing them from river channels and implies that gullies can develop from rills. Gullies are usually characterized by the following:

(a) Development when the hydrological cycle is disturbed by man, directly or indirectly.
(b) A distribution restricted by the occurrence of erodible sediment, regolith or soils.
(c) By only being intermittently occupied by flowing water. Furthermore, unlike river channels, gullies are only partially filled with water at high stages.
(d) There is no simple relationship between the slopes above a gully and the gully itself. A gully resembles a river valley in miniature rather than a river channel.
(e) Gully growth is regulated by the rate of release of very easily entrained material from the gully sides and floor so that water discharge and sediment outputs are frequently poorly related.

A classification of gully types based on Imeson and Kwaad (1980) is given in Table 13.1. Gully erosion is most extensive in areas where deep weathering profiles have developed or where slope or pediment deposits have accumulated over long periods of time (Fig. 13.4). In some areas gully erosion is accompanied by the process of tunnel erosion (Heede, 1971; Faber and Imeson, 1982; Stocking, 1981), whereby clay in dispersion is transported to gullies along tunnels which develop in the BC horizons of certain soils. This type of subsurface erosion is associated with soil having relatively high proportions of exchangeable sodium and low electrolyte concentrations (Imeson *et al.*, 1982). Rates of gully growth have been reported by many authors and these are usually in the order of 2–4 m year^{-1} for gullies with actively advancing headcuts (Ireland *et al.*, 1939; Graf, 1977, 1979; Imeson and Kwaad, 1980).

13.4　WIND EROSION

Wind erosion occurs in environments characterized by a poor or incomplete vegetation cover, notably in semiarid areas where rainfall is less than about 300 mm year^{-1}, or where, through drought or human activity, an incomplete vegetation cover is present. Wind erosion is favoured by high wind speeds, drought, low surface roughness and a low stability of soil aggregates. Human activity extends the risk of wind erosion into more humid areas, so that, for example, wind erosion is of frequent occurrence on sandy and peaty soils throughout northwest Europe.

TABLE 13.1   Sets of conditions characterizing particular gully types (after Imeson and Kwaad, 1980)

Gully type	Gully cross-section	Position in landscape	Principal source of runoff	Material in which gully is developed	Favourable conditions
Type 1	V-shaped	Anywhere, except valley bottoms, where runoff becomes concentrated	Overland flow	Relatively resistant weathering products of impermeable parent materials, B-horizons of deep soil profiles. The resistance of the material increases with depth	Intense rainfall, poor soil structure, steep slopes, poorly built terraces and tracks
Type 2	U-shaped	Anywhere in landscape except valley bottoms	Overland flow with a contribution of subsurface water of lesser importance; occasional seep caves at headcut	Relatively little resistant weathering products or slope deposits which do not increase in resistance with depth	Subhumid climate. Relatively old soils and slope deposits
Type 3	U-shaped	Anywhere, but usually on pediments and gentle lower slopes	Subsurface flow predominates, as is apparent from piping	Weathering products and slope deposits as for type 2	Dispersive soil materials. Subhumid climate with pronounced wet and dry seasons
Type 4 (arroyos)	U-shaped	Valley bottoms	Overland flow mainly from tributary gullies, and subsurface flow	Alluvium and slope deposits	Semiarid climate, lack of valley bottom vegetation, dispersive materials

FIGURE 13.4   Gully erosion on pediment deposits near Roma, Lesotho (a), and on old colluvial soils in northern Morocco (b)

For wind erosion to take place wind speeds of about $5\,m\,s^{-1}$ are needed (Eppink, 1983). Above this wind speed, sand grains are transported by the process of saltation. Grains are set into motion by the momentum they receive from grains already in transport and by lift and drag forces. Saltating particles, usually between $50-500\,\mu m$ in diameter, tend to rise sharply into the air at an angle of more than 70°, rising to heights mostly less than 30 cm, before falling back to the ground surface at a shallow angle of between 6° and 12° (Eppink, 1983). Material finer than $100\,\mu m$ can be transported in suspension and is present in dust storms. Coarser particles creep or roll over the surface. About 50–75 per cent of wind erosion takes place by saltation, 3–40 per cent by suspension and 5–25 per cent by creep (Bagnold, 1973; Chepil, 1945).

(b)

(caption opposite)

The mechanics of wind erosion have been treated by Bagnold (1973) but it is difficult to apply theoretical formulae to conditions other than those existing for well-sorted soil material. Woodruff and Siddoway (1965) developed an empirical equation to predict wind erosion based on the variables listed in Table 13.2. Using the nomograms and relationships given by the authors, or computer programs (Skidmore *et al.*, 1970), the equation can be used either to estimate the annual average soil loss or to determine the effect of changes in soil erodibility, roughness, equivalent field length and equivalent vegetation cover.

13.5 THE RESPONSE OF EROSIONAL SYSTEMS TO HUMAN IMPACT

The relationships between vegetation and erosion and between cultivation and soil loss are well known (Penman, 1963). It has long been realized that forest soils become more erodible when cultivated and that soils under cultivation become less erodible when put under grass. Furthermore, it is known that continuous cultivation tends to increase soil erodibility. The link between vegetation or land use and erosion is provided by the interrelationships which exist between vegetation and soil structure. Changes in the vegetation cover or cultivation bring about direct and indirect changes in the soil environment which lead to an alteration in the structural organization of the soil. These changes

TABLE 13.2 Variables used to predict wind and water erosion in the wind erosion equation (Woodruff and Siddoway, 1965) and the Universal Soil Loss Equation (Wischmeier and Smith, 1978)

The Wind Erosion Equation
The Wind Erosion Equation is used to estimate the potential average annual soil loss in tons/acre/year (E) as a function of I', C', K', L' and V.

I' : soil erodibility index. This is the potential soil loss in tons per acre from a wide unsheltered field with a bare smooth surface, corrected for knoll erodibility in the case of short windward slopes.

K' : soil ridge roughness factor.

C' : local wind erosion climatic factor: this is computed by dividing the cube of the mean annual wind velocity dividing the square of Thornthwaite's P–E ratio and multiplying this by a constant.

L' : the equivalent field length. This is the unsheltered distance across the field along the prevailing wind erosion direction.

V : The equivalent vegetative cover. This is obtained from information on the quantity, kind and orientation of the vegetation.

The Universal Soil Loss Equation
The USLE is used to predict the average rate of soil loss under a range of crop systems and management practises under specified physical conditions. The computed soil loss A is calculated as a function of $R\ K\ L\ S\ C\ P$.

R : The rainfall and runoff factor. This is obtained from the rainfall erosion index and a subfactor R_s used to account for snowmelt runoff. The rainfall erosion index is obtained from precipitation data.

K : The soil erodibility factor: this is the soil loss rate per erosion index unit for a given soil on a unit plot (72.6 ft long, 9 per cent slope continuously in clean-tilled fallow). Frequently estimated from soil parameters.

L : The slope length factor. This is the ratio of the soil loss from the field slope length being considered to that from the unit plot (72.6 ft) under identical conditions.

S : The slope steepness factor. This is the ratio of soil loss from the field slope gradient being considered to that from the unit (9 per cent) slope under identical conditions.

C : The cover and management factor. This is the ratio of soil loss from an area with a specified plant cover and management to that from an identical plot under continuous fallow.

P : The support practice factor. This is the ratio of soil loss when conservation practices are used to that occurring with straight row farming up- and downslope.

---

are important because the pathways that water can follow through or over the soil may be affected. The structural organization of the soil is particularly important in regulating the infiltration process and the amount and frequency of overland flow. It is useful to understand how soil structure and erosion are related, because this relationship can be manipulated by modifying land use or tillage in a way that minimizes the erosion hazard.

These points are illustrated by considering some of the changes which occur in a silty soil when a hypothetical forest ecosystem is cleared for cultivation. Such an example is hypothetical because in practice very much depends on the general soil and climatic environment. In the forest soil large water-stable aggregates are produced and there is a high macroporosity and correspondingly low bulk density. The larger soil aggregates are stabilized by fine roots and fungal hyphae (Tisdall and Oades, 1982). These aggregates remain stable under raindrop impact, or when moistened from a dry state, and they impart to a soil a high permeability or infiltration capacity. The corresponding soil under cultivation tends to produce aggregates which are less water-stable and which break down into very fine micro-aggregates ($< 50\,\mu$m). This is directly due to the effect of a reduction in numbers of roots and fungal hyphae, which stabilize large aggregates, and indirectly due to the effect that a reduction in the amount of organic material added to the soil has on the soil fauna. This results in the hypothetical cultivated soil having finer pores, a higher bulk density and a lower permeability than the corresponding forest soil.

A situation whereby land use is changed from forest to cultivation is extreme. Within a forest the soil structure will vary according to the composition of the vegetation and associated soil fauna. In the tropics, removing individual trees can increase soil temperatures dramatically and indirectly lead to a greater susceptibility of soils to erosion or to another type of soil aggregation, as is the case with certain andosols which irreversibly dry (Imeson and Vis, 1982). Under cultivation it has been shown that different soil aggregation characteristics develop under different field crops (Reid and Goss, 1981) and that soil aggregation is dynamic in time (Imeson and Vis, 1984).

The breakdown of soil aggregates into constituent micro-aggregates which occurs under wetting is a process called slaking. Since slaked materials are easily transported and deposited, they tend to develop surface crusts or seals. Crusts can be formed by deposition or by raindrop impact and they are extremely serious since they result in a quantitatively large reduction in the infiltration rate. Even if crusts are relatively resistant to raindrop impact, they lead to greater amounts of overland flow and to an increased rill and gully erosion hazard. Crusts are more effective in reducing infiltration if accompanied by the chemical dispersion of the clay fraction in the absence of organic matter (see Shainberg and Letey, 1984). They also influence seedling emergence directly by offering mechanical resistance and indirectly by affecting the transfer of gases between the soil and the free atmosphere.

In a natural forest ecosystem macropores which enable the rapid drainage of the surface soil are continually being formed and destroyed. Under cultivation, ploughing creates a high macroporosity which is subsequently reduced due to compaction and internal slaking. The initial roughness (and storage potential) is subsequently reduced and the potential erosion hazard increases. If the soil fauna and rooting characteristics can create new macropores, the effects of

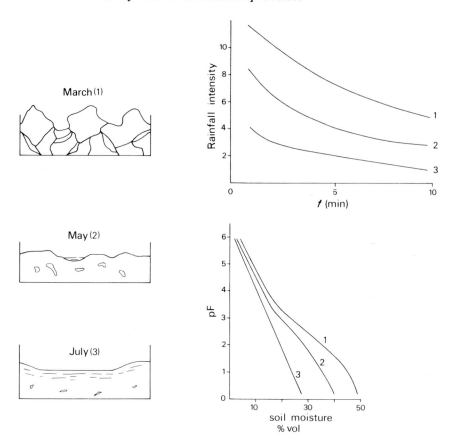

FIGURE 13.5   The effects of slaking. The ploughed surface of the loess soils in South Limbourg, The Netherlands, becomes increasingly smooth as soil aggregates slake under the influence of rainfall. There is a decrease in the macroporosity and a crust develops. The decrease in the infiltration capacity is indicated by the infiltration envelopes and the decrease in the porosity by the soil moisture characteristics (based on van Eijsden and Imeson, 1985)

slaking are less dangerous. If there is only a poor soil fauna (and a low biological activity) then the increase in erodibility is far greater.

   The effect of slaking on porosity and infiltration is indicated by the changes which occur in the Ap horizon of a loess soil in the Netherlands between March and July (Fig. 13.5). Ploughing in March (1) creates a high initial macroporosity and very high rainfall intensities are required to produce ponding. Rainfall between March and May (2) produces slaking and a crust extends over about 20 per cent of the surface. By July (3) the crust has increased to cover the greater part of the surface and, as the soil moisture characteristic indicates, there has been a large reduction in the number of coarse pores. Very little rainfall is required to pond the soil. The rate at which the above changes in the cultivated soil occur is dependent upon the aggregate stability of the soil and the amount or intensity of the rainfall.

The negative impact of man on the structure of cultivated soils has led to an increase during the last half century in what is often termed soil degradation. Soil degradation is expressed in a loss of soil structure and in an accompanying decrease in the resistance offered to slaking, dispersion and erosion. It is associated with a reduction in the amount of organic matter in the soil, a decline in the soil fauna and compaction by increasingly heavy agricultural machinery. With respect to compaction, also of importance is the increase in soil erodibility which occurs if soils are puddled as a result of harvesting or cultivation under conditions that are too moist. This can be seen in Western Europe where it may be economically sound to postpone the harvesting of sugarbeet and maize into the late autumn, but if wet weather sets in, and soils are puddled during harvesting, the erodibility of the soil during the subsequent growing season may by considerably increased. Soil degradation also results from the negative effect of fertilizer on soil structure, directly as in the case of phosphates which encourage soil dispersion, and possibly indirectly as in the case of copper-containing liquid manures which might harm the soil fauna.

13.6 ASSESSING THE EROSION HAZARD

The likelihood that a particular form of erosion will occur and present a hazard, should a particular human activity occur, can be assessed in several ways. In general, the geomorphological and climatological setting and the soil environment control the potential rate of soil loss and the geographical distribution of this problem. This is reflected in the rainfall, soil and slope factors used in the Universal Soil Loss Equation described below and in the FAO Methodology for Soil Degradation Assessment (FAO, 1979). It must be stressed that flooding and sedimentation, the offsite effects of erosion may present a greater problem than onsite erosion (Fig. 13.6).

A particular difficulty in assessing the erosion hazard is the relatively few measurements of erosion which are available to test predictions. Field evidence of erosion can provide useful information but it can be misleading. The erosion hazard is not necessarily reflected in the distribution of erosional phenomena nor indicated by rates of erosion measured over short periods of time. In evaluating the significance of erosional phenomena, information is required on first, the relationship between the phenomena and the materials in which they occur, secondly, the rate at which the phenomena are formed and obliterated, and thirdly, the importance of extreme meteorological events. Furthermore, erosion may have been triggered off by past human activity of a type no longer occurring, or ecological changes may have reduced or accelerated the risk of

FIGURE 13.6   Damage to crops by flooding and sedimentation following rainfall in the
northern Rif Mountains of Morocco

erosion to an extent not obvious from existing erosional phenomena. These
problems are apparent in Mediterranean or semiarid areas where it is not
immediately obvious if the many 'truncated' soil profiles and gullies reflect past
high rates of erosion or a relatively long survival time of these phenomena (see
Yair *et al.*, 1980).

For many years soil erosion assessment has been dominated by applications
of the Universal Soil Loss Equation (USLE), developed by Wischmeier and
associates using the results of plot experiments from experimental stations
operated by the US Soil Conservation Service. This equation predicts the rate
of soil loss, using factors which influence erosion by rainwash under a clearly
specified range of conditions. The factors included in the equation are indicated
in Table 13.2. To apply the formulae it is necessary to refer to the tables and
procedures outlined in Wischmeier and Smith (1978). Limitations of the
formulae are described by Wischmeier and Smith and a knowledge of these is
essential in evaluating the predictions which are obtained. Relatively recent
advances in the application of the USLE are described in SCSA (1976).

Several mathematical models have been developed for predicting soil erosion
during the last ten years. One of the most frequently reported is the CREAMS
model (Knisel, 1980), which can be used to estimate soil loss from field size
plots. This model has recently been applied to areas outside the United States

where it was developed, and also enables runoff, nutrient and pesticide losses to be predicted.

13.7 CONSERVATION PRACTICES

A great number of conservation practices have been developed to control erosion by water and wind. These techniques involve the management of plants and soil and the use of artificial constructions. The different techniques required for controlling gullying, erosion by wash processes and wind are described fully in many manuals and textbooks.

(a) Erosion by water

In the case of erosion by water, Bryan (1980) listed the objectives of field techniques as:

(1) Protecting the soil surface from raindrop impact.
(2) Maintaining or increasing infiltration capacity.
(3) Reduction of overland flow velocity.
(4) Reduction of soil erodibility.

Bryan also points out that soil and water conservation measures need to be considered together, since, in areas of high erosion risk, water shortage is common. It is usually recommended to develop a soil and water conservation plan at the scale of the small drainage basin, rather than to simply build structures to contain erosion at sites of erosion or sedimentation problems. Ecologically sound methods of manipulating the vegetation cover to provide a more active soil fauna and better soil structure can in the long run be more efficient than expensive terracing or dam construction. It is easier to persuade a farmer to reduce the erosion risk by preparing a coarser seed bed than to change his crop rotation or land use. In preparing a soil and water conservation plan a prerequisite is an inventory of the actual and potential soil erosion hazard. Central to such an inventory is knowledge of man's impact on the soil and hydrological cycle.

Several hundreds of measures are listed in most conservation textbooks and many of these provide excellent examples of how man can positively counteract the negative effects of accelerated erosion. Flannery (1977) distinguishes between (1) soil management (for example, contour cropping, chiselling and subsoiling and mulching), (2) plant management (for example, burning, livestock exclusion and field border planting), and (3) water management and conservation practices (for example contour furrowing, parallel terraces and grass waterways). Specifications for implementing such measures need to be developed locally, allowing for geographical variations in physical conditions. Failure to take account of local conditions can lead to conservation measures having a negative effect. However, intensive field trials of proposed conservation techniques are not always budgeted for in developing countries.

Soil and plant management techniques have as their objectives minimizing the exposure of the soil to erosive rainfall and improving soil structure. 'Conservation tillage' is a broad category of tillage practices aimed at either reducing cultivation or at retaining stubble. Charman (1982) gives a summary of reduced cultivation techniques adopted in New South Wales, Australia. He describes the ultimate in reduced cultivation as 'no tillage' or direct drilling, where the only soil disturbance is by seeding and herbicides are used to control weeds. Seeds are sown directly through the stubble and plant residues of previous crops. Although 'no tillage' techniques have been demonstrated to significantly reduce runoff and erosion at many locations, other problems of plant disease and nitrogen retention in stubble may occur (Charman, 1982). Stubble retention can be practised with normal machinery and can take the form of stubble incorporation or stubble mulching. Aveyard *et al.* (1983) and Packer *et al.* (1984) have compared different cropping systems during trials at the Wagga Wagga and Ginniderra Research stations in New South Wales. Runoff coefficients for example were reduced from 0.3–0.5 under conventional tillage to 0–0.2 under direct drilling with respect to a 40 mm simulated storm, having a recurrence interval of 1.5 years. Ponding times were also improved and sediment loss reduced to about 25 per cent of that under conventional tillage at Wagga Wagga and to about 2 per cent at Ginniderra.

Tillage and crop management techniques which maintain a good tilth that does not slake and which protect the soil from splash erosion are not always sufficient to reduce the erosion hazard. Mechanical protection techniques are then necessary, in order to prevent overland flow from achieving threshold entrainment velocities (Bryan, 1980). Such techniques range from contour tillage, which produces water-retaining furrows along the slope, to bench terraces. Terraces reduce the length and angle of the cultivated slope and are extremely effective. Various types of terrace are described, for example, by Sheng (1977) who gives specifications for construction and maintenance. In semiarid areas terraces usually have additional functions of providing level sites for cultivation and of retaining water for infiltration. In planning terraces it is essential to make allowances for the infiltration characteristics of the local soils, to construct adequate drainage systems and to reserve funds for maintenance. Failure can result in terraces producing more negative than positive benefits (Fig. 13.7).

Gully erosion control usually involves stabilizing gully side slopes and headcuts and reducing the amount and velocity of gully runoff. Many examples exist of successful gully stabilization programmes (Ireland *et al.*, 1939; Gray and Leiser, 1982). Gully control and reclamation practises are reviewed by Heede (1977). Reducing gully runoff amounts and velocities is achieved by check dams or fences, by reducing the runoff from gully sides and by preventing water from entering the gully via the headcut. In the long term, general improvements in the vegetation cover and soil structure in the drainage basin in which the gully occurs are beneficial. Stabilizing steep gully sides with vegetation is difficult when soil conditions inhibit plant growth. Examples of this problem are described by Schouten (1976) for acid sulphate soils in New Zealand and by Enright (1984) for mine waste dumps in New South Wales, Australia. In many cases stabilization of steep gully slopes is achieved by a combination of mechanical grading and biological control.

(b) Wind erosion

As with water erosion, conservation techniques for wind erosion control are aimed at increasing the resistance of the soil to erosion and at reducing the erosive force of the transporting medium. The objectives of measures designed to reduce wind erosion listed by Bryan (1980) are:

(1) Maintaining a surface cover.
(2) The reduction of wind velocity.
(3) The entrapment of moving soil.
(4) Increasing the resistance of the soil to erosion.

Reducing the wind velocity can be achieved by increasing the roughness of the soil or microrelief by means of rough ploughing, plants or vegetation debris

FIGURE 13.7   Terraces breached by gullies and tunnels in Northern Morocco

(mulching). Alternatively, the macro-roughness of the landscape can be increased by the use of shelterbelts or plantations (Eppink, 1983).

Ploughing ridges can be very effective in reducing wind speeds provided that the ridges are between 0.05 and 0.10 m high. Lower ridges are ineffective and higher ridges increase turbulence. A review of the function and use of shelterbelts is given by Bhimaya (1976). Planning and constructing shelterbelts requires great care. The shelterbelts have to have the correct form, height and density (permeability). The trees and shrubs used have to be selected to suit local conditions and require management and maintenance if they are to be effective. Although the positive effect of shelterbelts is great, they have the disadvantage of competing with other vegetation for land, water, light and nutrients.

Increasing the resistance of the soil to entrainment usually involves improving the soil structure by increasing the organic matter or clay contents. The clay content of the surface soil can sometimes be increased by deep ploughing which brings clayey B horizon material to the surface. Alternatively, at great expense, clayey material can be brought in from other areas. Organic matter can be added to the soil in many ways, for example as compost or liquid manure, or by the incorporation of plant residues.

13.8 CONCLUSIONS

Soil erosion is caused by human activity. The prevention of soil erosion requires as a prerequisite the solution of the social and economic problems which makes the implementation of technical soil conservation measures effective. Soil conservation techniques can control erosion, but in developing countries new techniques are required which minimize the amount of disruption to the agricultural community. To develop less disruptive conservation measures, more research into the mechanisms of soil erosion is required. However, new conservation methods will be no more effective than old ones, unless they are developed within the framework of the decision-making environment of the peasant farmer (Jungerius *et al.*, 1985).

The soil erosion and conservation problem requires a multidisciplinary approach. The physical geographer has many roles to play. Priority should, however, be given to establishing the extent of the soil erosion problem, so that its urgency can be evaluated by political decision-makers. Unless erosion can be shown to be economically serious or socially unacceptable, soil conservation will be given low priority. This is the case, for example, in the EEC where there has been no inventory of soil or land degradation and both the soil erosion hazard and the amount of damage which has already occurred, are simply not known. Obtaining this information, will present the physical geographer with a major challenge.

ACKNOWLEDGEMENTS

I wish to thank R. B. Bryan, F. J. P. M. Kwaad, G. van Eijsden, J. M. Verstraten and M. Vis for the interesting discussions on soil erosion. Mrs M. C. G. Keijzer-v. d. Lubbe is thanked for preparing the manuscript and Mrs O. de Vré for drawing the figures.

REFERENCES

Aveyard, J. M., Hamilton, G. J., Packer, I. J., Barker, P. J., 1983, 'Soil conservation in cropping systems in Southern New South Wales', *Journ. of Soil Conserv., New South Wales*, **39**, 113–120.

Bagnold, R. A., 1973, *The Physics of Blown Sand and Desert Dunes*, Chapman & Hall, London.

Bhimaya, C. P., 1976, 'Shelterbelts — functions and uses', *FAO Conservation Guide 3*, FAO, Rome, pp. 17–40.

Boon, W., and Savat, J., 1981, 'A nomogram for the prediction of rill erosion', in Morgan, R. P. C. (Ed.), *Soil Conservation, Problems and Prospects*, Wiley, Chichester, pp. 303–319.

Bryan, R. B., 1980, 'Soil erosion and conservation', in Gregory, K. J., and Walling, D. E. (Eds), *Man and Environmental Processes*, Dawson, London, pp. 207–221.

Bryan, R. B., Imeson, A. C., and Campbell, I. A., 1984, 'Solute release and sediment entrainment on microcatchments in the Dinosaur Park badlands, Alberta, Canada', *J. Hydrol.*, **71**, 79–106.

Charman, P. E. V., 1982, 'Managing soil for future profit', *Journ. Soil Cons. New South Wales*, **38**, 44–48.

Chepil, W. S., 1945, 'Dynamics of wind erosion', *Soil Sci.*, **60**, 305–320, 397–411.

Chu, S. T., 1978, 'Infiltration during an unsteady rain', *Water Res. Research*, **14**, 461–466.

Croley, T. E. II, 1982, 'Unsteady overland flow sedimentation', *J. Hydrol.*, **56**, 325–346.

Croley, T. E. II, 1983, 'Sedimentation dynamics in unsteady non-prismatic rills', *Proc. Conf. on Frontiers in Hydraulic Engineering ASCE/MIT*.

Croley, T. E., and Foster, G. R., 1984, 'Unsteady sedimentation in non-uniform rills', *J. Hydrol.*, **70**, 101–122.

Emmett, W. W., 1978, 'The hydraulics of overland flow on hillslopes', *US Geol. Surv. Professional Paper 662-A*.

Enright, N. F., 1984, 'Aluminium toxicity problems in mine waste', *Journ. Soil Cons. New South Wales*, **70**, 108–115.

Eppink, L. A. A. J., 1983, *Winderosie en Winderosiebestrijding en Bodembescherming*, Deel 2, L. H. Afd. Cultuurtechn., Wageningen.

Eijsden, G. van, and Imeson, A. C., 1985, 'De relatie tussen erosie en enkele landbouwgewassen in het Ransdalerveld, Zuid-Limburg', *Landschap* (in press).

Faber, Th., and Imeson, A. C., 1982, 'Gully hydrology and related soil properties in Lesotho', *IAHS Publ. 137*, pp. 135–144.

FAO, 1976, *Conservation in Arid and Semi-arid Zones*, FAO Conservation Guide 3, FAO, Rome.

FAO, 1977, *Guidelines for Watershed Management*. FAO Conservation Guide 1, FAO, Rome.

FAO, 1979, *A Provisional Methodology for Soil Degradation Assessment*, FAO, Rome.

Flannery, R. D., 1977, *A Handbook on Resource Conservation*, Lesotho Agricultural College, Maseru, Lesotho.

Foster, G. R., and Meyer, L. D., 1975, 'Mathematical simulation of upland erosion by fundamental erosion mechanics', in *Present and Prospective Technology for Predicting Sediment Yields and Sources*, USDA-ARS Rep. ARS-S-40, pp. 190–207.

Graf, W. L., 1977, 'The rate law in fluvial geomorphology', *Amer. Journ. of Sci.*, **277**, 178–191.

Graf, W. L., 1979, 'The development of montane arroyos and gullies', *Earth Surf. Proc.*, **4**, 1–14.

Gray, D. H., and Leiser, A. T., 1982, *Biotechnical Slope Protection and Erosion Control*, Van Nostrand Reinhold, New York.

Hamilton, G. J., White, I., Clothier, B. E., Smiles, D. E., and Packer, I. J., 1983, 'The prediction of time to ponding of constant intensity rainfall', *Journ. of Soil Cons. New South Wales*, **39**, 188–198.

Heede, B. H., 1971, 'Characteristics and processes of soil piping in gullies', *USDA Forest Res. Paper RM-68*, 15 pp.

Heede, B. H., 1977, 'Gully control structures and systems', FAO Conservation Guide 1, *Guidelines for Watershed Management*, FAO, Rome, pp. 181–219.

Hodges, W. K., 1982, 'Hydraulic characteristics of a badland pseudo-pediment slope system during simulated rainstorm experiments', in Bryan, R. B., and Yair, A. (Eds), *Badland Geomorphology and Piping*, GeoBooks, Norwich, pp. 127–151.

Horton, R. B., 1945, 'Erosional development of streams and their drainage basins: hydrophysical approach to quantitative morphology', *Bull. Geol. Soc. Amer.*, **56**, 275–370.

Hudson, N., 1971, *Soil Conservation*, Batsford, London.

Imeson, A. C., and Kwaad, F. J. P. M., 1980, 'Gully types and gully prediction', *Geogr. Tijdschr.*, **14**, 430–441.

Imeson, A. C., and Kwaad, F. J. P. M., 1982, 'Field measurements of infiltration in the Rif Mountains of Northern Morocco', *Studia Geomorph. Carp.-Balc.*, **15**, 19–30.

Imeson, A. C., and Verstraten, J. M., 1981, 'Suspended solids concentrations and river water chemistry', *Earth Surf. Proc. and Landf.*, **6**, 251–263.

Imeson, A. C., and Verstraten, J. M., 1985, 'The erodibility of highly calcareous material from southern Spain', (in press).

Imeson, A. C., and Vis, M., 1982, 'A survey of erosion processes in tropical forested ecosystems on volcanic soils in the Central Andean Cordillera, Colombia', *Geografiska Annaler*, **64A**, 3/4, 181–198.

Imeson, A. C., and Vis, M., 1984, 'Seasonal variation in soil erodibility under different land-use types in Luxembourg', *J. Soil Sci.*, **35**, 323–331.

Imeson, A. C., Kwaad, F. J. P. M., and Verstraten, J. M., 1982, 'The relationship of soil physical and chemical properties to the development of badlands in Morocco', in Bryan, R., and Yair, A., (Eds), *Badland Geomorphology and Piping*, GeoBooks, Norwich, pp. 47–70.

Ireland, H. A., Sharpe, C. F. S., and Eargle, D. H., 1939, 'Principles of gully erosion in the Piedmont of South Carolina', *US Dep. Agric. Techn. Bull. 63*, 143 pp.

Jungerius, P. D., Vis, M., and van der Wusten, H. H., 1985, 'The relationship between human settlement and natural environment in Beni Boufrah (Central Rif, Morocco) and its relevance to resource management (manuscript).

Kirkby, M. J., and Chorley, R. J., 1967, 'Throughflow, overland flow and erosion', *IASH Bull.*, **12**, 5–21.

Knisel, W. G., (Ed.) 1980, 'CREAMS. A field-scale model for chemicals, runoff and erosion from agricultural management systems', *US Dept. Agric. Cons. Res. Rep. no. 26*.

Mein, R. G., and Larson, C. C., 1971, 'Modelling the infiltration component of the rainfall–runoff process', *Water Resource Research Center, University of Minnesota, Bull. 43*.

Moore, I. D., Larson, C. L., and Slack, D. C., 1980, 'Predicting infiltration and micro-relief surface storage for cultivated soils', *Water Resource Research Center, University of Minnesota, Bull. 102*.

Mutchler, C. K., and Young, R. A., 1975, 'Soil detachment by raindrops', in *Present and Prospective Technology for Predicting Sediment Yields and Sources*, US Dept. Agric., ARS-S-40, pp. 113–117.

Packer, I. J., Hamilton, G. J., and White, I., 1984, 'Tillage practices to conserve soil and improve soil conditions', *Journ. Soil Cons. New South Wales*, **40**, 78–87.

Palmer, R. S., 1963, 'Influence of a thin water layer on waterdrop impact forces', *IASH Bull.*, **65**, 141–148.

Parlange, J. Y., Rose, C. W., and Sander, G., 1981, 'Kinematic flow approximation of runoff on a plane: an exact analytical solution', *J. Hydrol.*, **52**, 171–176.

Penman, H. L., 1963, *Vegetation and Hydrology*, Techn. Comm. No. 53, Commonwealth Bureau of Soils, Harpenden.

Reid, J. B., and Goss, M. J., 1981, 'Effect of living roots of different plant species on the aggregate stability of two arable soils', *Journ. of Soil Sci.*, **32**, 521–541.

Rubin, J., 1968, 'Numerical analysis of ponded rainfall infiltration', *IAHS-UNESCO Publ., Water in the unsaturated zone*, pp. 440–451.

Savat, J., 1977, 'The hydraulics of sheet flow on a smooth surface and the effect of simulated rainfall', *Earth Surf. Proc.*, **2**, 125–140.

Schouten, C. J. J. H., 1976, *Origin and output of suspended and dissolved material from a catchment in Northland (New Zealand), with particular reference to man-induced changes*, Publ. no. 23, Fys. Geogr. Bodemk. Lab., Amsterdam.

Scoging, H., 1982, 'Spatial variations in infiltration, runoff and erosion on hillslopes in semi-arid Spain', in Bryan, R. B., and Yair, A. (Eds), *Badland Geomorphology and Piping*. GeoBooks, Norwich, pp. 89–112.

SCSA, 1976, *Soil Erosion Prediction and Control*, Special Publ. no. 21. Soil Conservation Society of America.

Shainberg, I., and Letey, J., 1984, 'Response of soils to sodic and saline conditions', *Hilgardia*, **52**, 1–57.

Sheng, T. C., 1977, 'Protection of cultivated slopes-terracing steep slopes in humid regions', *FAO Soil Conservation Guide 1*, FAO, Rome, pp. 147–179.

Skidmore, E. L., Fischer, P. S., and Woodruff, N. P., 1970, 'Wind erosion equation: computer solution and application', *Soil Sci. Soc. Amer. Proc.*, **34**, 931–935.

Smith, R. E., 1972, 'The infiltration envelope: results from a theoretical infiltrometer', *J. Hydrol.*, **17**, 1–21.

Smith, R. E., and Parlange, J. Y., 1978, 'A parameter efficient hydrologic infiltration model', *Water Resources Res.*, **14**, 533–538.

Stocking, M., 1981, 'A model of piping in soils', *Transact. of Japanese Geomorph. Union*, **2**, 263–278.

Tanaka, T., 1982, 'The role of subsurface water exploration in soil erosion processes', *IASH Publ. 137*, pp. 73–87.

Tisdall, J. M., and Oades, J. M., 1982, 'Organic matter and water-stable aggregates in soils', *Journ. Soil Sci.*, **33**, 141–163.

US Soil Conservation Service, 1966, 'Procedure for determining rates of land damage, land degradation and volume of sediment produced by gully erosion', *Techn. Release no. 32*, Soil Conservation Service, US Department of Agriculture.

Walling, D. E., 1983, 'The sediment delivery problem', *J. Hydrol.*, **65**, 209–237.

Wischmeier, W. H., and Smith, D. D., 1978, 'Predicting rainfall erosion losses — a guide to conservation planning', *US Dept. Agric., Agriculture Handbook no. 537.*

Woodruff, N. P., and Siddoway, F. H., 1965, 'A wind erosion equation', *Soil Sci. Soc. Amer. Proc.*, **29**, 602–608.

Woolhiser, D. A., and Liggett, J. A., 1967, 'Unsteady, one-dimensional flow over a plane — the rising hydrograph', *Water Resources Res.*, **3**, 753–771.

Yair, A., Lavee, H., Bryan, R. B., and Adar, E., 1980, 'Runoff and erosion processes and rates in the Zin Valley Badlands, Northern Negev, Israel', *Earth Surf. Proc.*, **5**, 205–225.

PART V
BIOSPHERE

INTRODUCTION

Perhaps the most extensively studied realm of the earth's environment in relation to human activity is the sphere which embraces plants and animals. Indeed the effect that human activity had upon vegetation was one of the major things in the mind of George Perkins Marsh in 1864 when he wrote *Man and Nature* (Chapter 1). The modification of vegetation cover and of fauna and flora has been a continuing implication of man's occupation of different parts of the earth's surface. This was therefore one of the major reasons for the symposium which gave rise to the publication of *Man's Role in Changing the Face of the Earth* (Thomas, 1956). However, although the way in which human activity modifies vegetation communities is of paramount significance in understanding the development of ecological concepts such as the modifications of climax theory, nevertheless the tendency was for biogeography to focus more upon what was left rather than upon what was modified. There were some notable exceptions to this but it is comparatively recently that there has been a focus upon what has been described in the USA as cultural biogeography. In his textbook Simmons (1979), separated biogeography into two parts. The first shorter part was natural biogeography and the second and larger part was styled cultural biogeography. In such cultural biogeography there is concern with the effects of man in changing the genetic make-up of plants and animals, in redistributing them over the earth's surface and in altering the structure of many ecosystems. This therefore acknowledges one of the most recent trends in biogeography namely the concern with ecological processes that have been modified by human action and in some cases have been introduced by such action. In addition, there is a further trend which arises from such an interest in cultural biogeography and that is the interest in conservation. During the period since 1970, biogeographers have been very much aware of the way in which natural biogeography is a necessary basis for understanding cultural biogeography, and that cultural biogeography is an essential prerequisite to a sound understanding of the principles of conservation.

To do justice to the range and depth of studies which make up biogeography in three chapters is extremely difficult. However, they have been selected to show first of all how natural biogeography has been modified and, although it is not possible to do this for the whole world, Africa provides an excellent illustration, as shown by R. Whitlow in Chapter 14. The implications for the investigation of biogeography and the significance of genetic manipulation are subject of contemporary interest and these are reviewed by C. Harrison in Chapter 15. To complete the trio it is necessary to review the significance of human activity in relation to biogeography as a basis for conservation and this is done by I. G. Simmons in Chapter 16 in a way which leads naturally towards the final section of the book which is concerned with the general human impact.

REFERENCES

Simmons, I. G., 1979, *Biogeography: Natural and Cultural*, Edward Arnold, London.
Thomas, W. L., (Ed.), 1956, *Man's Role in Changing the Face of the Earth*, University of Chicago Press, Chicago.

Human Activity and Environmental Processes
Edited by K. J. Gregory and D. E. Walling
©1987 John Wiley & Sons Ltd.

14

Man's Impact on Vegetation: the African Experience

R. WHITLOW

14.1 INTRODUCTION

There are three main reasons for using Africa as an example to demonstrate man's impact on vegetation. Firstly, increasing human and livestock pressures are resulting in widespread and rapid denudation of the plant cover in Africa. This is having adverse environmental and socioeconomic consequences in the interplay of man–land relationships. Secondly, Africa provides examples of changes that are operative to varying degrees throughout the tropical world. It is hoped that a more coherent picture of man's disturbance of the vegetation will be achieved through concentrating on African examples. Thirdly, whilst there is a vast literature on African vegetation (Whitlow, 1984), there are relatively few accounts of biotic impacts on the plant cover which provide an overview of the African scene. This discussion is directed mainly at the destructive influences of man's activities; the effects of fire are outlined (14.2), followed by accounts of deforestation (14.3), cultivation (14.4), livestock impacts (14.5) and the consequences of concentrating wild animals in game reserves (14.6), and conservation issues are examined in the final section (14.7).

Of all elements of the physical environment the plant cover is possibly the most vulnerable to changes brought about by man's activities (Wilkinson, 1963; Goudie, 1981). Modification of the plant cover invariably has repercussions on processes such as nutrient and hydrological cycles as well as soil erosion, and also because plants are a vital resource for man providing foodstuffs, fuel and other materials, it is clear that the question of man's impact on vegetation is a subject of primary interest. In Africa the main processes of change have been clearance, burning and grazing (Keay, 1974), but the dissemination of plants to new areas has also resulted in locally important changes in the vegetation (Wild, 1968; De Vos, 1975; Stirton, 1978; Harlan et al., 1976).

Through the use of fire and tools, and more recently of chemicals, man has changed the character of the earth's plant cover to such an extent that few areas today have vegetation that is in a pristine state. Even the tropical rainforests, once thought to have been largely unaffected by human activity, have proved upon closer inspection to be of secondary origin in many areas (Flenley, 1979). This is especially true of African forests where human impact has been more severe and of longer duration than other forests in the tropics (Fontaine *et al.*, 1978). Taking natural vegetation to be that which is undisturbed by man, it is obvious that truly natural plant communities are rare. Localized remnants of such vegetation may persist only in remote or inaccessible sites. John Booth (1905, quoted by West, 1971) provided a useful summary of the extent of man's impact on African vegetation when he wrote

> Virgin I maintain, is only that which is bad: in Africa the swamp, the steppe, the stone land. One always detects, clear up to the mountain heights, the hand of passed generations. One finds even in the high-trunked 'primeval forest' of East Usambara (north east Tanzania) the sherds of big-bellied sugar-cane-beer pots and what here passes with the uninformed as 'high-meadow' is generally not original but derived from felled and burned forest.

Disturbance of vegetation is ubiquitous in Africa but is not always easy to detect. This is partly due to the ability of the vegetation to regenerate following the abandonment of croplands or cessation of grazing by livestock. For example, many savanna trees and shrubs regenerate readily by suckering and coppicing. Consequently secondary succession on abandoned or neglected farms is a relatively rapid process. Within a few years the regrowth may be indistinguishable from the surrounding woodlands, being composed of similar species. In contrast, regeneration of forest species following clearance is effected mainly by plants invading open sites in the form of seeds derived from trees adjacent to such areas. Furthermore, occasional fires and more extreme microclimates in abandoned clearings tend to inhibit the re-establishment of closed forest species. It may take several decades before the secondary forest resembles the original forest, assuming that no further disturbance takes place to prevent or set-back the regeneration processes. The ability of the vegetation to recover from disturbance depends not only on the modes of plant reproduction but also the changes in habitat conditions brought about by modification of the vegetation and a review of West African vegetation is given by Hopkins (1965).

In general terms, the extent of man's disturbance of the vegetation depends on population numbers, duration of settlement and levels of technology. Africa has a long history of human occupation with the East African rift valley region being a major hearth of early hominid evolution some two and a half million years ago (Butzer, 1977; Clark, 1980). Small dispersed groups of hunters and gatherers with their primitive technology had little impact on the vegetation for much of this period. Locally, the use of fire may have resulted in major changes in the plant cover, but the earliest evidence of man's use of fire in Africa only dates back some 60 000 years (Pullan, 1974a). With the development and spread of pastoralism and cultivation, along with the use of iron tools, so the impact on the vegetation increased. Keay (1974, p. 387) argues, however, that until the present century the vegetation of Africa has held 'a rough balance with the influences of traditional land use'. Although some areas do not support this viewpoint, it is evident for example that shifting cultivation systems, which operated under conditions of low population density allowed for long fallow periods during which the vegetation could regenerate. Because disease, periodic famines and tribal warfare acted as constraints on human and livestock populations, environmental impacts tended to be localized and of short duration. Obvious exceptions to this include the populous regions of North Africa (Thirgood, 1981) and the environs of the ancient cities in West Africa and on the East African coast.

This situation of apparent ecological harmony changed dramatically during and following the colonial era in Africa. A major reason for this had been a massive increase in human and livestock populations due to improvements in health care and disease control. For example, the pre-1900 human population of Africa was just over 100 million people but by 1950 this had doubled and in 1985 the population may reach 530 million (Udo, 1982). Although population densities are still low in relation to southeast Asia, they have increased markedly in recent years and settlement has been extended into previously sparsely populated or uninhabited regions. Inevitably, this has resulted in more widespread modification of the vegetation as trees have been cleared to make way for farms and fires used to destroy shrublands and encourage the spread of grasses for livestock. Continued population growth, despite the local existence of Malthusian-type checks, is likely to result in further denudation of the plant cover and De Vos (1975, p. 73) observed that 'the African environment today is deteriorating at an unprecedented and accelerating rate'.

Given the diversity of Africa's vegetation and the long and varied history of human activity on this continent, it is only possible to describe the major biotic impacts on the plant cover in very general terms. Various impacts are discussed separately but in reality several processes may be operative and interact in a given area at the same time as demonstrated by Rapp (1974, and Chapter 17) for the denudation of the plant cover in semiarid regions in Africa and as indicated by the modification of vegetation on large termite mounds in Central Africa (Pullan, 1979).

14.2 IMPACT OF FIRE

Fire has been described as the first great force used by man (Stewart, 1956). In Africa people have used fire for a variety of purposes including: the preparation of agricultural lands as in the *citimene* system in Zambia (Pullan, 1974a); removal of moribund grasses to encourage growth of new herbage for livestock; driving away bees during honey-gathering; and assisting in hunting of wild animals. The availability of safety matches has promoted burning in recent years (Glover, 1968) and man-made fires are recognized as having been important in the extension and maintenance of savannas, in combination with shifting cultivation and grazing (Whyte, 1974). However, naturally occurring fires started, for example, by lightning were an ecological force in African habitats long before man came on the scene (Komarek, 1971). One indication of this is the development and widespread distribution of fire-resistant or fire-tolerant grass and woody species, especially in savannas and Mediterranean regions (Naveh, 1975; Stuart-Hill and Mentis, 1982). For example, *Themeda triandra*, a common savanna grass in East Africa, germinates better following fires than without burning and it may be eliminated from areas where burning is prevented (Lock and Milburn, 1970).

The incidence and severity of fires are governed in part by climatic conditions and the nature of the vegetation. Africa has been called the *fire continent* because of the extent of regions which experience seasonal rainfall and prolonged dry periods which favour fires (Komarek, 1971). As a general rule, the wetter the growing season and the hotter and longer the dry season then the greater the chance of fires. Fires are prevalent, therefore, in savanna and Mediterranean regions, especially in the former where grasses are dominant. Grasses produce large quantities of combustible fuel in the rainy season and this burns readily in the dry season when the grasses die back (Rose-Innes, 1971). In arid regions burning is rare and localized because the vegetation cover is sparse and patchy. In more humid regions where forests occur, there is greater biomass than in the savannas, but the absence of a dry season during which the vegetation can dry out tends to inhibit burning. The relationships between climate, plant cover and fire are demonstrated in Fig. 14.1(a). The climatic association of fires has been used by Batchelder (1967) as a basis for delimiting the spatial and temporal patterns of burning in tropical areas. The results of this survey demonstrate the extent of areas that are prone to burning in Africa (Fig. 14.1(b)).

A crucial factor influencing the impact of fires on vegetation is the timing of the burning. Observations on experimental plots in savanna woodlands have demonstrated common features with respect to the effects of burning in the early and late dry season and in prevention of fires (Trapnell, 1959; Rose-Innes, 1971). In the Olokemeji Forest Reserve, in the derived savanna zone of Nigeria (Hopkins, 1965), plots within the forest were cleared of trees and burnt in 1929. Subsequently, they were subjected to three different treatments, namely protection from fire, early dry season burning, and late dry season burning. The results of this exercise, after maintenance of the regime for nearly 30 years, are summarized in Fig. 14.1(c). On the protected plot the forest regenerated well with numerous fire-sensitive species managing to establish themselves. The

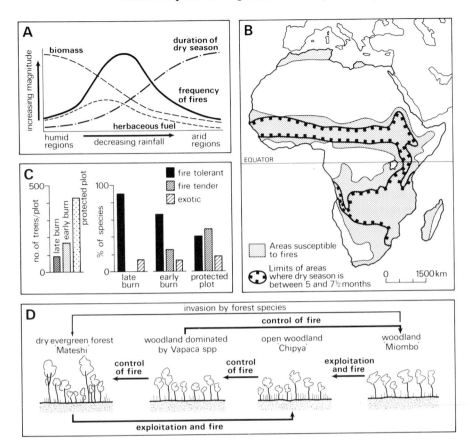

FIGURE 14.1 Fire ecology in Africa. (a) Hypothetical relationship of vegetation, climate and fires; (b) Areas prone to burning in Africa (based upon Batchelder, 1967); (c) Effects of fire on woody vegetation in Olokemeji Reserve, Nigeria (after Rose-Innes, 1971); (d) Woodland trends influenced by fires in northern Zambia (derived from Lawton 1972, 1978)

vegetation on the late season plot, in contrast, comprised an open tree savanna within which most of the woody species were fire-tolerant, having, for example, thick corky bark. The early burn plot was characterized by a woodland community with localized clumps of fire-sensitive forest species. These contrasts were attributed to the greater intensity, and hence damaging effects, of fires in the late dry season. This is due to the fact that the ground vegetation has dried out more completely by the end of this season thus providing abundant and more flammable fuel. The process is a reinforcing one since fires keep woody growth in check and favour grass cover; in turn, the grasses promote fires.

In his review of the use of fire in land management in Africa Phillips (1965, 1974) regards fire as 'a poor master but a good servant'. This statement is a warning against the destructive effects of uncontrolled or indiscriminate burning. For example, late season fires in savannas followed by intensive grazing by livestock and high-energy rainstorms can result in excessive erosion. Accumulation of plant material takes place in the absence of regular burning so that in the event of a fire occurring it tends to be very intensive. Fire can be, and is, employed successfully to prevent bush encroachment, a common problem on rangelands in Africa (Strang, 1973). Certainly, fire can tip the balance in favour of either grasses or woody species depending on its frequency of occurrence and intensity. In a detailed study of woodlands in northern Zambia, Lawton (1978), p. 177) considered that 'the vegetation on any one site at any one time depends more on its history of burning than on any other single factor'. The role of fire in determining plant succession in these woodlands is indicated in Fig. 14.1(d) and the ecological significance of fire has been reviewed by, for East Africa, Langlands (1967) and, for South Africa, Joubert (1977), while Pullan (1974a) provides an account of fire in African savannas. In general terms, one can concur with Keay (1974, p. 388) that fires, mainly man-made, in tropical Africa have 'prevented the vegetation from reaching its climatic climax . . .'.

14.3 IMPACT OF DEFORESTATION

Causes of deforestation in Africa as in other continents have included clearance for cultivation, gathering of firewood and supply of timber for construction purposes (Eckholm, 1977). Removal of woody vegetation may be a necessary precursor to productive use of land in farming, so deforestation, *per se*, should not be condemned out of hand. Problems arise where deforestation occurs in areas of marginal terrain where removal of protective forest cover results in accelerated erosion, for example, or where shortages of woodfuel arise in the absence of controlled exploitation or replanting of trees (Whitlow, 1979a, 1980a). Under such conditions regeneration of forests is difficult, if not impossible, and trees can no longer be regarded as a renewable resource.

Although extension of croplands, partly with the assistance of fire, and woodfuel gathering are primary causes of contemporary deforestation in Africa, commercial exploitation has contributed to the reduction of forest cover in the past. For example, exports of mahogany and other hardwoods from Nigeria began in the early nineteenth century. Extensive areas of forest were felled, sometimes indiscriminately, to supply timber for this lucrative trade (Adeyoju, 1965). Similarly, the mangrove forests in East Africa were exploited to provide poles for Arab merchants many centuries ago (Lind and Morrison, 1974). However, present-day forestry activities generally make provision for regeneration or planting of trees. The real problems lie with those cultivation and wood-gathering activities that, for practical reasons, result in widespread and irreversible loss of woody vegetation. The seriousness of deforestation depends on the extent of destruction, the nature of the sites cleared and the subsequent use of the land.

Data on the extent and rate of deforestation in Africa are often based on general estimates derived from crude surveys or vegetation maps. Since ideas as to what constitutes closed forest and open forest, for example, vary greatly, it is only possible to provide approximate values. Phillips (1974) cites data which show that, of a potential forest area of 235.6 million hectares, only 146.2 million hectares or 62 per cent remains, much of this being in a degraded state. A more generous estimate is given by Dasmann (1979), suggesting that there are around 190 million hectares of closed forest left in Africa; but the potential forest area, based mainly on climatic criteria, is nearly double this! The difference between these estimates is a reflection of the paucity of accurate data on forests. However, it seems likely that over one-third of the African forests have succumbed to the axe or fire. In countries such as Ethiopia only about 4 million hectares of forest remain out of an estimated original forest area of 45 million hectares (Dasmann, 1979). In contrast, Zaire still has substantial areas under forest. The extent of encroachment of derived savanna (Keay, 1959) into the major forest regions in Africa is shown in Fig. 14.2(a).

Of greater concern perhaps is the alarming rate of deforestation. For example, forest loss in the Ivory Coast is reported at 0.4 million hectares per year (Stott, 1978) and given an estimated forest area of about 5 million hectares in the mid-1970s (ENDA, 1981), it is clear that by now very little of this forest remains. Madagascar had some 12.5 million hectares of forest in 1960 (Chauvet, 1972) but Hedberg (1974) reports that forest destruction is now more intense than ever before, with some 0.3 million hectares being lost each year. At current rates of deforestation much of this forest will be destroyed by the end of this century. Even allowing for inaccuracies in the limited data available, it is obvious that extensive and rapid deforestation is taking place in Africa (Aubreville, 1974). Moreover, in the face of growing population pressures this destruction is likely to make considerable inroads into the surviving forests over the next few decades.

The forests of the Mediterranean littoral were an early victim of man's activities. In northern Morocco, the remnants of forests surviving deforestation (Mikesell, 1960) are indicated in Fig. 14.1(b). These forests, dominated by species of oak, cedar and fir, are restricted now to localized patches in mountainous areas with much of the present-day vegetation comprising sparse scrub. Denudation of the formerly extensive forests dates back to the Roman Empire period some 2000 years ago. Demands for timber for shipbuilding, temple construction and fuel resulted in considerable depletion of the forests at that time. Continued exploitation over the centuries has produced the barren landscapes of today. In particular, the forests have suffered from the effects of fire and depredations by livestock, especially goats. Mikesell (1960) estimates

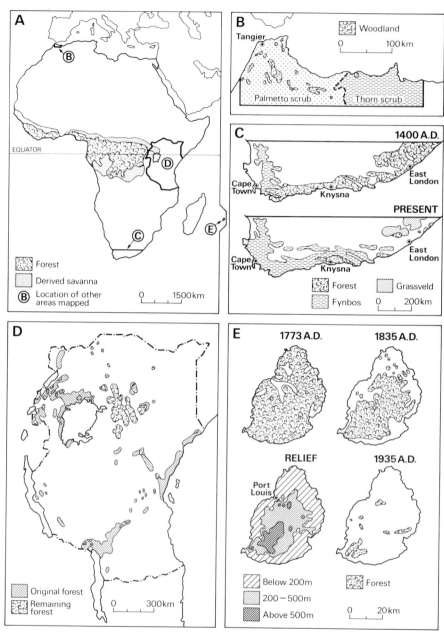

FIGURE 14.2   Case studies of deforestation in Africa. (a) Derived savannas in Africa (Pullan, 1974a); (b) Woodland remains in north Morocco (Mikesell, 1960); (c) Deforestation in the Cape region (Acocks, 1975); (d) Deforestation in East Africa (Kingdom, 1982); (e) Deforestation in Mauritius (Vaughan and Wiehe, 1937)

that the present wooded area constitutes about one-tenth of the potential forest area, the main agents of contemporary tree suppression being cutting, burning and browsing.

Forests in the south of the continent have suffered a similar fate, although destruction occurred at a much later date (Fig. 14.2(c)). Acocks (1975), on the basis of relic patches of vegetation, has reconstructed the vegetation patterns in the Cape area prior to European settlement. At this time the coastal mountain ranges and plains were covered in evergreen forests dominated by species such as the yellow-woods (*Podocarpus* spp.). Following the founding of the Cape settlement in 1652 and the subsequent eastward migration of the European farmers, these forests were subjected to exploitation of their valuable timber trees. For example, Graaff (1981) notes that there was a heavy demand for pitprops in the Witwatersrand gold mines in the late 1900s, much of the timber being obtained from the Cape forests. The remaining forests around Knysna are still exploited for their valuable timber, but this is done on a more rational scientific basis with certain areas being set aside for conservation. As the forests were destroyed, so the hardy shrub species of the Cape *fynbos* (comparable to the Mediterranean macchia) extended their range from the western Cape, whilst in areas north of East London the forests were replaced mainly by grasslands. The increased incidence of fires in these vegetation types served to inhibit forest regeneration, a situation similar to that in northern Morocco.

Forest regression has occurred in East Africa on a similar scale, especially in the long-settled coastal regions, the Udzungwa ranges and the areas northwest of Lake Victoria (Fig. 14.2(d)). The major causes of deforestation here have been clearance to make way for peasant farms and plantations, although there is evidence of forest clearance for cultivation some 3000 years ago (Morrison and Hamilton, 1974). Two points need emphasis with respect to this forest denudation. Firstly, removal of forests in mountain catchments has resulted in serious erosion problems, including widespread landsliding, in some areas (Temple and Rapp, 1972). Secondly, destruction of montane forests is threatening the survival of a unique flora and fauna about which little is known (White, 1981). In contrast, the lowland forests in parts of Uganda are able to regenerate and spread readily if biotic pressures are removed (Dale, 1954). Mauritius furnishes an even more dramatic and ecologically devastating history of deforestation (Fig. 14.2(e)). The onslaught on these forests began in 1598 following Dutch settlement on the island (Vaughan and Wiehe, 1937). The ebony trees were a valuable source of timber for the early settlers but much of the later forest clearance was carried out to make way for plantations, notably of sugar-cane. By 1880 the forests were described as being 'a picture of doleful ruin' and only a few patches of forest remained in largely unproductive mountain sites by 1935.

These examples demonstrate that deforestation has been taking place in Africa for a long time and has culminated in virtual elimination of forests in some areas.

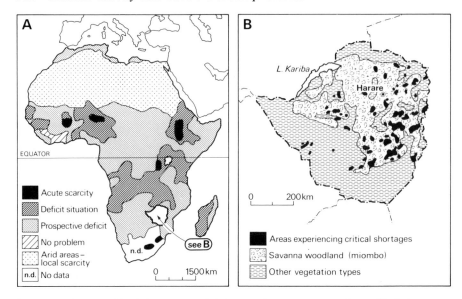

FIGURE 14.3   Woodfuel shortages and deforestation. (a) Degree of woodfuel shortages in Africa (based upon FAO, 1981); (b) Woodfuel shortages in Zimbabwe (Whitlow, 1980a)

Equally serious is the problem of destruction of woody vegetation to supply fuel for both rural and urban populations. Firewood is a staple energy source throughout Africa, but foraging for wood is becoming increasingly more difficult and time-consuming notably in more densely settled regions. The degree of woodfuel shortages in Africa is indicated in Fig. 14.3(a). Consumption of woodfuel in the early 1970s was estimated to be around 180 million cubic metres per year (Arnold and Jongma, 1978). Africa's population at that time was about 350 million people giving an average per capita consumption of 0.8 m³ per year. Since then, steady increases in population, in combination with further extensions of croplands have taken a further toll on the woody vegetation so that shortages of firewood are now quite common *in densely populated regions.*

Even in regions which FAO (1981) regards as potential future problem zones, there may already be local scarcity of woodfuel. For example, in Zimbabwe the densely settled peasant farming areas already experience acute shortages of woodfuel, to the extent that increasing use is being made of crop residues and cow dung for cooking and heating (Fig. 14.3(b)). Where there are adequate supplies of wood near settlements there is a tendency for selective exploitation. Favoured species are plundered so that the composition of woodlands is changed. As woodland resources dwindle people become less selective and any available trees or shrubs may be felled for firewood (Whitlow, 1979b). Of particular concern in Africa is the fact that reafforestation has been neglected in the past and to meet current and future demands for woodfuel and construction timber will require massive inputs of funds and manpower. There is little evidence of this taking place at present.

14.4 IMPACT OF CULTIVATION

In Africa it has been estimated that barely 5 per cent of the land has been cultivated out of a total area of some 3010 million hectares (Bridges, 1978). In densely populated regions the proportion of cropland may be considerably higher. For example, in the peasant farming areas of Zimbabwe, about 14 per cent of the land is cultivated, but in densely settled areas this value may increase to well over 50 per cent (Whitlow, 1980b). Similarly, in parts of West Africa extensive areas of croplands occur giving rise to a distinctive agricultural landscape which Pullan (1974b) has termed 'farmed parkland' (see Fig. 14.4(d)). In such areas very little indigenous plant cover remains and what does exist has been disturbed by woodfuel gathering and burning. There are, however, substantial areas where little or no cultivation takes place and Bridges (1978) estimates that there are over 700 million hectares of potential arable land in Africa. Much of this land is located in marginal areas where drought, nutrient deficiencies and pests are major constraints on crop production. Since Africa is a 'hungry continent' with malnutrition being commonplace (May, 1974) and is faced with the prospect of massive population increases, it is clear that further extension of cultivation must take place. This will require further destruction of the indigenous vegetation. In marginal situations this will probably result in accelerated land degradation, particularly where conservation measures are inadequate, and in an impoverishment of the surviving flora.

Cultivation necessarily involves partial or complete clearance of the indigenous plant cover, the nature of the ecological impact being dependent upon the land use system concerned (Puzo, 1978). Throughout much of tropical Africa the traditional agricultural systems are based upon shifting cultivation or bush fallowing, the latter being more characteristic of densely populated areas (Allan, 1965). The response of peasant farmers to increasing population pressure is usually to extend the area under cultivation, initially in more accessible and better quality lands, but as demand for food increases so marginal areas are brought under the hoe or plough. A common scenario depicting the changes in subsistence land use systems placed under increasing population pressure has been described by Whitlow (1979c) and is shown diagrammatically in Fig. 14.4(a).

Stage 1 in this model represents a period when cultivation is localized and extensive areas of 'natural vegetation' are largely unaffected by human activity, being utilized for hunting and gathering. As the human population increases the area under cropping is extended to produce more food as shown in Stage 2 of the model. Shifting cultivation systems rely on long fallow periods to maintain soil fertility and ensure reasonable crop yields. However, with prolonged

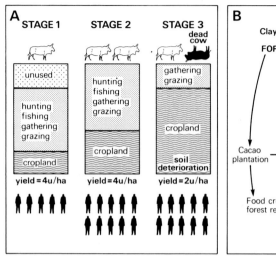

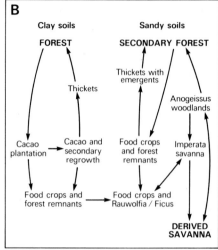

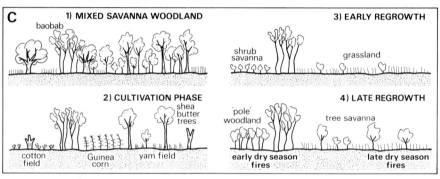

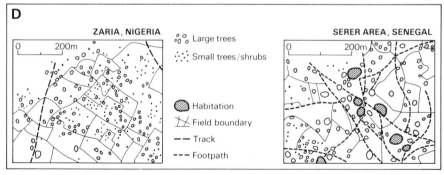

FIGURE 14.4   Impacts of cultivation on vegetation. (a) Changes in subsistence land use (Kay, 1975); (b) Successional trends in vegetation near Ibadan Nigeria (simplified from Clayton, 1978); (c) Changes in vegetation under bush fallowing in savanna areas (Pullan, 1974a); (d) Farmed parkland landscapes in West Africa (Pullan, 1974b)

cultivation and shortening of fallow periods, a cycle of degenerative changes is set in motion involving progressive extension of cropping into marginal sites with steep slopes or shallow soils for example (Allan, 1965). This is shown by Stage 3 of the model where ecological impacts are most obvious in rangelands. The situation is one of more and more animals being forced to survive on a diminishing area of land as cultivation encroaches into the grazing areas. Denudation of plant cover and erosion are most widespread in the rangelands and poor food supplies result in high mortality rates amongst the livestock, especially during stress periods such as droughts. It is important to realize, therefore, that cultivation activities may have an indirect and adverse impact on vegetation surrounding or adjacent to croplands.

In forest or woodland areas where shifting cultivation is practised the vegetation may present a mosaic of stands in various stages of growth, as discussed by Hopkins (1965) in relation to West Africa. In the derived savanna zone near Ibadan, Nigeria (Fig. 14.4(b)), Clayton (1958, p. 217) has pointed out that 'there is no clear distinction between natural vegetation and farmland; for the peasant farmer never completely clears his land before planting crops, and when it is exhausted he restores its fertility by letting it revert to bush for several years'. The result is a patchwork of stands of varying composition and structure (Fig. 14.4(b)). Three main trends in plant succession resulting from cultivation in this area have been outlined by Clayton (1958). Firstly, there is a sequence of cultivation–thicket–forest–recultivation; secondly, there may be degeneration in the regrowth vegetation as a result of repeated cultivation; and finally, food crop cultivation extends into cacao plantations affected by die-back and opening up of the tree canopy. Further information, particularly on the nature of regenerating vegetation on fallow plots, is given by Moss and Morgan (1970) and Aweto (1981).

Where population densities are high, cultivation is based on bush fallowing or permanent plots. This involves more complete clearance of the indigenous vegetation than in shifting cultivation, but large trees such as baobabs (*Adansonia digitata*) may be left intact amongst croplands. A sequence of vegetation development under bush fallowing has been outlined by Pullan (1974a) and is shown in Fig. 14.4(c). The nature of the regrowth vegetation depends on several factors such as the composition of the remnant flora which persists during the cultivation phase and the effects of fires as noted earlier. In West Africa and Zambia the 'farmed parkland' (e.g. Fig. 14.4(d)) is defined as a 'landscape in which well grown trees occur scattered through cultivated or recently fallow fields' (Pullan, 1974b, p. 119). Several types of farmed parkland can be differentiated on the basis of the species present and the size and spacing of trees. The baobab is a conspicuous element of farmed parkland in drier regions, whilst *Acacia* species and fig trees (*Ficus* spp.), which provide fodder for livestock or fruit, characterize more humid areas. In Zimbabwe a common tree in farmed parklands is the mbola plum (*Parinari curatellifolia*), favoured for its fleshy fruit and providing shade for cattle. A distinctive variation on farmed parkland is found in parts of Central Africa, where dense thickets survive on large termite mounds located within cultivated lands.

Cultivation usually requires at least partial destruction of the indigenous plant cover, but one interesting exception is provided by the growth of *Acacia senegal* in the Sudan (Obeid and Seif el Din, 1970). During the cultivation of millet, groundnuts and other staple foodstuffs, the acacia shrubs are cut back but not uprooted. When the lands are left to fallow the acacias regenerate forming uniform, dense stands. These can be exploited to supply a resinous substance called gum arabic, this being a valuable export commodity for the Sudan. Harvesting of the resin begins after the shrubs are about four years old and may continue for some ten years thereafter. The shrubs are cut back then to allow another phase of cultivation. The system ensures maintenance of soil fertility and a useful cash crop is produced during the fallow period. However, on cultivated lands throughout much of Africa there is a problem of degradation and declining yields. This is reinforcing the trend of extension of croplands and the associated reduction in the area, diversity and stature of the indigenous vegetation.

14.5   IMPACT OF LIVESTOCK

Semiarid to arid climatic conditions prevail over about half of Africa. Such areas are characterized by a sparse, patchy plant cover which varies in abundance and vitality depending on the amount and variability of rainfall. Livestock husbandry is an important human activity in these regions and in the dry savannas. Brown (1971) estimates that there are some 50 million pastoralists in sub-Saharan Africa alone, including tribes such as the Fulani and Masai. These people depend almost totally upon their animals for food, and human population growth has necessitated a concomitant increase in livestock numbers. This, combined with improvements in disease control and water supplies and the emphasis on quantity rather than quality of animals, inevitably has resulted in widespread overstocking of rangelands well beyond their carrying capacity (Dasmann *et al.*, 1973). The effect has been extensive denudation of the plant cover with only the more hardy drought- and 'goat-' resistant species being able to survive. This has been a major cause of desert encroachment, a process which becomes more acute during drought phases such as occurred in the Sahel zone of West Africa during the late 1960s and early 1970s (Warren and Maizels, 1977). Although livestock mortalities are high during stress periods such as prolonged droughts, this does reduce pressure on the land, allowing time for the vegetation to recover. There tends, therefore, to be a cyclic pattern of livestock increase–plant destruction–drought–livestock decrease–plant regeneration–livestock increase.

Overgrazing of African rangelands is as much a function of poor management as it is of too many animals. Any area which is not cultivated may be used for

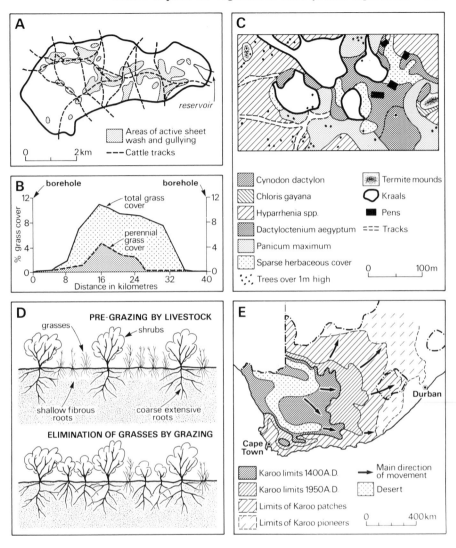

FIGURE 14.5 Impacts of livestock on vegetation. (a) Degradation along cattle trails in the Kisongo catchment, Tanzania (Murray-Rust, 1972); (b) Depletion of grass cover around boreholes in Botswana (Zumer-Linden, 1976); (c) Vegetation patterns around cattle kraals in part of Uganda (Turner, 1976); (d) Bush encroachment (compiled from various sources); (e) Advance of Karoo vegetation in South Africa (Acocks, 1975)

grazing and, typically, this is regarded as communal land. That is it is used by everyone and cared for by few. A good example of the consequences of such attitudes is the erosion which occurs along frequented tracks, especially where cattle and other animals go for water in streams or small dams (Fig. 14.5(a)).

Similarly, livestock tend to congregate in areas surrounding boreholes. This results in depletion of the plant cover around the boreholes and under-utilization of vegetation more distant from these sites. This is demonstrated in Fig. 14.5(b) for a semiarid area in Botswana (Zumer-Linden, 1976). One method of spreading grazing pressures more evenly is to develop water supply points in a given area. However, if this is not combined with controls on livestock numbers and rotational grazing, it may simply serve to increase the extent of denuded rangelands. A good example of this is given by Talbot (1973) who describes the disastrous ecological and social consequences of rangeland development in the Masai areas of Kenya and Tanzania. Cattle kraals (or enclosures) provides a contrast where concentration of livestock may have beneficial effects on the plant cover. This is a result of the increased fertility of soils around kraals due to inputs of manure, although some of this may be removed for use on cultivated plots. Trampling favours the survival of stoloniferous grasses in the vicinity of kraals. Distinctive vegetation pattern occurs around such enclosures as shown in Fig. 14.5(c) (Turner, 1967).

Selective grazing is often part of the problem of overgrazing. Plant communities in semiarid areas comprise two major components, the woody species (mainly shrubs) and the herbaceous species (mainly grasses). These compete for moisture in the surface soil horizons, but the deeper-rooting shrubs are able to utilize water in the subsoil as well (Fig. 14.5(d)). Selective removal of the grasses by livestock disturbs the balance between these two groups of plants. The palatable perennial grasses tend to be eaten first followed by annual species, gradually reducing the cover and vigour of the grass sward. This provides a competitive advantage to the shrubs, especially the more 'aggressive' species like *Acacia karoo*, which may increase in density to form impenetrable stands beneath which little else grows. A critical factor in this process of bush encroachment (Walker, *et al.*, 1981) is the reduction of plant litter. Infiltration rates on bare soils within overgrazed rangelands are considerably lower than those in areas where a good grass cover is maintained (Kelly and Walker, 1976). With less water infiltrating into the soil and increased runoff there is a progressive desiccation of the habitat which, in turn, adversely affects the composition and productivity of the vegetation from the point of view of livestock owners. Periodic burning has been employed to eradicate shrubs but has not always proved successful (Strang, 1973).

It has been estimated that over 6.5 million square kilometres of Africa are faced with high-to-very-high risks of desertification hazards (FAO, 1977). Much of the desert encroachment that has taken place may be attributed to overstocking and overgrazing (Chapter 17). Attention has tended to focus on regions in the Sahel zone of West Africa and the Horn of Africa, mainly because of the large numbers of peasant families adversely affected by droughts in recent years. A less well-known example is the spread of desert-like conditions in southern Africa. A substantial area of South Africa has been affected by the spread of xerophytic shrubs and succulents of what is known in that country as the Karoo veld (Acocks, 1964). This has been brought about as a result of selective grazing and overgrazing, mainly by sheep. The extent and rate of advance of the Karoo veld at the expense of grassland and other vegetation types is shown in Fig. 14.5(e). There is a very real danger that, with continued grazing pressures, the Karoo patches beyond the limits of the main Karoo areas may, in due course, coalesce. Once established these hardy plants are virtually impossible to eradicate. It is clear, therefore, that livestock in Africa have brought about major changes in the nature of the plant cover and are a key element in suppressing the regeneration of vegetation in some areas (De Vos, 1975).

The environmental and socioeconomic constraints on improvements in the standards of range management and increases in livestock production are immense, albeit not entirely insurmountable (Blair Rains, 1982). A plan which is receiving serious attention is an FAO proposal to eradicate tsetse fly (*Glossina* spp.) from large areas of the infested African savannas and forests (Fig. 14.6(a)). Tsetse flies are the vector of a debilitating, sometimes fatal, disease called trypanosomiasis, more commonly known as sleeping sickness. This affects both people and livestock, especially cattle. A study by Bourn (1978) has demonstrated that livestock populations, and hence production, could be increased markedly if tsetse fly could be eliminated from rangelands in Africa (Fig. 14.6(b)). Linear (1982, p. 2) has evaluated the ecological and other implications

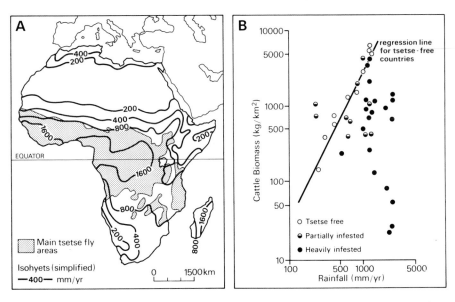

FIGURE 14.6   Tsetse fly distribution and livestock in Africa. (a) Tsetse fly distribution and rainfall (based upon Clark, 1980); (b) Cattle distribution, rainfall and tsetse infestation (based upon Bourn, 1978)

of the FAO proposal and concluded that the 'dreaded tsetse fly may be the only thing standing between Africa's remaining tropical forests (and woodlands) and the developer's axe'. Bourn (1978) and Ormerod (1976) have expressed similar reservations about the ecological effects of controlling and reducing trypanosomiasis. A project such as the FAO tsetse eradication scheme could result in a massive increase in human and livestock populations and the extension of settlement into hitherto sparsely populated or marginal regions. The destruction of vegetation and land degradation which would be associated with such changes may prove disastrous in the long term. This does not imply that all development should be prevented. Rather it is a case for careful evaluation of large-scale, technology-based projects so that maximum benefits can be derived and negative effects avoided (Dasmann *et al.*, 1973).

14.6 IMPACT OF WILD ANIMALS

Africa has an extremely diverse fauna, rich in large mammals (Coe, 1980). These animals were an integral part of African ecosystems well before the evolution and spread of man. Close, sometimes symbiotic, relationships exist between animals and the flora as shown by Lamprey (1963) for the Tarangire Game Reserve in Tanzania (Fig. 14.7(a)). Defoliation of grasses by herbivores has influenced the nature of evolution in some plant species (Stuart-Hill and Mentis, 1982) and seed dispersal by animals has been an important element in the spread of plants, in certain cases over quite considerable distances, as shown by Wickens (1976) for the Jebel Marra in the Sudan. Different animal species frequent different habitats, sometimes on a seasonal basis (Jewell, 1980). Consequently, as a group, wild animals utilize a wider variety of plants than do cattle or goats, for example. Their ecological impact under 'normal conditions' is far less damaging than that of domestic livestock. This does of course depend upon the size and number of animals involved, with large mammals have a significant impact on the vegetation in some areas (Cumming, 1979).

The ecological balance of wildlife and vegetation has been disrupted in recent decades by a human and livestock population explosion which has involved 'man's progressive occupation of land and the consequent restriction of the ranges of wild animals' (Laws, 1970, p. 2). As cultivation and grazing have been extended and intensified so there has been reduction in area of wildlife habitats as well as a deterioration in the quality of these habitats. This in turn has resulted in a decline in wildlife populations, a trend that has been exacerbated by ruthless poaching in some parts of Africa (Jackson, 1982; Kingdon, 1982). Very few animals, notably the larger mammals, are able to survive outside game reserves for these reasons. As human population pressures increase, so man and wildlife come into increasing conflict (e.g. Myers, 1972; Swift, 1972) and both animals and habitats suffer as a result of this conflict. Moreover, whereas 'bush-meat' was a valuable source of food for people in the past, with few

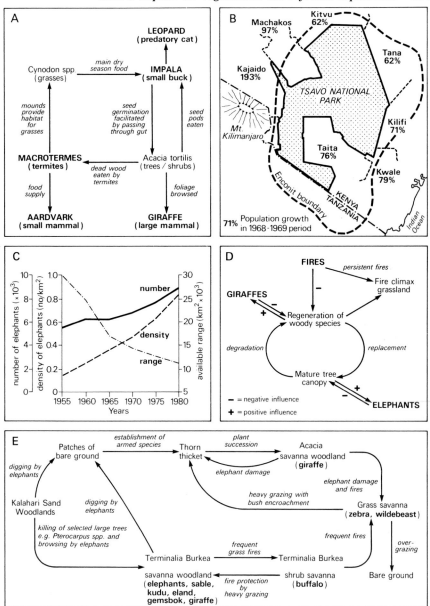

FIGURE 14.7   Impacts of wild animals on vegetation. (a) Interrelationships of plants and animals in Tarangire Game Reserve, Tanzania (Lamprey, 1963); (b) Tsavo National Park, Kenya (Myers, 1973); (c) Change in elephant population and available range in the Sebungwe region, Zimbabwe (Cumming, 1981); (d) Impacts on *Acacia tortilis* woodlands in Serengeti, Tanzania (Pellew, 1983); (e) Vegetation changes in Wankie National Park, Zimbabwe (modified extensively from Boughey, 1963 and using personal communication from Rushworth, 1984)

exceptions, where game ranching is carried out, for example, this is no longer the case.

Major impacts on the vegetation have resulted from effectively restricting animals to game reserves where they are protected more easily. Some of the African game reserves are quite large and Tsavo National Park in Kenya is over 20 000 km² in extent whilst Kruger National Park in South Africa is of similar size. Nevertheless, existing parks are often inadequate to support their resident wildlife populations. Indeed some scientists predict that this will result in 'faunal collapse' in East African reserves in the long term (Soulé *et al.*, 1979). Part of this problem relates to the fact that the park areas are smaller than the eco-units within which animals would move around under normal circumstances, that is in the absence of man. This is illustrated in Fig. 14.7(b) for Tsavo National Park (Myers, 1973). The area that is required to meet the year round needs of the animals in this park, especially the elephants, is estimated to be about 44 000 square kilometres. This is more than double the area of the park and, moreover, transcends the national boundaries by extending into Tanzania! The problem is compounded by growing human population pressures in the peasant farming areas adjacent to the park. Between 1948 and 1969, population densities increased by about 75 per cent in most of these areas (Myers, 1973). There has been growing pressure to allow livestock to graze in game parks as has happened in the Masai Amboseli Game Reserve (Western and Van Praet, 1973). Under these conditions it is not surprising that major changes have occurred in the nature of the vegetation cover within many game reserves.

The most dramatic changes have been brought about by the African bush elephant (*Loxodonta a. africana*) which Laws (1970) regards as having a greater impact on African habitats than any other animal, with the exception of man. The major problem relates to the great size and immense appetites of elephants who are both browsers and grazers, depending on the availability of food. Where animals are confined to game reserves through the encroachment of human settlement, so the density of animals in the reserves increases and their ranges decrease (Fig. 14.7(c)). The result in many areas has been the widespread destruction of forest and woodland, with quite sizeable trees including large baobabs being stripped of their bark or pushed over by foraging elephants. The general trend is one of woody vegetation giving way to open grasslands and the onset of desert-like conditions in drier regions (Laws, 1970). The increase in grass cover favours more frequent and intensive fires, so periodic burning may suppress the regrowth of the woody species. Pellew (1983) has shown that giraffe browsing also has contributed to the suppression of regenerating *Acacia tortilis* in parts of Serengeti in Tanzania (Fig. 14.7(d)). Although there are varying views on the causes and consequences of vegetation changes in African game reserves it is clear that elephants have been a major agent of rapid deforestation in recent years. The habitat changes initiated by elephants necessarily have had an affect on the relative abundance of other animals, with grazers increasing and browsers decreasing as the woody plant cover has given way to herbaceous vegetation. This is demonstrated by Boughey (1963) for secondary plant communities derived from the destruction of teak (*Baikiaea plurijuga*) woodlands and other woodland types on Kalahari Sands in southwest Zimbabwe (Fig. 14.7(e)).

Wildlife is the mainstay of tourism in some African countries. In addition, it is a potentially valuable source of food, especially in semiarid regions which are unsuited to crop production (Coe, 1980). The ecological basis of 'game ranching' relates to the use of a wide variety of woody and herbaceous species by different wild animals and the fact that these animals are generally better adapted to their environment than domestic livestock (Walker, 1979). Game ranching has been carried out on a commercial level in Zimbabwe for some twenty years now, but very few other African countries have managed to pursue this activity successfully. Part of the reason for this is the need for a sophisticated, rather different type of management to that used for domestic livestock (Pullan, 1981). However, exploitation of wildlife could prove more rewarding economically and less damaging ecologically than attempts to increase livestock production in the drier savanna regions (Jewell, 1980).

14.7 CONCLUSION

It is evident that human activity, directly or indirectly, has brought about massive changes in the character of the African vegetation. What emerges from this chapter is the increasing rate and destructive nature of changes in the plant cover of Africa and it is pertinent to comment on conservation issues in Africa.

Two aspects need to be stressed with respect to the African flora. Firstly, knowledge of the plant species, their distribution and ecology is far from complete. Extensive areas of the continent have barely been explored let alone subjected to systematic collection and description of their flora or mapping of the main plant communities (Fig. 14.8(a)). Secondly, man's activities are posing a major and increasing threat to plant species in certain parts of Africa. For example, there are large numbers of endangered or vulnerable species in the Cape region of South Africa (Fig. 14.8(b)), whilst White (1981) reports that many isolated remnants of montane forest species in East and Central Africa are faced with destruction. The essence of the problem is summarized by Hepper (1979, p. 41) who states that throughout Africa 'detailed taxonomic research is being overtaken by events which are exterminating the vegetation with its constituent species before thorough exploration has been undertaken'.

In the face of growing population pressures and the demands of development, the prospects for the African flora are not good. Hedberg (1979), for example, regards knowledge of the flora and organizations to implement conservation measures as vital prerequisites for effective protection of African plants. The first of these, as noted above, is incomplete or inadequate in many areas. The

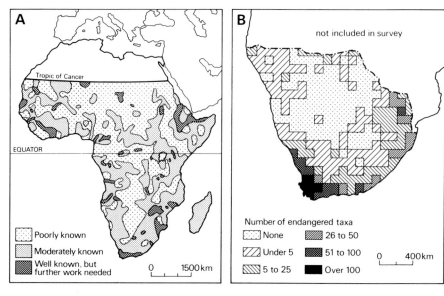

FIGURE 14.8   Knowledge of African flora and threatened plants. (a) Extent of floristic exploration in Africa (based upon Hepper, 1979); (b) Endangered plants in southern Africa (following Hall *et al.*, 1984)

second prerequisite exists in some countries but not in others. A major problem here is that plants are often treated as secondary to animals and soils in conservation legislation and programmes. The conservation effort as a whole is hampered by the fact that 'conservationists in Africa are often working against immense difficulties. Staff shortages, lack of training, limited financial resources, and above all, lack of tradition, make conservation of nature very difficult' (Hedberg, 1974, p. 441). Moreover, conservation is often viewed as anti-development, rather than being the rational exploitation of resources. Consequently, conservation activities lack the political support needed to ensure their effectiveness (see Chapter 16).

What arguments, then, can be put forward to protect vegetation in Africa? Verdcourt (1968), in a symposium on 'Conservation of Vegetation in Africa South of the Sahara', describes direct and indirect uses of plants as being valid reasons for conservation. Plants are used directly for a variety of purposes including foodstuffs, drugs, fibres. The potential of many African plants is yet to be established, so that to destroy them before their benefits can be assessed and exploited would be foolish. The indirect uses, if anything, offer more forceful reasons for conservation. Such uses include the prevention of soil erosion and maintenance of water supplies, aspects which are clearly of practical relevance in Africa (Whitlow, 1983). Indiscriminate denudation of the plant cover often marks the early stages of land degradation and a progressive lowering of productivity. This, in turn, has a negative feedback on human and livestock populations. The most compelling reason for conserving vegetation is *survival*.

The ecological principles for effective long-term utilization of the plant cover have been established for a long time (Dasmann *et al.*, 1973). The problem lies in translating these into practice within different environmental and socioeconomic settings in Africa. A multidisciplinary approach to this problem has been developed by Walker *et al.* (1978) with particular reference to the needs and circumstances of Africa. Conservation must become an integral part of development rather than being tacked on as an afterthought or, even worse, being ignored completely. This conservation should, however, be orientated towards rational exploitation rather than preservation, since it is only through its utilitarian value that conservation can be justified in poverty-stricken societies as occur in Africa. Nevertheless, it takes time to change people's attitudes and to introduce practical measures to manage the plant cover more effectively. Unfortunately, Africa does not have time to spare. The burgeoning human and livestock populations are placing increasing stresses on African habitats. The short-term prognosis must be continued degradation of the plant cover to the detriment of man and his animals.

REFERENCES

Acocks, J. P. H., 1964, 'Karoo vegetation in relation to the development of deserts', in Davies, D. H. S. (Ed.), *Ecological Studies in Southern Africa*, Dr. W. Junk, The Hague, pp. 100–112.

Acocks, J. P. H., 1975, 'Veld types of South Africa', *Memoirs of the Botanical Survey of South Africa*, no. 40.

Adeyoju, S. K., 1965, 'The forest resources of Nigeria', *Nigerian Geogr. Jnl.*, **8**(2), 115–126.

Allan, W., 1965, *The African Husbandman*, Oliver & Boyd, Edinburgh.

Arnold, J. E. M., and Jongma, J., 1978, 'Fuelwood and charcoal in developing countries', *Unasylva*, **29**, 2–9.

Aubreville, A. M. A., 1974, 'The disappearance of the tropical forests of Africa', *Unasylva*, **1**(1), 5–11.

Aweto, A. O., 1981, 'An ecological study of forest fallow communities in the Ijebu-Ode/Shagamu area, southwestern Nigeria', *Singapore Jnl. Trop. Geog.*, **2**(1), 1–8.

Batchelder, R. B., 1967, 'Spatial and temporal patterns of fire in the tropical world', *Proc. Tall Timbers Fire Ecol. Conf.*, Tallahassee, Florida, vol. 6, 171–206.

Blair Rains, A., 1982, 'African pastoralism', *Outlook on Agriculture*, **11**(3), 96–103.

Boughey, A. S., 1963, 'Interaction between animals, vegetation and fire in Southern Rhodesia', *Ohio Jnl. Sci.*, **63**(5), 193–209.

Bourn, D., 1978, 'Cattle, rainfall and tsetse in Africa', *Jnl. Arid Env.*, **1**, 49–61.

Bridges, E. M., 1978, 'Soil, the vital skin of the earth', *Geography*, **63**(4), 354–361.

Brown, L. H., 1971, 'The biology of pastoral man as a factor in conservation', *Biol. Cons.*, **3**(2), 93–100.

Butzer, K. W., 1977, 'Environment, culture and human evolution', *Amer. Scientist*, **65**, 572–584.

Chauvet, B., 1972, 'The forests of Madagascar', in Battistini R., and Richard-Vindard, G. (Eds), *Biogeography and Ecology of Madagascar*, Dr. W. Junk, The Hague, pp. 191–199.

Clark, J. D., 1980, 'Early human occupation of African savanna environments', in Harris, D. (Ed.), *Human Ecology in Savanna Environments*, Academic Press, London, pp. 41–71.

Clayton, W. D., 1958, 'Secondary vegetation and the transition to savannah near Ibadan, Nigeria', *Jnl. Ecol.*, **46**, 217–238.

Coe, M., 1980, 'African wildlife resources', in Soule, M. E., and Wilcox, B. A. (Ed.), *Conservation Biology–An Evolutionary : Ecological Perspective*, Sinauer, Sunderland, pp. 273–302.

Cumming, D. H. M., 1979, 'The influence of large herbivores on savanna structure in Africa', S. A. Savanna Ecosystems Workshop, CSIR, Pretoria, May (1979).

Cumming, D. H. M., 1981, 'The management of elephant and other large mammals in Zimbabwe', in Jewell, P. A., Holt, S., and Hart, D. (Eds), *Problems in Management of Locally Abundant Wild Mammals*, Academic Press, New York, pp. 91–118.

Dale, I. R., 1954, 'Forest spread and climatic change in Uganda during the Christian era', *Empire For. Rev.*, **33**(1), 23–29.

Dasmann, R. F., 1979, 'Biosphere exploitation and conservation: a world picture' in Hedberg, I. (Ed.), *Systematic Botany, Plant Utilization and Biosphere Conservation*, Almqvist & Wiksell, Stockholm, pp. 155–157.

Dasmann, R. F., Milton, J. P., and Freeman, P. H., 1973. *Ecological Principles for Economic Development*, John Wiley, London.

De Vos, A., 1975, *Africa, The Devastated Continent? Man's Impact on the Ecology of Africa*, Dr. W. Junk, The Hague.

Eckholm, E., 1977, 'The shrinking forests', *Focus*, **28**(1), 12–16.

ENDA, 1981, *Environment and Development in Africa*, United Nations Environmental Program, Pergamon Press, Oxford.

FAO, 1977, *World Map of Desertification: Explanatory Note*. FAO and UNESCO, Paris.

FAO, 1981, 'Fuelwood supplies in the Third World', *Ambio.* **10**(5), 236–237.

Flenley, J. R., 1979, *The Equatorial Rain Forest: A Geological History*, Butterworths, London.

Fontaine, R. G., Gomez-Pompa, A., and Ludlow, B., 1978, 'Secondary successions', in *Humid Tropical Ecosystems*, Natural Resources Research, 14, UNESCO, Paris.

Glover, P. E., 1968, 'The role of fire and other influences on the savannah habitat, with suggestions for further research', *East Afr. Wildlife Jnl.*, **6**, 131–137.

Goudie, A., 1981, *The Human Impact: Man's Role in Environmental Change*, Blackwell, Oxford.

Graaff, G. De, 1981, 'Republic of South Africa', in Kormondy, B. J., and McCormick, J. F. (Eds), *Handbook of Contemporary Developments in World Ecology*, pp. 485–510.

Hall, A. V., De Winter, B., Fourie, S. P., and Arnold, T. H., 1984, 'Threatened plants in Southern Africa', *Biol. Cons.*, **28**, 5–20.

Harlan, J. R., De Wet, J. M. J., and Stemler, A. B. L., 1976, *Origins of African Plant Domestication*, Mouton, The Hague.

Hedberg, I., 1974, 'Follow-up of the AETFAT meeting at Uppsala in 1966 on "Conservation of vegetation in Africa south of the Sahara"', *Boissiera*, **24**, 437–441.

Hedberg, I., 1979, 'Possibilities and needs for conservation of plant species and vegetation in Africa', in Hedberg, I. (Ed.), *Systematic Botany, Plant Utilization and Biosphere Conservation*, Almqvist & Wiksell, Stockholm, pp. 83–104.

Hepper, F. N., 1979, 'Africa', in Hedberg, I. (Ed.), *Systematic Botany, Plant Utilization and Biosphere Conservation*, Almqvist & Wiksell, Stockholm, pp. 41–46.

Hopkins, B., 1965, *Forest and Savanna*, Heinemann, London.

Jackson, P., 1982, 'The future of elephants and rhinos in Africa', *Ambio* 9(4), 202–205.

Jewell, P. A., 1980, 'Ecology and management of game animals and domestic livestock in African savannas', in Harris, D. R. (Ed.), *Human Ecology in Savanna Environments*, Academic Press, London, pp. 353–381.

Joubert, D. M., 1977, 'Ecological effects of fire: an overview', *S. Afr. Jnl. Science*, **73**, 166–169.

Kay, G., 1975, 'Population pressure and development prospects in Rhodesia', *Rhodesia Science News*, **9**(1), 7–13.

Keay, R. W. J., 1959, *Vegetation Map of Africa: Explanatory Notes*, Oxford University Press, Oxford.

Keay, R. W. J., 1974, 'Changes in African vegetation', *Env. and Change*, **2**, 387–394.

Kelly, R. D., and Walker, B. H., 1976, 'The effects of different forms of land use on the ecology of a semi-arid region in south-eastern Rhodesia', *Jnl. Ecol.*, **64**, 553–576.

Kingdon, J., 1982, *East African Mammals: An Atlas of Evolution in Africa*, Volume III D (Borids), Appendix I — Conservation, Academic Press, London.

Komarek, E. V., 1971, 'Lightning and fire ecology in Africa', *Proc. Tall Timbers Fire Ecol. Conf.*, Tallahassee, Florida, **11**, pp. 473–512.

Lamprey, H. F., 1963, 'Ecological separation of the large mammal species in the Tarangire game reserve, Tanzania', *East African Wildlife Jnl.*, **1**, 63–92.

Langlands, B. W., 1967, 'Burning in East Africa, with particular reference to Uganda', *East Afr. Geogr. Rev.*, **5**, 21–37.

Laws, R. M., 1970, 'Elephants as agents of habitat and landscape change in East Africa', *Oikos*, **21**(1), 1–15.

Lawton, R. M., 1972, 'A vegetation survey of Northern Zambia', *Palaeoecology of Africa*, **6**, 253–256.

Lawton, R. M., 1978, 'A study of the dynamic ecology of Zambian vegetation', *Jnl. Ecol.*, **66**, 175–198.

Lind, E. M., and Morrison, M. E. S., 1974, *East African Vegetation*, Longman, London.

Linear, M., 1982, 'Gift of poison — the unacceptable face of development aid', *Ambio*, **11**(1), 2–8.

Lock, J. M., and Milburn, T. R., 1970, 'The seed biology of Themeda triandra Forsk in relation to fire', in Duffey, E., and Watt, A. S. (Eds), *The Scientific Management of Animal and Plant Communities for Conservation*, Blackwell, Oxford, pp. 337–349.

May, J. M., 1974, 'The geography of malnutrition in Africa south of the Sahara', *Focus*, **25**(1), 1–10.

Mikesell, M. W., 1960, 'Deforestation in Northern Morocco', *Science*, **132**(3425), 441–448.

Mikesell, M. W., 1969, 'The deforestation of Mount Lebanon', *Geogr. Review*, **59**(1), 1–28.

Morrison, M. E. S., and Hamilton, A. C., 1974, 'Vegetation and climate in the uplands of south-western Uganda during the later Pleistocene period. II — forest clearance and other vegetational changes in the Rukiga Highlands during the last 8000 years', *Jnl. Ecol.*, **62**, 1–32.

Moss, R. P., and Morgan, W. B., 1970, 'Soils, plants and farmers in West Africa', in Garlick, J. P., and Keay, R. W. J. (Eds), *Human Ecology in the Tropics*, Pergamon Press, Oxford, pp. 1–31.

Murray-Rust, D. H., 1972, 'Soil erosion and reservoir sedimentation in a grazing area west of Arusha, Northern Tanzania', *Geografiska Annaler*, **54A**, 325–343.

Myers, N., 1972, 'National parks in savannah Africa', *Science*, **178**, 1255–1263.

Myers, N., 1973, 'Tsavo National Park, Kenya, and its elephants: an interim appraisal', *Biol. Cons.*, **5**(2), 123–132.

Naveh, Z., 1975, 'The evolutionary significance of fire in the Mediterranean region', *Vegetatio*, **29**(3), 199–208.

Obeid, M., and Seif El Din, A., 1970, 'Ecological studies of the vegetation of the Sudan. I. *Acacia senegal* (L.) Wild. and its natural regeneration', *Jnl. Appl. Ecol.*, **7**, 507–518.

Ormerod, W. E., 1976, 'Ecological effect of control of African trypanosomiasis', *Science*, **191**, 815–821.

Pellew, R. A. P., 1983, 'The impacts of elephant, giraffe and fire upon the *Acacia tortilis* woodlands of the Serengeti', *Afr. Jnl. Ecol.*, **21**, 41–74.

Phillips, J. F. V., 1965, 'Fire as master and servant: its influence in the bioclimatic regions of Trans-Saharan Africa', *Proc. Tall Timbers Fire Ecol. Conf.*, Vol 5, 7–109.

Phillips, J., 1974, 'Effects of fire in forest and savanna ecosystems of sub-Saharan Africa', in Kozlowski, T. T., and Ahlgren, C. E. (Eds), *Fire and Ecosystems*, Academic Press, New York.

Pullan, R. A., 1974a, 'Burning impact on African savannas', *Geogr. Mag.*, **47**(7), 432–438.

Pullan, R. A., 1974b, 'Farmed parkland in West Africa', *Savanna*, **3**(2), 119–151.

Pullan, R. A., 1974c, 'Farmed parkland in Zambia', *Zambia Geogr. Assoc. Mag.*, **26**, 1–17.

Pullan, R. A., 1979, 'Termite hills in Africa: their characteristics and evolution', *Catena*, **6**, 267–291.

Pullan, R. A., 1981, 'The utilization of wildlife for food in Africa: The Zambian experience', *Singapore Jnl. Trop. Geog.*, **2**(2), 101–113.

Puzo, B., 1978, 'Patterns of man–land relations', in Werger, M. J. A. (Ed.), *Biogeography and Ecology of Southern Africa*, Dr. W. Junk, The Hague, pp. 1049–1112.

Rapp, A., 1974, 'A review of desertization in Africa—water, vegetation and man', Secretariat for International Ecology, Stockholm, Report no. 1.

Rose-Innes, R., 1971, 'Fire in West African vegetation', *Proc. Tall Timbers Fire Ecol. Conf.*, Tallahassee, Florida, Vol. 11, pp. 147–173.

Soulé, M. E., Wilcox, B. A., and Holtby, C., 1979, 'Benign neglect: a model of faunal collapse in the game reserves of East Africa', *Biol. Cons.*, **15**, 259–273.

Stewart, O. C., 1956, 'Fire as the first great force employed by man', in Thomas, W. L. (Ed.), *Man's Role in Changing the Face of the Earth*, University of Chicago Press, Chicago, pp. 115–133.

Stirton, C. H. (Ed.), 1978, *Plant Invaders: Beautiful but Dangerous*, Department of Nature and Environmental Conservation of the Cape Province Administration, Cape Town.

Stott, P. A., 1978, 'Tropical rain forest in recent ecological thought: the reassessment of a non-renewable resource', *Progress in Phys. Geog.*, **2**(1), 80–98.

Strang, R. M., 1973, 'Bush encroachment and veld management in South-central Africa: the need for a reappraisal', *Biol. Cons.*, **5**(2), 96–104.

Stuart-Hill, G. C., and Mentis, M. T., 1982, 'Coevolution of African grasses and large herbivores', *Proc. Grassld. Soc. South Afr.*, **17**, 122–128.

Swift, J., 1972, 'What future for African national parks', *New Scientist*, July, 192–194.

Talbot, L. M., 1973, 'Ecological consequences of rangeland development in Masailand, East Africa', in Farvar, M. T. and Milton, J. P. (Eds), *The Careless Technology*, Tom Stacy, London, pp. 694–711.

Temple, P. H., and Rapp, A., 1972, 'Landslides in the Mgeta area, Western Uluguru Mountains, Tanzania', *Geografiska Annaler*, **54A**, 157–193.

Thirgood, J. V., 1981, *Man and the Mediterranean Forest: A History of Resource Depletion*, Academic Press, London.

Trapnell, C. G., 1959, 'Ecological results of woodland burning experiments in Northern Rhodesia', *Jnl. Ecol.*, **47**, 129–168.

Turner, B. J., 1967, 'Ecological problems of cattle ranching in Combretum savanna woodland in Uganda', *E. Afr. Geogr. Rev.*, **5**, 9–19.

Udo, R. K., 1982, *The Human Geography of Tropical Africa*, Heinemann, London.

Vaughan, R. E., and Wiehe, P. O., 1937, 'Studies on the vegetation of Mauritius. I—A preliminary survey of the plant communities', *Jnl. Ecol.*, **25**(2), 289–343.

Verdcourt, B., 1968, 'Why conserve natural vegetation?' *Acta Phytogeographica Suecica*, **54**, 1–9.

Walker B. H., 1979, 'Game ranching in Africa', in Walker, B. H. (Ed.), *Management of Semi-Arid Ecosystems*, Elsevier, Amsterdam, pp. 55–81.

Walker, B. H., Norton, G. A., Conway, G. R., Comins, H. N., and Birley, M., 1978, 'A procedure for multi-disciplinary ecosystem research: with reference to the South African savanna ecosystem project', *Jnl. Appl. Ecol.*, **15**, 481–501.

Walker, B. H., Ludwig, D., Holling, C. S. and Peterman, R. M., 1981, 'Stability of semi-arid savanna grazing systems', *Jnl. Ecol.*, **69**, 473–498.

Warren, A., and Maizels, J. K., 1977, 'Ecological change and desertification', in Secretariat of UN Conference on Desertification (Ed.), *Desertification: its Causes and Consequences*, Pergamon Press, Oxford, pp. 169–259.

West, O., 1971, 'Fire, man and wildlife as interacting factors limiting the development of climax vegetation in Rhodesia', *Proc. Tall Timbers Fire Ecol. Conf.*, **11**, 121–146.

Western, D., and Van Praet, C., 1973, 'Cyclical changes in the habitat and climate of an East African ecosystem', *Nature*, **241**, 104–106.

White, F., 1981, 'The history of the Afromontane archipelago and the scientific need for its conservation', *Afr. Jnl. Ecol.*, **19**, 33–54.

Whitlow, J. R., 1979a, 'Deforestation—some global and national perspectives', *Proceedings of Geog. Assoc. of Zimbabwe*, **12**, 13–30.

Whitlow, J. R., 1979b, *The Household Use of Woodland Resources in Rural Areas*, Natural Resources Board, Harare.

Whitlow, J. R., 1979c, 'A scenario of changes in subsistence land use and its relevance to the tribal areas of Zimbabwe', *Zambezia*, **7**(2), 171–190.

Whitlow, J. R., 1980a, 'Deforestation in Zimbabwe: problems and prospects', *Supplement to Zambezia*, University of Zimbabwe.

Whitlow, J. R., 1980b, 'Land use, population pressure and rock outcrops in the tribal areas of Zimbabwe Rhodesia', *Zim. Rho. Agric. Jnl.*, **77**(1), 3–11.

Whitlow, J. R., 1983, 'Hydrological implications of land use in Africa, with particular reference to Zimbabwe', *Zim. Agric. Jnl.*, **80**(5), 193–212.

Whitlow, J. R., 1984, 'The study of vegetation in Africa: a historical review of problems and progress', *Singapore Jnl. Trop. Geog.*, **5**(1) (in press).

Whyte, R. O., 1974, *Tropical Grazing Lands—Communities and Constituent Species*, Dr. W. Junk, The Hague.

Wickens, G. E., 1976, 'Speculations on long-distance dispersal and the flora of Jebel Marra, Sudan Republic', *Kew Bull.*, **31**, 105–150.

Wild, H., 1968, 'Weeds and Aliens in Africa: The American Immigrant', Inaugural lecture, University of College of Rhodesia, Salisbury.

Wilkinson, H., 1963, 'Man and the Natural Environment', *Occasional Papers in Geography*, no. 1. Department of Geography, University of Hull.

Zumer-Linden, M., 1976, 'Botswana', *Ecol. Bull.*, **24**, 171–187.

Human Activity and Environmental Processes
Edited by K. J. Gregory and D. E. Walling
©1987 John Wiley & Sons Ltd.

15

Human Activity and
Species Response

C. M. HARRISON

15.1 INTRODUCTION

One of the most intractable problems facing ecologists and biogeographers is the difficulty of reconciling what is known to be an immense potential amongst natural organisms for generating new species, with the apparent failure of species to adapt to those environmental changes associated with human activity. So, for example, the list of species which have become extinct in historic time is a long one and the list of endangered species is lengthening (IUCN, 1978). By contrast, there are comparatively few species whose ranges have expanded to occupy habitats associated with human disturbance and even fewer new species have evolved to take advantage of these areas. And yet, in certain environments some species are able to acquire immunity and resistance to synthetic pesticides faster than chemists can manufacture new formulations (Bull, 1982), and recent developments in the field of genetic engineering have opened up all kinds of new possibilities for creating novel species in the laboratory. These conflicting trends, the one suggesting the vulnerability of species to environmental change associated with modern civilization, and the other suggesting the ability of certain natural and artificial processes to promote novelty, prompt a reappraisal of the role of natural variation and natural selection in the evolutionary process and, in particular, of the role of human disturbance in the creation of new fit forms and the extinction of others. The aim of this chapter is to illustrate how a combination of different methods of study drawn from the closely related disciplines of ecology, geography and genetics can provide a more comprehensive understanding of how organisms respond to human impact than is the case when each discipline is studied in isolation.

The chapter starts with a brief introduction to natural variation and to the role of natural selection in promoting novelty. It moves on to examine how various studies of environmental change provide support for the view that large-scale disturbances are a repeated feature of the environment to which plant and animal populations and assemblages have to respond. It also shows how many ecologists and geneticists have come to accept that, at least in the middle latitudes, species respond to environmental disturbance in an individualistic manner, such that some individuals appear to be preadapted to change and take advantage of those new opportunities for colonization that emerge after disturbance. Such findings suggest that, if natural selection favours those individuals who transmit copies of their genes to future generations most abundantly, individuals in a changing environment are faced with the task of allocating limited resources between survival and reproduction. The result of this allocation, is the life history. The third section of the chapter looks more closely at the concept of bionomic strategies and the theory of adaptive life histories. It is suggested that these concepts provide a new means of approaching ecology from an evolutionary perspective which may have particular merit for those concerned with both the long-term future of species survival and for those concerned with managing and predicting the outcome of particular disturbance regimes associated with human activity. A final section illustrates how these new approaches may be applied to the ecology and biogeography of organisms which inhabit urban areas. The urban area is used because, of all the environments modified by human activity, city areas must rank among the most strongly modified and disturbed. Organisms which live in the urban area might be expected to exhibit particular life-histories indicative of these conditions and an understanding of how these species adapt to the human impact might be applicable to other habitats subject to change.

15.2　NATURAL VARIATION AND NATURAL SELECTION

All plants and animals are subject to the process of natural selection through competition with neighbours, other species or predation. The outcome of natural selection is measured by reproductive success or failure, or more correctly as the differential change in the relative frequency of genotypes due to differences in the ability of their phenotypes to obtain expression in the next generation (Berry, 1982). Natural selection will only serve to promote new fit forms when accompanied by environmental change, for in practice, natural selection acts as a stabilizing influence by removing inefficient offspring. In this context many human activities, such as intensive cultivation or urbanization, can be regarded as increasing selection pressure over and above that normally experienced by wild organisms. Once it is appreciated that even small selection pressures can account for major evolutionary changes, as for example with the doubling of the human skull capacity between *Australopithecus* and modern man, which required a mean selection pressure of only 0.04 per cent per generation, then the high selective pressures encountered in habitats strongly modified by human activities may be expected to promote novelty at the expense of less-fit forms.

Novelty is introduced when some character that is preadapted to new environmental circumstances achieves a higher fitness than its neighbours. In other words, new forms are opportunists and take advantage of favourable conditions available for colonization. Forms that are not preadapted may become locally extinct. There are now numerous examples to illustrate this form of directional selection. Bradshaw (1978) for example has shown that heavy metal-tolerance is widespread amongst natural populations of grasses such as *Agrostis* species, but tolerant phenotypes come to predominate in derelict areas of toxic waste, where the selective pressure is measured at 45–65 per cent. The poor survival and abundance of these metal-tolerant forms in normal swards can be attributed to their low tolerance of crowding. Similarly, the well-known phenomenon of industrial melanism in the peppered-moth (*Biston betularia*) can be explained in part by selective predation by birds on the non-melanic forms. Cryptic colouration provides protection from predation and melanic forms which normally occur infrequently thoughout Britain come to predominate in industrial areas. Both these examples illustrate how selection in disturbed environments can promote extreme phenotypic forms at the expense of average forms and in this sense, selection can be said to be directional. Normally, selection favours intermediate forms and hence serves to preserve the status quo. Both processes work on populations, but environmental change and disturbance may be expected to favour directional selection rather than stabilization selection. Indeed, the fossil record seems to support the view expressed by Gould and Eldredge (1977) which states that evolution proceeds as series of punctuated equilibria. According to this view, periods of rapid speciation accompany phases of environmental disturbance and are followed by periods of stabilization. However, it is not clear whether the processes of directional selection which operate at the micro-level, that is amongst local populations of a single species, can also lead to the formation of new species during periods of rapid speciation. New species have evolved in short periods of tens of years, as for example on the Hawaiian islands where new species of *Drosophila* evolved in association with the recent introduction of the banana, and this island group provides remarkable evidence of the ability of a few founding individuals to give rise to hundreds of new species over longer periods of thousands of years (Carson and Templeton, 1984). But the processes at work on isolated, oceanic islands are likely to be very different from those at work on large, continental areas and the case of the Hawaiian islands cannot be taken as the general rule (Loveless and Hamrick, 1984).

Such examples leave unanswered the question of how new species arise during periods of rapid speciation, but they do serve to reinforce two basic tenets of Neo-Darwinism, namely that wild populations exhibit immense natural variation and that natural selection appears to be the strongest agency which changes gene frequency. Geneticists have not been slow to explore the application of this knowledge to the development of new fit forms in the laboratory, and biotechnology, based on recombinant DNA, has opened up new opportunities for improving upon natural selection (Warr, 1984). Gene cloning, whereby desirable genes of a donor are inserted into the DNA of plasmid vectors and are replicated inside host bacteria, has been achieved at the cellular level, but advances at the level of whole organisms, especially amongst the higher organisms, are compounded by the very real difficulties encountered in identifying the complex of genes which interact to produce desirable traits. Many important traits are not the product of single genes, unlike pesticide resistance, for example, which is determined by a single gene locus. However, progress in engineering new plant strains has been rapid in the last ten years largely because whole plants can be regenerated from single cells grown in culture (Cross, 1984). Callus cultures, which use tumour-inducing vectors to infect plant cells with desirable genes, have proved to be very effective. Like their wild relatives, regenerated plants exhibit a high frequency of variation in their phenotypes, and the chances of encountering new variants which possess desirable traits are enormously enhanced when artificial selection is made from many thousands of donor cells. In this way, disease-resistant properties can be selected for and combined with high-yielding properties. So far, it has proved difficult to engineer new crop varieties amongst monocotyledons (Chilton, 1983), but it is only a matter of time before real benefits will accrue in this area too. Nevertheless, because even these advanced techniques depend upon the initial natural variation exhibited by the parent cells, genetic engineering has to be supported by a programme of genetic conservation. Only by preserving natural variation in gene banks which include representatives of old crop varieties and their wild ancestors, will genetic engineers have an adequate source of novelty with which to work.

15.3   ENVIRONMENTAL CHANGE

During the 1970s and 1980s, the realization that any population of plants or animals possessed a high degree of phenotypic and genotypic variation was coupled with an increased awareness of the variability of environmental conditions. This reappraisal of the assumed constancy of environmental conditions suggested that organisms and environment were not necessarily in equilibrium. Moreover, the high variability of local populations of organisms was likely to be matched by an equally variable, local environment. Ecologists who accepted these premises also came to challenge the conventional wisdom of co-evolution, whereby organisms with a common evolutionary history were thought to have become organized into functionally linked communities through the prevalence of inter-specific interactions. Rather than accepting the existence of plant and animal communities which behaved as if they were super-organisms, many ecologists emphasized the individualistic nature of species response and began to challenge the significance of competition and species interactions as determinants of the structure and composition of assemblages. According to these ideas, species assemblages were viewed as the product of stochastic processes and, over time, their species composition and structure changed greatly. Such assemblages exhibited no tendency of persistence or resilience. Stochasticity, it should be emphasized, does not imply lack of causation, merely a lack of consistent patterning which would be required were the concept of community organization to be invoked. Not all ecologists have come to accept the individualistic nature of the species association (see, for example, Roughgarden (1984)) but many of its proponents suggest that the burden of proof now lies with those favouring the concept of community organization (Strong, 1984). In other words, there is a requirement to demonstrate interaction and mutual dependency amongst species before the concept of community organization can be accepted uncritically.

The individualistic hypothesis was first advanced by Gleason in the 1920s (Gleason, 1926) and renewed interest in its usefulness as a theory capable of explaining species change and vegetation dynamics has been expressed both by ecologists working in a variety of different habitats subject to known disturbance regimes (Sousa, 1984) and by palaeoecologists concerned with the reconstruction of former environments (Walker, 1982).

(a)   Palaeoecological studies of forest history

A notable example of how the pollen record has been used to provide evidence in support of the individualistic hypothesis, is found in the work of Davis (1981) in North America. Her comprehensive studies of the Quaternary history of North America reveal that forests seldom maintain a constant species composition for more than two to three hundred years at a time. Forest assemblages, as revealed by the palynological record for the temperate areas, are most readily interpreted as chance combination of species without an evolutionary history. In New Hampshire for example, there is no simple expansion northwards of the forest in parallel with the movement of isotherms after the withdrawal of the ice front, because the migration routes of different species differ greatly. In practice, the composition of vegetation at any one point in time or space appears to be largely controlled by species' immigration rates during the Quaternary. Boreal species such as spruce (*Picea species*), larch (*Larix laricina*), balsam fir (*Abies balsamea*) and Jack/Red pine (*Pinus banksiana/resinosa*), which today have overlapping ranges (see Fig. 15.1), exhibited markedly different patterns and rates of migration. For example, the Appalachian Mountains proved to be a barrier to some species and not to others, whilst the availability of seed presumably controlled rates of disperal as well. Davis's conclusion that during much of the Quaternary period (1.7 million years) forest species adjusted their ranges individually to alternating glacial and interglacial conditions lends support to the view that species occurred in loose associations and not as tight complexes that responded as units to environmental change. They also serve

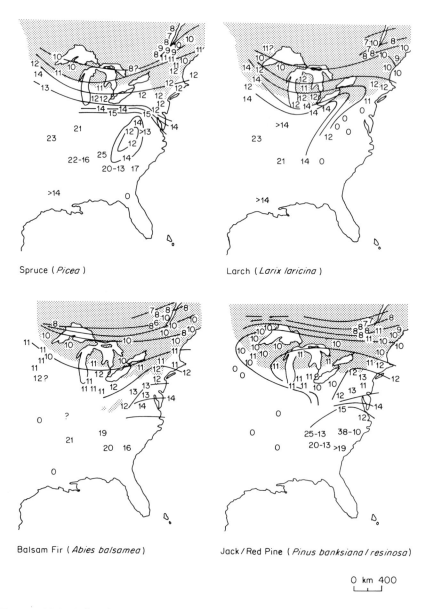

FIGURE 15.1 Migration maps of forest species. The numbers refer to the radio-carbon age (in thousands of years) of the first appearance of each tree after 15 000 years ago. Isopleths are drawn to connect points of similar age and represent the leading edge of the expanding tree population. The stippled area represents the modern range of each tree species. (*From West* et al., *1981. Reproduced by permission of Springer-Verlag*)

to suggest that the distribution of many species were in disequilibrium with climate. Factors other than climate appear to have influenced dispersal and colonization rates and additional knowledge about the mechanisms which enable species to disperse and invade an established forest are required before a complete understanding of the fluctuating nature of forest histories will be gained.

(b)   Modern studies of forest ecology

Palaeoecological studies such as those of Davis provide a much needed temporal dimension to studies that attempt to assess the significance of human disturbance and its relationship to environmental change (see also Battarbee, 1984, and Battarbee *et al.*, 1985). Other work on modern forests which has sought to determine the successional status of different forest types (West *et al.*, 1981) also lends support to an interpretation of forest dynamics that is based on the chance occurrence of disturbances factors and the known ecological limits of particular species. Indeed, the weight of current opinion suggests that most modern forests are subject to chronic, synchronous, large-scale disturbances during their life histories. Disturbances such as fire, hurricanes, volcanic eruptions, severe frost and disease all play their part. Horn (1981) suggests that on this basis the pattern of species colonization which succeeds each disturbance can be interpreted, in some instances, as accidents of history while, for other successions, a pattern can be interpreted in terms of the ecology of the species present. Horn's observations made as a result of several studies of temperate forests of known disturbance histories, show that the progress of secondary succession is dependent upon several processes — processes that include the type and frequency of the disturbance regime, regeneration and replacement opportunities, competitive relationships between species and the facilitation of the invasion of one species by another. All of these processes can be observed in modern forests and are likely to have also played their role in ancient forests, as a result the simple expectation of convergent secondary succession, in which adjacent stands develop towards a recognizable, stable state of similar species composition, is clearly a remote possibility.

Horn's conclusions have been reinforced by others who have examined the nature of succession, for example Connell and Slatyer (1977) and Miles (1979), and the growing acceptance of these ideas amongst ecologists has spawned a host of simulation models of succession that incorporate stochastic processes and the individualistic nature of species' responses (Table 15.1). Many of these models handle the ecological characteristics of upwards of thirty species in respect of dispersal properties, establishment opportunities, growth rates and probabilities of mortality. For example, Shugart *et al.* (1981) discuss four sets of forest simulation models that have been devised for use with forests as diverse as those of the subtropics of Australia and the mixed deciduous and coniferous forests of North America. One of the main attractions of such models is that

TABLE 15.1   Brief descriptions and testing of the four forest simulation models

Model	Verification
FORET (Shugart and West, 1977) Model simulates dynamics of 33 tree species typical of Southern Appalachian deciduous forests. Nominal locations of simulated stands are on moist sites in Anderson Co., Tennessee.	1. Prediction of composition of post-chestnut-blight forests.
BRIND (Noble *et al.*, 1980 Shugart and Noble, 1981) Model simulates dynamics of 18 arborescent species found on southeast-facing slopes from 900- to 2400-m elevations in the Brindabella Range, Australian Capital Territory. Nominal locations of simulated stands are in the Cotter catchment near Canberra, A.C.T. All forests are dominated by *Eucalyptus*.	1. Prediction of qualitative pattern of tree species replacement at 1600-m elevations in the Brindabella Range under different wildfire frequencies.
FORAR (Mielke *et al.*, 1977; Mielke *et al.*, 1978) Model simulates 33 species found on upland sites in Union Co., Arkansas. Forest are of a mixed oak-pine type.	1. Prediction of forest composition based on an 1859 reconnaissance (Owen, 1860) of southern Arkansas. 2. Prediction of yield tables for loblolly pine (*Pinus taeda*)
KIAMBRAM (Shugart *et al.*, 1980 Shugart *et al.*, 1981b) Model simulates the dynamics of 125 species found in the subtropical rain forest in the vicinity of the New South Wales–Queensland border. Nominal locations of the simulated stands are Wiangaree State Forest, New South Wales.	1. Comparisons with composition of stands of known age in Lamington National Park, Queensland.

Validation	Application
1. Prediction of composition of pre-chestnut-blight forests.	1. Prediction of chronic atmospheric pollutant effects on forests of Southeastern U.S. (West *et al.*, 1980; McLaughlin *et al.*, 1978). 2. Reconstruction of 16 000-year pollen chronology under changing climatic conditions (Solomon *et al.*, 1980). 3. Prediction of habitat dynamics for non-game birds following harvest of select hardwood trees. (Smith *et al.*, 1981).
1. Prediction of fire-succession response of communities at different altitudes in the Brindabella Range. 2. Prediction of altitudinal zonation of forests from 900- to 2400-m in the Brindabella Range. 3. Prediction of mean dbh, basal area and stocking density for alpine ash (*Eucalyptus delegatensis*) stands.	1. Model is currently being tested for its utility in managing the watersheds supplying Canberra, A.C.T.
1. Prediction of abundances of trees by species on uplands in southern Arkansas. 2. Prediction of density by diameter class of upland forests in southern Arkansas.	1. Design of tree management schemes for providing habitat for the rare and endangered bird, the redcockaded woodpecker (*Dendrocopus borealis*). (Mielke *et al.*, 1978;
1. Prediction of abundances and basal areas of trees in mature forests at Wiangaree State Forest, N.S.W.	1. Evaluation of timber harvest schemes for Australian subtropical rainforest (Shugart *et al.*, 1981b).

(after Shugart *et al.* 1981a).

they can predict the outcome of change in forests dominated by long-lived species. Normally, the longevity of the canopy species would prevent direct observation of the changes that would accompany different disturbance regimes. Moreover, the fact that many modern forests originated after disturbance, as for example is the case with the forests of the Boundary Waters Canoe Area (BWCA) in North Minnesota discussed by Heinselman (1981), suggests that management policies which alter pre-settlement disturbance regimes will have far-reaching repercussions on the structure and diversity of future forests. In his study of the BWCA, Heinselman shows that the pre-colonial fire regime differed for each of the major forest types with respect to the intensity of the fire (crown or ground), to its periodicity (100–400 years) and to the extent of the burn (small–large). Forest management policies and practices can obviously benefit from the application of models that are designed to simulate these fire regimes. Cattelino *et al.*, (1979) discuss the wider application of simulation models of post-disturbance succession. Such models are not without their limitations, however, and Kessell (1981) draws attention to some of them. Not least is the need to validate each model for a new habitat and species complement for, by adopting the individualistic response of species to disturbance and by recognizing that disturbance regimes can vary widely in terms of their frequency, intensity and complexity, each model ceases to have general validity. Notwithstanding this limitation, the application of such models in the field of management is likely to prove rewarding and their foundation on the individualistic response does appear to have greater validity than previous models of succession based on the organismic analogy.

One of the other developments in ecological theory that has been promoted by these studies of post-disturbance change has been an interest in the concept of life history (Horn, 1981), for, in confronting disturbance, an organism has to make an allocative decision, namely whether to invest in self-preservation or in regeneration through reproduction. The outcome of this allocative decision is the life history and when similar or analogous genetic characteristics recur widely among species or populations so as to cause them to illustrate similar ecologies, different life strategies may be recognized (Grime, 1979).

15.4   BIONOMIC STRATEGIES

The growing acceptance amongst ecologists and biogeographers of the need to understand autecological facets of existence before making tenuous inferences about synecological influences or properties, has generated a new interest based on the life history of individuals and the populations to which they belong. In essence this approach focuses on reproductive success and failure and provides a means of analysing the relationship between the ecology of animals and plants and its evolutionary significance. Southwood (1981), for example, suggests that if natural selection operates successfully, it does so most efficiently by acting on the reproductive ability of the individual, hence the demographic characteristics of organisms and populations may be regarded as a function of selection processes. Building on this approach, ecologists have observed that two alternative reproductive strategies are common amongst animals: the one which invests energy in rapid growth and massive reproductive output at the expense of a short lifespan (r-selection) and the other which invests energy in a longer lifespan, slow growth and very low recruitment (K-selection). Both strategies are potentially efficient but are favoured by different environmental circumstances. According to Southwood, environments that are subject to sudden and drastic change provide organisms with numerous opportunities for colonization but exert heavy mortality. They favour a short life-cycle and maximum reproductive output. Conversely, in stable and persistent environments, mortality losses are low, crowded conditions develop and longevity and the ability to delay reproduction over a long period, are favoured. Whole taxonomic groups of animals tend to be largely 'r-selected' or 'K-selected'. Viewed in this way the bionomic strategy exhibited by a population is seen as the product of both the ecological requirements of an organism and its evolutionary success. Moreover, 'disturbance' assumes an ecological and evolutionary significance (Sousa, 1984).

Amongst plants, Grime (1977) has suggested that at least three common strategies may be observed. He proposes that plants have evolved different strategies for coping with varying intensities of both stress and disturbance. Grime defines stress as phenomena which restrict photosynthetic production, such as shortages of light, water or mineral nutrients or suboptimal temperatures. Disturbance is defined as the partial or total destruction of biomass and arises from the activities of grazing animals, pathogens or people. The three strategies identified in this way are competitors, ruderals and stress-tolerators. Competitors occupy habitats which are favourable in terms of available resources and low disturbance, whilst ruderals successfully occupy these favourable habitats when disturbance is increased. The most substantial way in which this three-strategy model differs from that of the 'K-selected' or 'r-selected' model of animal ecologists is in its recognition of stress-tolerance as a distinctive strategy which has evolved in intrinsically unproductive habitats or under conditions of extreme resource depletion induced by the vegetation itself. Stress-tolerators exhibit conservative physiologies which enable them to survive under conditions of low resource availability and occupy habitats that could not sustain the high growth rates exhibited by competitors. Moreover, by uncoupling the juvenile and established phases of growth, Grime's model accommodates those cases in which a plant may be subject to 'r-selection' early on its life-history and to 'K-selection' at a later stage, as, for example, when a plant colonizes a newly vacant area that remains abandoned for some considerable length of time.

The advantage that this life history approach has over traditional approaches is that plants and animals and the assemblages to which they belong are described and analysed, not in terms of their emergent properties, but in terms of their dynamic attributes. So, for example, a comparison of organisms and assemblages requires information about average mortality, fecundity and dispersal and their age-specific schedules (Horn and Rubenstein, 1984) as well as information about dominance, abundance, physiognomy, feeding habits etc. By focusing on these demographic and dispersal properties, a more comprehensive explanation of how and why particular assemblages come to occupy a particular habitat may be achieved than has been the case, and the role of human disturbance as a selection process may be more critically examined. The following examples drawn from the urban area illustrate how ecological and biogeographical studies may benefit from this approach.

15.5 SPECIES RESPONSE TO URBANIZATION

Observers have often commented upon the impoverished nature of urban wildlife, such that increasing urbanization leads to a graded reduction in the diversity of plants and animals (Davis, 1976). This is certainly the case with wildlife in the London area, moreover it is also noticeable that few of the organisms which inhabit the urban area can be regarded as truly urban. At one and the same time, human activities in the urban area have led to significant local extinctions and they have failed to provide new opportunities which favour the evolution of new urban species. However, although the assemblages of organisms which are present in the urban area are less diverse than those found in the surrounding countryside, they are supplemented by a number of alien and adventive organisms that have been introduced by human agency. So in total, the flora and fauna of urban areas is often an impressive one (Teagle, 1982). For example, the London area includes some 2055 different plant species (Burton, 1983) and 100 species of breeding birds (GLC, 1984). An explanation of why some species survive in the urban area and others do not has to involve several factors, such as the level of disturbance experienced in city areas, the reduction in the extent of available habitat and in the number of niches that can be supported, and the influence of urban-related factors such as the lowered humidity of city centres. All of these factors interact to produce a complex environment that is heterogeneous both in space and time (Douglas, 1983). The studies introduced below provide some examples of how different species have responded to this remarkable environment.

(a) Response to the development cycle

Disturbance can be defined as a 'discrete, punctuated killing, displacement or damaging of one or more individuals (or colonies) that directly or indirectly creates an opportunity for new individuals (or colonies) to become established' (Sousa, 1984). In the urban area this involves high levels of atmospheric pollution

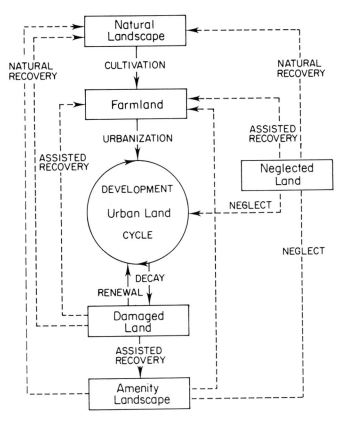

FIGURE 15.2   The development cycle of urban land. (*Reproduced by permission of J. F. Handley, The Groundwork Trust*)

as well as the pressure exerted by the presence and activities of people at work and play. Most of these kinds of disturbance provide constraints on some organisms and opportunities for others and the outcome of disturbance will often be determined by the position a plot or habitat patch occupies in the development cycle of urban land (see Fig. 15.2). Many habitat patches have a very short life cycle, as for example is the case with plots scheduled for urban renewal, other plots have a longer lifespan associated with their role as amenity areas, although each will be managed in a different way and each will have its own distinctive regime of disturbance. Moreover, the location and size of each patch in relation to possible sources of colonists will affect the rate and kind of the species response, whether or not colonization is assisted deliberately. Under these circumstances any succession of plant and animal assemblages which takes place on vacant land is likely to be highly individualistic. In practice very few studies have been conducted on the succession of assemblages that can be encountered in the urban area (Barrett, 1985) so that we are ignorant of the role that different strategies play in the colonization process but several studies have been undertaken on sites where recovery has been assisted (Bradshaw and Chadwick, 1980).

One of the unexpected findings of this work on the ecology of land reclamation in urban areas has been the realization that these areas present fewer difficulties for plant growth than many other types of derelict land where toxicity is a problem. The soils of derelict land in urban areas are often composed of brick rubble, cement and mortar and are deficient in nitrogen but, in many other respects, these soils are unexpectedly fertile. Their intrinsic fertility partly explains why natural plant colonization takes place quite rapidly and why landscape reclamation can be achieved very readily by planting legumes such as *Trifolium repens* (Bradshaw, 1982).

The most successful group of plants represented in the urban areas are the ruderals, together with aliens and ephemerophytes (species growing wild only temporarily). For example Kunick (1982) in a study of the flora of nine cities of central Europe found that on average, these two groups accounted for over 45 per cent of the urban flora and in some cities, for example Vienna, Warsaw, Lodz and Posnam, accounted for well over half the flora. Such remarkable success suggests that these species may have common characteristics. Indeed, such species do exhibit a number of characteristics in terms of their strategies of growth and dispersal which allow them to take advantage of the spatially and temporally disturbed conditions that prevail in the city. Of primary importance is their high investment in reproduction. As long ago as 1943, Salisbury pointed out that many of the plants encountered as 'weeds' in the city are annuals and produce enormous quantities of seeds (Salisbury, 1943). Moreover, some species of the genera such as *Polygonum*, *Atriplex* and *Chenopodium*, several members of which occur in urban areas, exhibit a rapid ripening of the seed, such that flowers and ripe seeds occur in the same inflorescence. Even in the absence of disturbance, many ruderals are short-lived and seed production is followed by death of the parent. The result of natural selection in most ruderals has been the development of early and lethal reproduction (Harper, 1977). In some ruderals seeds may be dispersed over a wide geographical area, in others they may lie buried in the soil and remain dormant for years. In either case, the ruderal strategy is pre-adapted to take advantage of disturbance and explains why it is so successful in the city.

The success of alien and ephemerophytes may also be attributed to the fact that several successful colonists possess a ruderal strategy. For example, an alien race of the perennial wall rocket (*Lactuca ferriola*) appeared in London about fifty years ago. Now it is abundant throughout the metropolis especially where motorway work and the like have bared the ground (Burton, 1983). Other successful aliens such as the rose-bay willow herb (*Chamaenerion angustifolium*) and Japanese knotweed (*Polygonom cuspidatum*) exhibit a similar dependency.

In part this ruderal strategy explains why Dony (1979) records that aliens and garden escapes prove to have been the single most successful group in the Hertfordshire flora between 1800 and 1979 during a period in which this county was progressively urbanized.

(b) Response to a reduction in habitat

Such examples can only provide a very superficial explanation of how species adapt to urban conditions, but an approach based on life-history and plant-strategies explicitly requires the observer to examine both the ecology and demography of a species and to enquire into their evolutionary significance as adaptations to disturbance. Amongst plants, the ruderal strategy has proved to be the most successful in urban areas but plants exhibiting other strategies can survive as well. They do so, however, often in those areas of encapsulated countryside that have survived the onslaught of urban growth and in those linear habitats associated with rivers, canals, communication lines and service corridors. None of these habitats are immune to the pressures of development that inflict any open area in the city, and their position in the development cycle affects the number and type of opportunities for colonization and establishment. Under these conditions the dispersal capabilities of a species are critical and for those species that cannot invest heavily in freely dispersed seeds, then local extinction is a real possibility. The likelihood of extinction is made all the more certain when such habitats are small in size and remote from a source of potential colonists (Davis and Glick, 1978). One study by Cousins (1982) examined the species/area effect for birds and snails in London. Small bird species are predicted to survive in urban centres because territory area is positively correlated with body size. Since the distribution and extent of favourable habitats is fragmented and reduced in the city centre, small birds only might be expected to survive. In the case of snails, small species might be expected because, for this group of relatively immobile species, fecundity might be expected to play an important role in lessening the chances of extinction. Since small species tend to have short generation times and high rates of reproduction, so might small species be expected to be favoured in the city centre.

Cousins's study revealed a decline in species density with increasing urbanization for both animal groups—the species/area effect—but whereas smaller birds were encountered in the city centre, the effect on species size was less clear for snails. Because snails depend upon the production of mucus as a surface on which to move, they are strongly affected by the moisture conditions of their surrounding habitat. Under conditions of moisture stress, a large surface to volume ratio is a disadvantage and paradoxically, large snails can be at an advantage in the drier city centre. This study also raises another paradox, namely that the most typical of all urban birds, the feral pigeon and increasingly the herring gull, are both quite large. An explanation of this phenomenon has to be sought in terms of their feeding ecology, since both depend upon the large supplies of waste food that abound throughout the urban area and their high mobility allows them to exploit these supplies. This study shows that a species/area effect may operate in the urban area but it cannot be fully explained for an individual species without a detailed understanding of the known ecology of the species and especially of its dispersal capabilities. It also suggests that no one life-strategy is likely to be optimal in the urban area, but by comparing species in terms of their life histories, the relative importance of fecundity, mortality and dispersal capabilities can be established. This approach is likely to prove more instructive than traditional ecological approaches, particularly in those environments subject to varying rates of disturbance and habitat fragmentation such as the urban area.

(c)   Response to atmospheric pollution

One further example of how a life-history approach has provided new insights is provided by Seaward (1982) in his studies of the recovery of lichens in city centres during a ten-year period of ameliorating atmospheric conditions. Lichens are able to survive in extremely harsh environments under conditions in which vascular plants may be totally excluded and for this reason have been used to monitor both the deteriorating (Hawksworth and Rose, 1976) and ameliorating conditions associated with atmospheric pollution in and around urban areas (Henderson-Sellers and Seaward, 1979). Lichens are able to tolerate what amount to some of the most 'stressed' conditions, such as those provided in the city centre. They do so by maintaining slow growth rates and by physiological acclimation whereby a high proportion of photosynthate is sequestered rather than expended in growth. Most lichens are long-lived and they are also classically regarded as opportunists by virtue of their ease of dispersal and early appearance in the colonization of a freshly exposed surface. In terms of the three plant strategies recognized by Grime they are 'stress-tolerators' *par excellence.*

In the urban area, species such as *Lecanora muralis*, which are capable of benefiting from a variety of saxicolous substrates, such as dressed stone and other artificial materials such as concrete, mortar, asbestos-cement and, to a lesser extent, tarmacadam and brick, are at an advantage (Seaward, 1976). On these substrates, the artificially high pH of the growth medium provides a buffering effect against the artificially low pH of urban rain water. In the West Yorkshire conurbation *Lecanora muralis* has spread into the city at the rate of 9 km^2 per annum, mainly through its ability to exploit calcareous substrates during a period of lowering levels of ambient pollution. But, detailed monitoring of the life-history of this species has revealed unexpected variations in the rate of growth of particular colonies. Early colonists, which became established soon after improvements in conditions of SO_2 show consistently higher growth rates than later colonists, even though the influx of propagules and amelioration remained constant. Seaward attributes this variation to the inhibiting effect exerted by the pioneers on secondary colonists. He suggests that pioneers exploit the more favourable sites and relegate secondary colonists to suboptimal sites; he also points to the possibility of chemical suppression of growth by the initial colonists. Without the detailed monitoring of sites and a consideration of the life-history of individual colonies, the depressed growth rates of late colonists might have been interpreted as indicative of a change in the level of pollution, when in practice this has remained constant.

(d) Response to recreation pressure

Disturbance that results from high levels of atmospheric pollution and from the development cycle associated with urban land are in many respects self-evident, although the species response is often a highly individualistic one. Other forms of disturbance are less readily studied and the impact of outdoor recreation on animals falls into this category. Numerous studies have been conducted on the ecological effects of recreation, but the burden of work has examined the effects on vegetation rather than animals. Moreover, few studies have attempted to measure the dose and response such that both the stimulus and its effect are presented in quantitative terms — the former as some variant of numbers of people engaged in different activities in a given area over a given time period, and the latter in terms of breeding densities for example. In practice too, the number of studies that have been conducted in the urban or urban-edge environment is also limited and as a result we know little of the response of urban birds to recreation. One study by Cooke (1980), compared the 'fly-away distances' of several bird species to an approaching person in suburban and rural areas. The study found that suburban birds were less easily disturbed than rural ones and Cooke interpreted this behavioural response as evidence that habituation had occurred. On this basis, rural birds might be deemed to be more vulnerable to disturbance associated with an approaching person than urban birds, but unless that behavioural response is reflected in lowered breeding densities for example, it is difficult to place a meaningful interpretation on the results.

One study that has successfully demonstrated a negative relationship between recreation pressure and breeding densities is provided by Van der Zande *et al.* (1984). They found that the recreational use of urban-fringe woodlands during a weekday, significantly affected the densities of eight out of thirteen common bird species. Some species were more vulnerable than others and relative vulnerability also depended upon the type of woodland (deciduous or coniferous). By examining the recreation intensity which depressed bird densities by 50 per cent, they were able to produce maps to indicate where 'unacceptable levels of disturbance' were encountered (see Fig. 15.3). In this study no attempt was made to determine why one species was more vulnerable than another, but now that

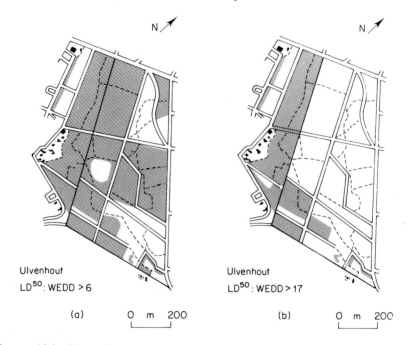

FIGURE 15.3   Maps of Ulvenhout Forest in Holland showing the LD_{50} parcels. These parcels are those with a recreation intensity above which 50 per cent of the maximum density of birds will have disappeared. (a) Two very susceptible birds (*Sylvia borin* in deciduous parcels and *Streptopelia turtur* in coniferous parcels). (b) Two moderately susceptible birds (*Phylloscopus collybita* in deciduous parcels and *Columba palumbus* in coniferous parcels) (After Van der Zande *et al.* 1984)

a significant relationship has been demonstrated between recreational use and breeding densities, the scene has been set for other more detailed studies such as those relating to life-history (Van der Zande, 1984).

15.6   CONCLUSION

The fact that few strictly urban species have evolved lends support to the view that most urban habitats appear to serve as analogues of naturally occurring ones. On the other hand, the nature of those species which are most successful, namely the ruderals, suggest that it is the intensity of disturbance which sets urban habitats apart from their rural analogues. If so, then amongst plants, ruderals as a group appear to be peculiarly resilient to both extinction and the genetic repatterning promoted by natural selection following upon disturbance. It may be that insufficient time has elapsed on an evolutionary time-scale for new species to have arisen, or that ruderal species possess an all-purpose genotype which equips them well for the disturbance regimes of the urban area (Baker,

1965). Whatever the explanation for the extraordinary success of these species in the face of some of the most intensive selection pressure exerted by human activity, it is clear that the study of urban wildlife offers numerous opportunities for ecologists and biogeographers to examine how organisms in different habitats respond to different disturbance regimes. Moreover, the development of the concept of life-history and its application to the study of urban wildlife appears to offer the prospect of gaining new insight into how some forms achieve greater fitness than others. New techniques of bioengineering may generate new, fit, urban-forms where natural selection has failed to, but in the absence of any new, truly urban species, and on the evidence of the failure of many of our native species to survive in urban areas, it behoves us all to gain a better understanding of urban organisms and to devise new disturbance regimes, new techniques of assisted recovery and new spatial patterns of habitats to ensure that all types of organisms have a permanent place amongst the wildlife of cities.

REFERENCES

Baker, H. G., 1965, 'Characteristics and modes of origin of weeds', in Baker, H. G., and Stebbins, G. L. (Eds), *The Genetics of Colonizing Species*, New York, Academic Press, pp. 147–172.

Barrett, I., 1985, *Research Needs in Urban Ecology*, Ecological Parks Trust, London, 43 pp.

Battarbee, R. W., 1984, 'Diatom analysis and the acidification of lakes', *Phil. Trans. R. Soc. Lond. B*, **305**, 451–477.

Battarbee, R. W., Flower, R. J., Stevenson, A. C., and Rippey, B., 1985, 'Lake acidification in Galloway: a palaeoecological test of competing hypotheses', *Nature*, **314**, 350–352.

Berry, R. J., 1982, *Neo-Darwinism*, Studies in Biology no. 144, Edward Arnold, London, 68 pp.

Bornkamm, R., Lee, J. A., and Seaward, M. R. D., 1982, *Urban Ecology*, Blackwell, Oxford, p. 370.

Bradshaw, A. D., 1978, 'Pollution and evolution', in Mansfield, T. A. (Ed.), *Effects of Air Pollution on Plants*, Cambridge University Press, Cambridge.

Bradshaw, A. D., 1982, 'The biology of land reclamation in urban areas', in Bornkamm, R. *et al.* (Eds), *Urban Ecology*, Blackwell, Oxford, pp. 293–304.

Bradshaw, A. D., and Chadwick, M. J., 1980, *The Restoration of Land*, Blackwell, Oxford.

Bull, D., 1982, *A Growing Problem*, Oxfam, Oxford.

Burton, R. M., 1983, *Flora of the London Area*, London Natural History Society.

Carson, H. L., and Templeton, A. R., 1984, 'Genetic revolutions in relation to speciation phenomena: the founding of new populations', *Ann. Rev. Ecol. Syst.*, **15**, 97–131.

Cattelino, P. J., Noble, I. R., Slatyer, R. A., and Kessell, S. R., 1979, 'Predicting the multiple pathways of plant succession', *Environmental Management*, **3**, 41–50.

Chilton, M., 1983, 'A vector for introducing new genes into plants', *Scientific American*, **248**, 36–45.

Connell, J. H., and Slatyer, R. O., 1977, 'Mechanisms of succession in natural communities and their role in community stability and organization', *Am. Nat.*, **111**, 1119–1144.

Cooke, A. S., 1980, 'Observations on how close certain passerine species will tolerate an approaching human in rural and suburban areas', *Biol. Conserv.*, **18**, 85–88.

Cousins, S. H., 1982, 'Species size distributions of birds and snails in an urban area', in Bornkamm, R., et al (Eds), *Urban Ecology*, Blackwell, Oxford, pp. 99–110.

Cross, M., 1984, 'Gene vector looks better and better', *New Scientist*, 1st November.

Davis, A. M., and Glick, T. F., 1978, 'Urban ecosystems and island biogeography', *Environ. Conserv.*, **5**, 299–304.

Davis, B. N. K., 1976, 'Wildlife, urbanization and industry', *Biological Conservation*, **10**, 249–291.

Davis, M. B., 1981, 'Quaternary history and the stability of forest communities', in West, D. C., Shugart, H. H., and Botkin, D. B. (Eds), *Forest Succession*, Springer Verlag, Heidelberg, pp. 132–153.

Dony, J. C., 1979, 'Changes in the flora of Hertfordshire', *Trans. Herts. Nat. Hist. Soc.*, **27**, 255–264.

Douglas, I., 1983, *The Urban Environment*, Edward Arnold, London, 229 pp.

Gleason, H. A., 1926, 'The individualistic concept of the plant association', *Bull. Torrey. Bot. Club.*, **53**, 7–26.

GLC (Greater London Council), 1984, *Ecology and Nature Conservation in London*, Ecology Handbook No. 1, GLC, London.

Grime, J. P., 1977, 'Evidence for the existence of three primary strategies in plants and its relevance to ecology and evolutionary theory', *Am. Nat.*, **111**, 1169–1194.

Grime, J. P., 1979, *Plant Strategies and Vegetation Processes*, Wiley, Chichester.

Gould, S. J., and Eldredge, N., 1977, 'Punctuated equilibria: The tempo and mode of evolution reconsidered', *Palaeobiology*, **6**, 119–130.

Harper, J. L., 1977, *Population Biology of Plants*, Academic Press, London, 892 pp.

Hawksworth, D. L., and Rose, F., 1976, *Lichens as Pollution Monitors*, Edward Arnold, London.

Heinselman, M. L., 1981, 'Fire and succession in the conifer forests of northern North America', in West, D. C., Shugart, H. H., and Botkin, D. B. (Eds), Springer Verlag, Heidelberg, pp. 374–405.

Henderson-Sellers, A., and Seaward, M. R. D., 1979, 'Monitoring lichen reinvasion of ameliorating environments', *Environ. Pollut.*, **19**, 207–213.

Horn, H. S., 1981, 'Causes of variety in patterns of secondary succession', in West, D. C., Shugart, H. H., and Botkin, D. B. (Eds), *Forest Succession*, Springer Verlag, New York, pp. 24–35.

Horn, H. S., and Rubenstein, D. I., 1984, 'Behavioural adaptations and life history', in Krebs, J. R., and Davies, N. B. (Eds), *Behavioural Ecology*, 2nd edn, Blackwell, Oxford, pp. 279–300.

IUCN (International Union for Nature and Natural Resources), 1978, *The IUCN Plant Red Data Book*, IUCN, Morges, Switzerland.

Kessell, S. R., 1981, 'The challenge of modelling post-disturbance plant succession', *Env. Mgt.*, **5**, 5–13.

Kunick, W., 1982, 'Comparison of the flora of some cities of the Central European lowlands', in Bornkamm, R. *et al.*, (Eds), *Urban Ecology*, Blackwell, Oxford, pp. 13–22.

Loveless, M. D., and Hamrick, J. L., 1984, 'Ecological determinants of genetic structure in plant populations', *Ann. Rev. Ecol. Syst.*, **15**, 65–95.

McLaughlin, S. B., West, D. C., Shugart, H. H., and Shriner, D. S., 1978, 'Air pollution effects on forest growth and succession: Application of a mathematical model'. Cooper, H. B. H. Jr. (Ed.), *Air Pollution Control Association*, Vol. 71, Houston, Texas, pp. 1–16.

Mielke, D. L., Shugart, H. H., and West, D. C., 1977, 'Users' manual for FORAR a stand model of upland forests of southern Arkansas', *ORNL/TN 5767*, Oak Ridge National Laboratory, Oak Ridge, Tennessee.

Mielke, D. L., Shugart, H. H., and West, D. C., 1978, 'A stand model for upland forests of southern Arkansas'. *ORNL/TM-6225*, Oak Ridge National Laboratory, Oak Ridge, Tennessee.

Miles, J., 1979, *Vegetation Dynamics*, Studies in Ecology, Chapman & Hall, London.

Noble, I. R., Shugart, H. H., and Schauer, J. S., 1980, 'A description of BRIND a computer model of succession and fire response of the high altitude Eucalyptus forests of the Brindabella Range, Australian Capital Territory'. *ORNL/TM-7041*. Oak Ridge National Laboratory, Oak Ridge, Tennessee.

Roughgarden, J., 1984, 'Competition and theory in community ecology', in Salt, G. W. (Ed.), *Ecology and Evolutionary Biology*, University of Chicago Press, London, pp. 3–21.

Salisbury, E. J., 1943, 'The flora of bombed areas', *Nature*, **151**, 462–466.

Seaward, M. R. D., 1976, 'Performances of *Leconora muralis* in an urban environment', in Brown, D. H., Hawksworth, D. L., and Bailey, R. H., *Lichenology: Progress and Problems*, (Eds), Academic Press, London, pp. 323–357.

Seaward, M. R. D., 1982, 'Lichen ecology of changing urban environments', in Bornkamm, R., *et al.* (Eds), *Urban Ecology*, Blackwell, Oxford, pp. 184–190.

Shugart, H. H., and West, D. C., 1977, 'Development of Appalachian deciduous forest succession model and its application to assessment of the impact of the chestnut blight'. *J. Environ. Management*, **5**, 161–179.

Shugart, H. H., and Noble, I. R., 1981, 'A computer model of succession and fire response of the high altitude eucalyptus forests of the Brindabella Range, Australia Capital Territory', *Aust. J. Ecol.*, **6**, 25–36.

Shugart, H. H., West, D. C., and Emanuel, W. R., 1981, 'Patterns and dynamics of forests: an application of simulation models', in West, D. C., Shugart, H. H., and Botkin, D. B. (Eds), *Forest Succession*, Springer Verlag, Berlin, pp. 74–94.

Shugart, H. H., Mortlock, A. T., Hopkins, M. S., and Burgess, I, P., 1980, 'A computer simulation model of ecological succession in Australian sub-tropical forest'. *ORNL/TM-7029*, Oak Ridge National Laboratory, Oak Ridge, Tennessee.

Shugart, H. H., Hopkins, M. S., Burgess, I. P., and Mortlock, A. T., 1981(b), 'The development of a succession model for sub-tropical rainforest and its application to assess the effects of timber harvest at Wiangaree State Forest, New South Wales'. *J. Environ. Management*, **11**, 243–265.

Smith, T. M., Shugart, H. H., and West, D. C., 1981, 'FORHAB. A forest simulation model to predict habitat structure for non-game bird species', in Capen D. E. (Ed.), *Symposium on the Use of Multivariate Statistics in Studies of Wildlife Habitat*. University of Vermont Press. Burlington.

Solomon, A. M., Decourt, H. R., West, D. C., and Blasing, T. J., 1980, 'Testing a simulation model for reconstruction of prehistoric forest-stand dynamics'. *Quat. Res.*, **14**, 275–293.

Southwood, T. R. E., 1981, 'Bionomic strategies and population parameters', in May, R. M. (Ed.), *Theoretical Ecology*, 2nd edn, Blackwell, Oxford, pp. 26–48.

Sousa, W. P., 1984, 'The role of disturbance in natural communities', *Ann. Rev. Ecol. Syst.*, **15**, 353–391.

Strong, D. R., 1984, 'Natural variability and the manifold mechanisms of ecological communities', in Salt, G. W. (Ed.), *Ecology and Evolutionary Biology*, University of Chicago Press, London, pp. 47–55.

Teagle, W. G., 1982, *The Endless Village*, Nature Conservancy Council, Peterborough.

Van der Zande, A. N., Berkhuizen, J. C., Van Latesteijn, H. C., ter Keurs W. J. and Poppelaars, A. J., 1984, 'Impact of outdoor recreation on the density of a number of breeding bird species in woods adjacent to urban residential areas', *Biol. Conserv.*, **30**, 1–39.

Van der Zande, A. N., 1984, *Outdoor Recreation and Birds: Conflict or Symbiosis!*, Offsetdrukkerij Kauters B. V., Alblasserdam, 269 pp.

Walker, D., 1982, 'Vegetation's fourth dimension', *New Phytol.*, **90**, 419–430.

Warr, R., 1984, *Genetic Engineering in Higher Organisms*, Studies in Biology no. 162, Edward Arnold, London, 58 pp.

West, D. C. McLaughlin, S. B., and Shugart, H. H., 1980, 'Simulated forest response to chronic air pollution stress'. *J. Environ. Qual.*, **9**, 43–49.

West, D. C., Shugart, H. H., and Botkin, D. B., 1981, *Forest Succession*, Springer Verlag, New York, 517 pp.

Human Activity and Environmental Processes
Edited by K. J. Gregory and D. E. Walling
©1987 John Wiley & Sons Ltd.

16

The Conservation of Plants, Animals and Ecosystems

I. G. SIMMONS

16.1 A MOSAIC OF ECOLOGICAL SYSTEMS

If we accept that one of the aspirations of geography is the presentation of an holistic perspective which enfolds both objective measurements of the physical environment and man's perception and uses of it, then in the present context we might note that man's economy is inextricably linked with that of nature: *Oikos* is the root of both ecology and economics. The major purpose of this essay is to examine some ideas about the functioning and relationships of different types of ecological systems and to see how they might have relevance for our attitudes to land use and other spatial patterns, including 'conservation'.

As a starting point, I propose to take the basic concepts of Eugene Odum's (1969) paper, 'The strategy of ecosystem development', which are to some extent amplified in his book *Ecology* (2nd edn, 1975). These are based on the characteristics of the different stages which occur during succession in the development of ecological systems towards the self-maintaining equilibrium which is sometimes called the 'climax' or 'mature' condition. The end point of successions, the mature stage, is especially notable for the high degree of internal symbiosis which is developed, for the closed nutrient cycles, and for the high degree of stability which is maintained, although piecemeal renewal is of course essential. In spatial terms, the trend is towards a mosaic of ecosystems moving towards maturity but with patches of early stages for example where new land is created, where catastrophic events have occurred and where the death of large organisms such as senescent trees has taken place.

We can erect a classification by designating four main types of ecosystem;

(1) Early succession systems with low biotic diversity and linear food chains and a high net community production.
(2) Mature systems which generally show the opposite characteristics from

type (1). Notably, a lot of biomass is supported by the energy flow, relative to type (1), and the food chains are predominantly web-like, with the detritus stage very important.

(3) A mixture of (1) and (2), at times of general environmental change, for example, or where natural fires bring about the juxtaposition of the two types of system.

(4) Inert systems, with little or no life: volcanoes are one example, ice-caps another.

If we can place spatial boundaries to each of these types of system we have the basis for a compartment model which can link each to the other. In the natural world the tendency is for type (1) to become type (2) but within the transition, some reversion to type (1) is always happening. Similarly, type (1) systems encroach upon areas of type (4) when they colonize cold volcanic lava flows or recently deglaciated terrain, for example. Volcanic activity or glacier advance increases the area of type (4). There is no doubt that some time during the present epoch, after the major post-glacial climatic adjustments but before both the intensive and extensive ecological impacts of agricultural societies, the world must have looked like a mosaic of those types. At present it clearly does not, but we are reminded by Odum that ecosystems manipulated by man can be described in the same terms as natural ecosystems. Thus early successional phases of natural ecosystems are equivalent to the simple systems of modern agriculture: the mixed systems of type (3) would apply to areas of mixed forest and farmland; the mature systems are the same in both, often being deliberately protected by man as parks or reserves, and the inert systems have their equivalents in cities (which like volcanoes give off SO_2 and particulate matter) and derelict land. Using the type and quantity of energy flow as a criterion, it is possible to combine both natural and man-manipulated ecological systems in one classification (Table 16.1). Here the source of the energy becomes a differentiating factor. Some systems (Type 1 in Table 16.1) are entirely solar-powered and may use only the incoming solar radiation which they fix as chemical energy. Others (Type 2) may receive a natural subsidy of organic matter brought in by natural processes, and estuaries are an obvious example of this type. Another major category of systems (Types 3 and 4) are those which are subsidized by man-procured supplies of fossil fuels (i.e. stored photosynthesis). Some systems are indeed subsidized by fossil fuels but remain fixers of solar energy (e.g. mechanized agriculture, short-cycle forestry); others, like the city, are entirely powered by fossil (and in some cases, nuclear) fuels. The absolute throughput of energy per unit area (energy intensity) increases through Types 1–4 (Table 16.1).

These ecosystem types can also be related to the compartment model. Successional systems belong to Type 3 and may have a very high energy intensity if powered by fossil fuel but little more than mature systems (which equate to

TABLE 16.1 Ecosystems classified according to source and level of energy

Ecosystem type	Annual energy flow kilocalories per sq metre per year (Kcal.m^{-2}.yr^{-1})	
	Range	Average (estimated)
1. Unsubsidized natural solar-powered ecosystems: e.g. open oceans, upland forests. Man's role: hunter-gatherer, shifting cultivation	1 000–10 000	2 000
2. Naturally subsidized solar-powered ecosystems: e.g. tidal estuary, lowland forests, coral reef. Natural processes aid solar energy input: e.g. tides, waves bring in organic matter or do recycling of nutrients so most energy from sun goes into production of organic matter. These are the most productive natural ecosystems on the earth. Man's role: fisherman, hunter–gatherer	10 000–50 000	20 000
3. Man-subsidized solar-powered ecosystems: Food and fibre producing ecosystems subsidized by human energy as in simple farming systems or by fossil fuel energy as in advanced mechanized farming systems: e.g. Green Revolution crops are bred to use not only solar energy but fossil energy as fertilizers, pesticides and often pumped water. Applies to some forms of aquaculture also.	10 000–50 000	20 000
4. Fuel-powered urban–industrial systems. Fuel has replaced the sun as the most important source of immediate energy. These are the wealth-generating systems of the economy and also the generators of environmental contamination: in cities, suburbs and industrial areas. They are parasitic upon Types 1–3 for life support (e.g. oxygen supply) and for food; possibly fuel also although this more likely comes from under the ground except in LDCs where wood is still an important domestic fuel.	100 000–3 000 000	2 000 000

The most productive natural ecosystems and the most productive agriculture seem to have upper limits of ca. 50 000 Kcal/m^2/yr.

Source: Odum (1975)

Type 1) if only human energy is based upon them. There is no explicit equivalent of the mixed systems. The fuel-powered industrial systems (Type 4) are clearly the inert category and are parasitic on the others, although we should not underestimate either the organic production of city suburbs or the value of the information created and transmitted in urban structures.

The major inference we can make from the compartment model and from Table 16.1 is that human strategies often run counter to those of nature. Whereas nature moves towards mature, self-maintaining systems, man moves towards one type of successional phase at the expense of the mature and mixed systems, and increasingly requires subsidiary energy to keep up productivity. He also increases the area of inert systems. In turn, many kinds of wastes reduce the diversity of organisms present in all types of system (Woodwell, 1970). The further implication, not perhaps openly stated by Odum but clearly intended, is that the matrix of ecosystem types produced by the imposition of human strategies on those of nature will in future be subject to unpredictable and uncontrollable fluctuations, that is they will not be stable enough as a habitat to sustain large numbers of our species.

The concept of stability in ecology is not, as is apparent from Chapter 15, an easy one. If it is used in two senses (Hill, 1975), then the first relates to a change of species composition over time, and an early successional stage is obviously much less stable in this respect than a mature ecosystem. The second refers to the ability of an ecosystem to return to its original successional pathway or self-maintaining state after a perturbation and is perhaps of most interest here. Jowett (1972) makes the analogy of a marble in a saucer: if hit, the marble will normally return to its position in the depression but it can be flipped so hard that it rolls out of the saucer entirely and finds an equilibrium position elsewhere. In man-manipulated systems, large quantities of energy, somatic and fossil based, are spent in keeping marbles in saucers. Without stability there can be no sensible prediction and hence no rational planning for the future, only an immediate single-purpose response to random events. The forms and scale that instability might take, resulting from unsuitable man-management of ecosystems are very varied and not easy to foresee in detail. Desertification, soil erosion, pest outbreaks, disease epidemics, famines and climatic change are all among the possibilities. Rapid fluctuations in the human environment will also inhibit resource uses which are sustainable in the long-term, i.e. uses which yield 'interest' rather than deplete 'capital' (WCS, 1980).

Viewed in this context 'conservation' becomes an attitude towards stability and sustainability on the continental and global scales in terms of ecosystem change and the economic development of human societies, especially the non-industrial nations. Those actions which tend to or seem likely to produce instability are seen as anti-conservative, whereas those which preserve or enhance stability are seen as part of the conservation effort. The general problem is to keep fluctuations of ecosystem parameters within manageable bounds, remembering the demands made by man upon the systems and the nature of the technology which he uses to manipulate them.

The bulk of this chapter will be concerned with brief accounts of the main trends in the human use of the earth which tend to maintain or to diminish the overall stability and sustainability of its ecosystems, both natural and managed. Readers will notice that few of the authors of the publications which are cited are geographers, because these are fields in which they have not generally been active, with the notable exception of desertification. In terms of a general synthesis by a geographer which strongly argues a particular case based on the sustained yield of plant production, S. R. Eyre's *The Real Wealth of Nations* should be consulted.

16.2 ANTI-CONSERVATION TRENDS

(a) Intensification of agriculture

Without attempting a complete list of destabilizing influences, a number of outstanding degradational processes must be discussed, and leading the list are the effects of agriculture throughout the world. In many places the trend is towards intensification of production, which has several unfortunate results. The most obvious of these is the complete breakdown of the soil when it is tilled or cropped at an intensity which it cannot tolerate, the erosional results of which are discussed in Chapter 13. Such processes are now most common in less-developed countries (LDCs) where human population pressure is severely pressing upon food production, but similar trends can be observed in some developed countries (DCs) where, for example, the absence of ley phases in tight cropping schedules on land in areas of medium to high rainfall brings about dangerously low levels of organic matter and threatens the stability of the soils. The drainage of soils may be affected by plastic deformation and smear caused by wheels and tillage implements, and so poorer germination and restricted root development may lead to more open soil and hence more erosion (Agricultural Advisory Council, 1970). The extension of agricultural ecosystems into areas formerly consisting of mature systems also leads to immense losses of soil, including its constituent organic matter and mineral nutrients. Traditional systems of shifting agriculture usually replaced these losses during the fallow period, but the curtailment of the rotation cycle and the forcing of peasant agriculture onto very steep slopes, makes such recovery barely possible. The loss of mature forest on mountains then contributes to a higher frequency and height of floods lower down the river basin so that floods can often be seen as a symptom of intensified land use within a watershed (see Chapter 4). Even in a soil conservation-conscious nation like the USA, an average of 26.9 tonnes ha^{-1} of topsoil is lost annually from agricultural land. This material also reduces the usefulness of lakes, reservoirs and rivers to the tune of $500 million per year not counting the interference with the fish populations caused by enhanced silt loads and eutrophication. To offset the losses caused by such erosion, Pimentel *et al.* (1976) calculate that 494 000 kcal ha^{-1} (50×10^6 bbl oil per year) of energy inputs have to be deployed on the land.

The crop components of agro-ecosystems can also be destabilized. In the past this was done mainly by weeds and pests, so that the application of a great battery of chemical agents has in general brought both ecological and economic stability to these systems. But, as is well known, the effect of some pesticides has been to bring about a deviation-amplifying condition. This has happened when an insect, for example, has built up resistance to a chemical biocide and has become a serious pest before new measures to combat it can be developed; or when the residues of a biocide have accumulated in a food chain and have proved lethal to a non-target organism which was an important structural member of a linked ecosystem. The breeding of plants for modern agriculture has meant a trend to genetic uniformity, in order to ensure an homogeneous response to water and fertilizers, a single cropping time, and undifferentiated response to industrial processing. In the USA 53 per cent of the cotton crop is in three varieties, 95 per cent of the peanut crop in nine varieties, and 65 per cent of the rice crop in four varieties (NAS, 1972).

Such practices increase the vulnerability of crops to failure from events like drought and insect attack, since in a stand with genetic variety some individuals would probably survive. The problem seems especially worrying in LDCs where the Green Revolution has brought a considerable dependence upon high-yield varieties with a large measure of genetic uniformity and most agricultural research and development stations are now trying to preserve genetic diversity in crop species. Intensification of agriculture anywhere in the world brings problems arising from runoff containing fertilizers rich in nitrogen and phosphorus. This may often lead to nutrient enrichment of fresh waters with consequent disruption of plant and animal life, an algal 'bloom' often being the first sign. This process is called eutrophication, and it is also one result of the discharge of untreated sewage from towns and cities (Holdgate, 1979).

The outcome of certain human uses of the land, allied to possible changes in climate, is seen in the set of phenomena called 'desertification' (see Chapter 17). This is defined as 'the spread of desert conditions for whatever reasons, desert being land with sparse vegetation and very low productivity associated with aridity, the degradation being persistent or in extreme cases irreversible' (Grove, 1977). The conversion of land into a virtually inert system and its associated susceptibility to rapid erosion, clearly puts the process into the destabilizing category and while not perhaps the only causative factor, the effects of grazing, cultivation, woodland clearance, and borehole development are clearly not to be ignored (Mabbutt, 1977; Warren, 1984).

(b) Urbanization and industrialization

One major process of the present time which is common to most parts of the world is urbanization and industrialization, which usually produce inert systems in place of all the other categories, either directly because of their replacement by a built environment, or indirectly because the wastes reduce or even eliminate many forms of life. The quantity of land in urban–industrial use is very variable from nation to nation, even within industrial economies. In the USA, urban uses plus highways account for ca. 29×10^6 ha of land (including 270 000 ha devoted to parking), which is 3.2 per cent of the total land area, including Alaska (Pimentel *et al.*, 1976). In England and Wales about 10 per cent was 'urban' in the 1960s (Anderson, 1977; Best, 1981). However, not all the urban–industrial land is built-up or inert and Best and Ward (1956) showed, for example, that in a London suburb the value of food output per unit area was close to that of the best farmland. Nevertheless, numerous examples of the more or less complete sterilization of land by dense building or by industrial heaps and holes, are easily brought to mind, with the presence of such wild life as starlings and pigeons being accounted for by the fact that they rely on energy imported from outside the city. Spin-off effects of cities and industries also destabilize ecosystems (Douglas, 1983). Raw sewage and calefacted water, and acid mine drainage, alike reduce the biological diversity and biomass of water bodies; fall-out of atmospheric contaminants may kill plants or reduce their growth rate; cumulative poisons like polychlorinated biphenyls (PCBs) or methyl mercury may result in the mortality of many organisms including, in the latter case, man himself. Floods too, may be exacerbated by urbanization and its resultant rapid runoff (Chapter 5, Fig. 5.3). Not all such developments must necessarily be viewed as negative. Industrial life has so far brought a much more comfortable, and culturally enriched, life than subsistence agriculture or pastoral nomadism, and it can be argued that it is the surpluses generated by industrial life that permit the development and transmission of knowledge that will enable their destabilizing effects to be overcome. Again, the cities are indeed parasitic upon other systems for food and are obviously powered by non-renewable fossil fuels, but perhaps in burning these sources of energy we transform them into the knowledge of how to manage without them.

(c) War

From being a localized 'phenomenon whose ecological effects were largely temporary, war has gradually become a destabilizer at virtually national scales with long-lasting effects. Just as the shell-holes of the First World War are preserved in a fossil form in Flanders, so much more freshly preserved are the immense numbers of craters which mark the Republic of Vietnam. The Vietnamese war of re-unification represents the first time that intentional anti-environmental actions were a major component of the strategy and tactics of one of the adversaries. Characterized by chemical warfare against mangroves, forests and crops, together with the effects of artillery on forests, with mechanized land clearing, and with attempts to wash out the Ho Chi Minh trail by artificial rain-making, industrialized war has perhaps never before been applied in such intensity to one country (Westing, 1984). Neither need it be forgotten that in the various forms of thermonuclear device, the potential for ecological destabilization and even sterilization is very high (SIPRI, 1977). This is especially so if the smoke and particulate matter resulting from a nuclear exchange are of sufficient quantity and reside long enough in the atmosphere to reduce temperatures and levels of photosynthesis to very low magnitudes for perhaps a year after the event. This would bring about the so-called 'nuclear winter' after which no ecosystem or human society would ever be the same again (Turco, 1983; Ehrlich *et al.*, 1983, Dotto, 1986).

(d)   Species extinction

Many of the conversions of system type discussed above produce one kind of result in the extinction of species of plants and animals either by direct extirpation or by alterations of habitats. This is no new event for it has been postulated that during the terminal Pleistocene of many parts of the globe, palaeolithic hunters administered a *coup de grâce* to many genera and species of large mammal herbivores which were at the time under stress because of environmental change; this is the phenomenon known as 'Pleistocene overkill' (Martin, 1973). But the expansion of empires in the seventeenth century and, above all, the coming of the industrial revolution have increased the numbers of animal extinctions to a considerable degree. Between 1600 and 1699, there were twelve extinctions of mammals, birds and marsupials, whereas from 1900 onwards there have been at least eighty-one (Ziswiler, 1967). The extinction rate of species and subspecies appears to be about one per year, compared with one every ten years from 1600 to 1950, and perhaps one every 1000 years during the period of the great dying of the dinosaurs (Myers, 1979). In spatial terms, the vulnerability of oceanic islands is outstanding, and the recently occupied continents of North America and Australia also contribute relatively high numbers of the total.

The extinction of plants must not be forgotten. Habitat changes have ousted populations of many plants and, as with animals, island floras have been particularly susceptible to introduced herbivores such as goats and sheep. For example, Phillip Island 1610 km east of Sydney, was thickly forested in 1774 but within 100 years grazing animals had helped to remove the forest and two endemic plants along with it. The extinction of the dodo, endemic to Mauritius, seems to have meant that a particular tree (*Calvaria major*) has not regenerated since the late seventeenth century. The hypothesis is that crushing in the dodo's gizzard was necessary to break the tough endocarp and allow the excreted but undigested seed to germinate. Now there are only about thirteen trees left, all over 300 years old (Temple, 1977). Globally, one estimate (Tinker, 1971) suggests that 20 000 plant species, representing about 10 per cent of the world's flora, are now threatened with extinction. Among such plants there are probably many that might be useful as well as beautiful or stabilizing components of ecosystems.

(e) Response to fluctuations

The destabilizing effects of human activity affect the world mosaic of ecosystems via several components of the environment. They may interact adversely with the soil and produce soil erosion or infertility; they may enter an ecosystem via the role of organisms in terms of extinctions of taxa, or via the introductions of exotic species which can 'explode' into new habitats, or via the direct and residual effects of pesticides. Organisms can also be affected by wastes of a toxic kind. The hydrological cycle can be destabilized by particular land use practices, producing floods, and we are ignorant of the possible future fluctuations of ecosystems which may result from the extinction of inert systems and the wastes they produce. Such processes are aptly described by Woodwell (1970), who brings together the evidence from deforestation, toxins, radioactive wastes, eutrophication and other similar processes. He talks in terms of such events as the loss of biotic diversity, of population instability, especially of small rapidly reproducing organisms such as rodents and insects, and of a movement towards a world which is not self-augmentive and homeostatic any more but requires constant tinkering to patch it up. Each act of tinkering, however, generates a need for further action by man.

The diverse suite of actions by human societies to manipulate ecological systems can be summed up in terms of the use of energy. Increasing amounts of energy (which at present mostly come from fossil fuels in the DCs) are being demanded in order to cope not only with the requirements for intensification and industrialization described above but with the aftermath of destabilized systems. Because fossil fuels are finite in amount, the alternative energy futures of man hold the key to the type of attitude we develop towards the world's mosaic of ecological systems: this is discussed further in a later section (16.3(c)) of this essay.

16.3 CONSERVATION TRENDS

(a) Protected landscapes and ecosystems

It is common in many nations to give protected status to areas of mature communities. So we have landscapes and ecosystems which have a special legal status and are called by a particular name: National Parks, Nature Reserves,

Game Parks and Protected Landscapes are all part of this complex (Simmons, 1981). They may in fact be relict in character since a great deal of their surroundings will have been converted by human activity to successional or inert systems. The formally protected lands are complemented by residual areas of 'unused' land where little manipulation of the ecosystems has taken place, e.g. parts of the Sahara and the North American tundra. The present network of parks and resources, excluding Greenland and Antarctica, covers about $1 \times 10^6 \, \text{km}^2$, or 1.1 per cent of the earth's land surface, and concentrates on the spectacular and unique rather than the representative. Of the world's 193 terrestrial biogeographical provinces, 35 have no national parks or equivalent reserves, and a further 38, although having at least one such area, are inadequately covered (WCS, 1980).

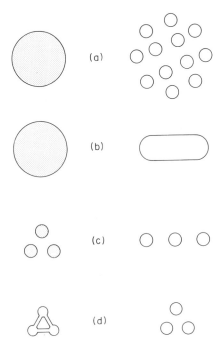

FIGURE 16.1 The geometrical rules of design of natural preserves, based on current biogeographic theory. The design on the left results in each case in a lower spontaneous extinction rate than the complementary one on the right. Both the left and the right figures have the same total area and represent preserves in a homogeneous environment. (a) A continuous preserve is better than a fragmented one, because of the distance and area effects. (b) A round design is best because of the peninsula effect. (c) Clumped fragments are better than those arranged linearly, because of the distance effect. (d) If the preserve must be divided, extinction will be lower when the fragments can be connected by corridors of natural habitat, no matter how thin the corridors. (Source: Wilson and Willis, 1975)

The size and configuration of a reserve is usually determined as much by politics as biology, but a spatially isolated population is also genetically isolated. If rates of dispersion across unprotected terrain are low, it may be essential to keep corridors of protected land connecting the main reserves. It is possible to design reserve systems whose shape and spatial relations appear to maximize the probability of survival of a threatened species (Fig. 16.1). It often looks as if a single large reserve is the most advantageous, but many smaller and scattered reserves are less susceptible to the spread of epidemic diseases or fire: edge-habitat species will thrive best where there is a high perimeter-to-area ratio; and while a single large reserve may result in fewer extinctions, smaller reserves may preserve species which otherwise would compete with each other.

Some reserves are large enough to protect whole sets of ecosystems. Nations with a lot of wild terrain can designate large zones, usually called National Parks (NP), for the conservation of biota and scenery. The parks of eastern and central Africa are perhaps the best known (Myers, 1972a, 1972b), apart from the spectacular landscapes of some of the western Cordillera of North America at places like Banff-Jasper NP, Greater Lake NP, and Yosemite NP. Even arid areas have their attractions: Australia has an Ayers Rock–Mount Olga National Heritage Area (Lacey and Sallaway, 1975). The great national parks grade into the concept of the wilderness, where large areas of land are set apart from all economic use and man's intrusion is allowed at only a low level for scientific purposes or low-impact recreations such as hiking. Wilderness zones may be designated within national parks or, as in the USA, special wilderness areas may be created within reserves of public land (Simmons, 1966). The outstanding instance is Antarctica where both the ecosystems and the ice-cap are protected by the Agreed Measures appended to the Antarctic Treaty of 1959. A wide measure of protection is given to native flora and fauna and steps are taken to prevent the introduction of alien species which might 'explode' in the biologically impoverished ecosystems of the edges of Antarctica. Whether such protection will persist in the face of increased economic interest in the resources of the Antarctic seas, its continental shelf and sub-ice rocks, and more tourism, will be a severe test of international cooperation in conservation.

(b) Outdoor recreation

Protection of landscapes and ecosystems is often carried out because areas of wild terrain are particularly sought after for recreation, whether of an active kind such as walking and camping or more inactive such as looking at scenery or cultural monuments (Simmons, 1975). Thus landscapes may be protected which also contain wild biota or even manipulated ecosystems as islands of less intensively used land within more productive landscapes. The national parks of England and Wales, and Japan, are of this type. However, management primarily for recreation may be antithetical to conservation aims (Hendee and

Stankey, 1973). In any event, large numbers of visitors usually lead to erosion around gathering places and paths, and to disturbance of the fauna. Scavengers will inevitably emerge from the fauna, and bighorn sheep, British hill sheep, blue jay, yellowhammer and brown bear have all been observed by the author filling this niche. During 1963–69, conflicts between people and grizzly bears in Yellowstone National Park had come to the level of 4.4 injury incidents per year (Cole, 1974). Nevertheless, multiple-purpose management can, if sophisticated, often keep people away from the sensitive ecosystems or biota and so the recreation areas can contribute to the total of protected systems.

(c)   Conservation of genetic variety

The tendency of modern resource development is to reduce biological diversity both by eliminating species and by breeding domesticated animals and plants from a very narrow range of genotypes within favoured taxa. The practice of cloning, which gives a uniform genetic inheritance to a crop and is therefore economically desirable, makes the organism an easy prey to climatic shifts, the introduction of new pests or indeed the resurgence of old ones. So the build-up of resources of genetic diversity becomes an important part of any conservation programme (Frankel and Hawkes, 1975; Rendel, 1975; WCS, 1980).

One place where this is done is within zoological and botanical gardens where the biota are kept in the live adult form and encouraged to breed, surplus young being sold off or possibly stored as seen in the case of plants. Apart from preserving threatened species and providing a pool of genetic material, it is argued that the presence of such places is a part of public education which will encourage the formation of biological awareness in individuals. Some zoos and gardens try harder than others in this aim. Outside such institutions, particular individuals and corporate bodies may be involved in special breeding and protection programmes. At one extreme is an idiosyncratic individual who breeds varieties of cows, pigs, or chickens which are ignored by the commercial producers and hence decline in numbers; at the other is the international research institute determined to collect and preserve viable seeds from every known variety and ancestor of a particular crop plant, for the whole world.

Gene banks may therefore take a variety of forms. Not to be underestimated are the wild places of the earth, small and large, though the latter are clearly the most important. Nobody knows when a wild plant or animal may come in useful, whether for introducing a new strain into an already domesticated crop, or as an agent of biological control, or as an entirely new crop. To take a single example, a desert plant of southwestern North America, the jojoba (*Simmondsia chinensis*) may provide the first acceptable substitute for sperm whale oil, as well as the basis of an economy for native peoples of the region (Hinman, 1984). The preservation of variety in agriculture also seems useful, as well as under experimental conditions at research institutes. These places may also have the facilities and skills to preserve parts of plants and animals containing genetic information, for example as viable seeds, as cell and tissue cultures, or as frozen semen, or perhaps even frozen embryos in the future. Human control over the nature and function of living organisms is likely to increase rapidly in the future. Much of this change is associated with the techniques of 'genetic engineering' which is applied at the subcellular level (see pp. 383). But there are numerous other developments in the field of biotechnology which are of significance, such as enhanced preservation of foods, the conversion of organic wastes to fuels, the scavenging of toxic substances and the transformation of common substances like cellulose to industrial feedstocks like ethanol and methane. Other new techniques include the cloning of cell lines to produce antibodies, and the breeding of plants tolerant of, for example, poor soils or saline conditions.

(d) Low-impact technology

The fundamental idea behind low-impact (or 'appropriate', 'alternative', 'soft') technology is access to a modest sufficiency of resources without destabilizing ecological systems: its slogan might be 'sustainability without subsidy', referring largely to fossil fuel-based energy. It often represents an attempt to live within the carrying capacity of existing systems rather than replace them with new systems requiring a higher energy subsidy to keep them stable. The ideas seem to have a particular relevance for LDCs, where the shortcomings of technology transfer from the DCs have in recent years gained a lot of exposure, not only in ecological terms (Taghi Farvar and Milton, 1972) but also socially, particularly in terms of rural unemployment and the drift to the cities. Thus 'ecodevelopment', relying more on equilibrium sources of energy together with local and traditional methods of self-sufficiency, and possibly the industrialization of renewable resources such as plant life, is the basis for exchange economies (Riddell, 1981; Tolba, 1982).

In terms of food, for example, the universalist approach of the Green Revolution is modified by a renewed valuation of traditional agriculture which must have had a survival value, since it lasted so long. Local knowledge and practical science is reflected in efficient and stable systems like shifting agriculture, wet rice terraces, and Polynesian gardens. There are many difficulties, such as how do the systems cope with rapid population growth; and can an ecologically sound way of using Amazonia be developed before it is all destroyed? The potential of forests can also be revalued since they can be sources of industrial materials, animal fodder, and human food; forest cultivation then becomes a more progressive idea than forest clearance. Protein from leaves and 'weeds' (both terrestrial and aquatic, such as water hyacinth) might add to food resources. The development of energy sources for LDcs would aim at freeing them from dependence on DC technology and prices for all except perhaps a highly technical finishing stage to an otherwise decentralized, labour-intensive process. But in rural areas, solar collection, wind power, and fermentation of organic wastes may all be applicable, and preferable to centralized sources of powers (Sachs, 1976). Such an approach has another long-term advantage: it promotes a diversity of cultural–economic ways of life, which in total has greater survival value than a reduction to a uniformity of lifestyle. It seems probable, too, that the greater social security brought about by modest but significant developments may also change attitudes to the frequency of childbearing, and so induce slower rates of population growth.

In developed countries, the scope for a change to low-impact technology might at first sight seem limited. Yet, spurred on by such eventualities as soil erosion, pesticide problems, and the spiralling price of energy, new approaches in agriculture are being brought forward. In intensive cultivation, for example, the practice of minimum tillage, a return to crop rotation, and the use of animal manure are advocated (Pimentel *et al.*, 1973) as ways of reducing both soil erosion and fossil energy input. Techniques to use lower quantities of inputs more effectively, such as fertilizer balls inserted near growing roots and trickle irrigation from punctured pipes, are also becoming more popular. Beyond these measures there are the ideas of 'radical agriculture' which transforms the consumers' demands; more self-sufficiency even in urban areas, less demand for a standardized product of a quasi-industrial nature, fewer convenience foods and, above all, less consumption of meat, underlie this approach (Wardle, 1977)

Both DCs and LDCs share the continuing interest in biological control of weeds and pests rather than the indiscriminate use of chemicals. True, new generations of toxins are appearing which are more target-selective and less persistent than, for example, the chlorinated hydrocarbons much used in the 1950s and 1960s, but where effective integrated bioenvironmental control (using techniques such as parasites, pathogens and predators of pests, habitat manipulation, crop spacing and rotation, breeding of host plant resistance, and sterile insect methods) can be maintained or introduced, it may often be cheaper and certainly pose fewer environmental problems such as the growth of resistant strains and the accumulation of poisonous residues (Huffaker, 1971).

(e)   Soft energy paths

Interest is growing in both DCs and LDCs in alternatives to the obvious development of nuclear fission power as a partial or total replacement for fossil fuels as they become too expensive to extract. Yet the advantages of alternatives to nuclear power are seen as both environmental and social. The former includes the avoidance of the waste storage problems associated with nuclear reactors, and the large quantities of waste heat per unit of electricity generated, and the latter, the turning away from possibilities of home-made plutonium bombs and the creation of a small priestly elite who would virtually control the country. Alternative energy sources, like geothermal and tidal power would, it is argued, be decentralized, would not add to the heat burden of the atmosphere, and would not pose waste-disposal difficulties. For DCs, the problem seems to be whether, even with energy conservation programmes of a high order, such alternative sources could, even with fossil fuels still available, provide sufficient power to obviate the breakdown of a basically industrial society. Public inquiries in England into the expansion of nuclear fuel reprocessing plant at Sellafield, Cumbria (1977), and the siting of a PWR at Sizewell, Suffolk (1983–84), highlighted many of these questions, and there is a lively debate between the protagonists of nuclear power (e.g. Hoyle, 1977) and those enthusiastic about alternative paths (e.g. Lovins, 1977).

16.4 ALTERNATIVE FUTURES

(a) Introduction

Implicit in most of this chapter has been the idea that different kinds of man–nature relationships are possible in the future. In the various alternatives, all the different ecological systems so far discussed might have modified roles, along lines which are not yet clear.

The chapter ends, then, with an outline of some of the emerging characterization of two of the most discussed of alternative man–nature relationships of the future. Some possible influences of these upon the ecosystem types of the front section of this chapter are given in Table 16.2.

(b) A technological future

This set of ideas is committed to the continued growth of material economies as measured by such indices as Gross National Product. The champions argue that not only will DCs continue to prosper, but LDCs will rise out of their poverty given such growth. The basis of the development will be the application of science and technology in the manipulation of nature, in which the provision of abundant industrial energy will be the key factor, made possible, eventually, by modern fusion power.

The power to change nature conferred by possession of such power might be used in two ways. The first would be to replace natural systems with totally man-made ones: given enough energy, we speculate that food can be made from granite, and artificial environments for all purposes, including the storage or processing of wastes, would replace all natural systems except presumably the atmosphere. At the extreme we can imagine with Fremlin (1964) the whole globe being covered with a building 2000 storeys high and housing $60\,000 \times 10^{12}$ people (890 years' growth from the present at 2 per cent per annum). The second alternative in this category would be the uncoupling of the ecosphere and the econosphere by, for example, putting industrial plant underground or offshore,

TABLE 16.2 Possible trends of ecosystem types in different man–nature relationships

Ecosystem type	Man–nature relationship			
	Equilibrium economy	High technology (coupled)	High technology (uncoupled)	
Successional (natural)	Retain place in natural and near-natural ecosystems.	Enlarged number in places where other systems disturbed by high energy impact, e.g. derelict land; energy affluence means reclamation of them not thought necessary.	Retain place in natural and near-natural systems.	
Successional (man-directed, e.g. agricultural)	Important but subject to limits.	Agriculture presumably of much diminished importance (except? for luxury prod.) ∴ of industrial production of food. Systems may revert to mature or be made inert.	If food production industrialized, then more natural area and role of these ecosystems much diminished; undergo succession to more mature system type.	
Mature	Important as stabilizing element in total system, also for recreation, aesthetic purposes etc.	Diminished role if wholesale manipulation of ecosystems made feasible by abundant energy: converted to inert or successional systems.	Enhanced area if pressure to convert to inert or successional states are lifted. Importance is highly valued.	
Inert (natural)	Balance achieved: nature creates but succession converts to productive systems. Major perturbations (e.g. ice-ages) may disrupt 'normal' balance.	Likely to diminish if adapted to inert but man-directed systems.	As in equilibrium economy: 'normal' place in nature.	
Inert (man-directed)	Compartment is restricted in area and where possible diminished by reclamation/conversion to productive systems.	Likely to increase as technology and/or built environment substitutes for all kinds of organic systems.	Specialized areas of inert systems may act as a decoupling buffer between ecosphere and econosphere.	
Mixed systems	Highly valued as diverse environments with capability of transforming successional mature.	Conversion to inert or purely mature systems likely.	Conversion to mature systems likely.	

and possibly enclosing cities under domes. If food could be made industrially, given cheap energy, then many ecosystems could be allowed to revert to a natural or mature condition. An initial step towards such a condition is described by Häfele (1974) who envisages power production from nuclear parks of about 30 GW capacity producing electricity and hydrogen (to replace hydrocarbon fuels) and containing all stages of the plutonium economy on one site. Waste heat would be initially led away to areas of the ocean which were thought to be ecologically and meteorologically insensitive to the release of the waste heat.

The outer limits of such a future system are perhaps dependent upon the dispersal of heat to space without disrupting the predictable functioning of the atmosphere. At present this is not generally a problem, although some cities do produce large quantities of heat. The Hung Hom district of Hong Kong produces a midwinter flux which is twice the daily incoming solar radiation in midwinter, and in Sydney, Australia, the equivalent figure is 49 per cent. Cities might therefore be the first places to have a limit imposed upon their growth by problems of waste-heat dispersal (Jaske, 1973; Kalma and Newcombe, 1976).

(c) An equilibrium future

This alternative view is based on a rejection of the idea of the replacement of nature and the substitution of a desire to live within the perceived constraints of the ecological envelope. Thus the soft energy paths and appropriate technology discussed above are seen as a mode of life for DCs and LDCs alike. The throwaway society is replaced by one devoted to recycling and thrift, whether it be food, energy or materials. In such a view, natural areas (i.e. mature systems) would lose any taint of being luxuries for the rich and be seen as essential parts of the environmental tesselation (Odum and Odum, 1972).

Transition to such an economy for the DCs might well be difficult (Daly, 1973; Pirages, 1977) but it is seen as having greater survival value, and greater cultural and ecologically diversity than the alternative (Dasmann, 1975). One other immediate problem would be whether it could cope with population growth in certain key LDCs in Asia and Latin America. In this economy, traditional economic growth would not be sought partly because the role of centralized authority necessary for such a trajectory would be greatly diminished. Neither would there be outer limits in the sense used: perception of the limits of the environment would feed back into a desire to live well within them, i.e. at a preferred level of resource use rather than an absolute level, rather like many hunter–gatherer peoples of the past (Lee and De Vore, 1968).

16.5 FINAL WORDS

It is clear that 'conservation' is intimately bound up with both economics and ecology. Both disciplines have in the past spoken with different languages and it is clear that they need to be brought together for a rational assessment of the futures available to the economics of both man and nature. One approach is that of Georgescu-Roegen (1971, 1976) who uses the concept of entropy (defined loosely as a measure of the unavailable energy in a thermodynamic system) as a linkage between the two conceptual systems. His argument summarized suggests that economic activity of an industrial kind is responsible for the rapid increase in entropy. The stock of low entropy on the globe is limited and solar energy alone represents a virtually free low-entropy addition. This is all perhaps a complicated way of saying that he favours an equilibrium man–nature relationship and in this he is joined by many others, for it has the additional advantage of not foreclosing any options for our descendants.

But the technological alternative has its attraction, not the least being a promise of material plenty for all. The responsibility for deciding which path is taken is decided by societies through their political institutions or through the transformation of the consciousness of individuals. The discipline of geography, in its desire to understand the connections of the natural and social systems of the world must have a key role to play in promoting an enhanced awareness, among a wider range of people, of their linkages with the earth.

REFERENCES

Agricultural Advisory Council (UK), 1970, *Modern farming and the Soil*, HMSO, London.
Anderson, M. A., 1977, 'A comparison of figures for the land use structure of England and Wales in the 1960s', *Area*, **9**, 43–45.
Best, R., 1981, *Land Use and Living Space*, Methuen, London.
Best, R., and Ward, D., 1956, *The Garden Controversy*, Wye College Papers in Agricultural Economics.
Cole, G. F., 1974, 'Management involving grizzly bears and humans in Yellowstone Park 1970-3, *BioScience*, **24**, 335–338.
Daly, H. (Ed.), 1973, *Toward a Steady-State Economy*, Freeman, San Francisco.
Dasmann, R. F., 1975, *The Conservation Alternative*, Wiley, Chichester.
Dotto, L., 1986, *Planet Earth in Jeopardy, Environmental Consequences of Nuclear War*, Wiley, Chichester.
Douglas, I., 1983, *The Urban Environment*, London, Edward Arnold.
Ehrlich, P. R., *et al.*, 1983, 'Long-term biological consequence of nuclear war', *Science*, **222**, 1293–1300.
Eyre, S. R., 1978, *The Real Wealth of Nations*, Edward Arnold, London.
Frankel, O. H., and Hawkes, J. G. (Eds), 1975, *Crop Genetic Resources for Today and Tomorrow*, IBP Studies 2, Cambridge University Press, Cambridge.
Fremlin, J. H., 1964, 'How many people can the world support?', *New Scientist*, **24**, 285–287.
Georgescu-Roegen, N., 1971, *The Entropy Law and the Economic Problem*, MIT Press, Cambridge, Mass.
Georgescu-Roegen, N., 1976, *Energy and Economic Myths*, Pergamon Press, Oxford.
Grove, A. T., 1977, 'Desertification', *Progress in Physical Geog.*, **I**, 296–310.
Häfele, W., 1974, 'A systems approach to energy', *American Scientist*, **62**, 438–447.
Hendee, J. A., and Stankey, G. H., 1973, 'Biocentricity in wilderness management', *Bioscience*, **23**, 535–538.

Hill, A. R., 1975, 'Ecosystem stability in relation to stresses caused by human activities', *Can. Geog.*, **19**, 206–220.

Hinman, C. W., 1984, 'New crops for arid lands', *Science*, **225**, 1445–1448.

Holdgate, M. W., 1979, *A Perspective of Environmental Pollution*, Cambridge University Press, Cambridge.

Hoyle, F., 1977, *Energy or Extinction?* Heinemann, London.

Huffaker, C. B. (Ed.), 1971, *Biological Control*, Plenum Press, New York.

Jaske, R. T., 1973, 'An evaluation of energy growth and use trends as a potential upper limit in metropolitan development', *The Science of the Total Environment*, **2**, 45–60.

Jowett, D., 1972, 'The quantitative assessment of environmental impacts', in Ditton, R. B., and Goodale, T. L. (Eds), *Environmental Impact Analysis: Philosophy and Methods*, University of Wisconsin, Madison, pp. 127–136.

Kalma, J. D., and Newcombe, K. J., 1976, 'Energy use in two large cities: a comparison of Hong Kong and Sydney, Australia', *Env. Studs.*, **9**, 53–64.

Lacey, J. A., and Sallaway, M. M., 1975, 'Some aspects of the formulation of a planning policy for the management of the Ayers Rock–Mount Olga National Heritage Area', *Proc. Ecol. Soc. Austral.*, **9** 256–266.

Lee, R. B., and De Vore, I. (Eds), 1968, *Man the Hunter*, Aldine Press, Chicago.

Lovins, A. B., 1977, *Soft Energy Paths: Towards a Durable Peace*, Penguin, Harmondsworth.

Mabbutt, J. A., 1977, 'Climatic and ecological effects of desertification', *Nature and Resources*, **13**, 3–9.

Martin, R. S., 1973, 'The discovery of America', *Science*, **179**, 969–974.

Myers, N., 1972a, 'National parks in savannah Africa', *Science*, **178**, 1255–1263.

Myers, N., 1972b, *The Long African Day*, Macmillan, New York.

Myers, N., 1979, *The Sinking Ark*, Pergamon, Oxford.

NAS (National Academy of Sciences, USA), 1972, *Genetic Vulnerability of Major Crops*, NAS, Washington D. C.

Odum, E. P., 1969, 'The strategy of ecosystem development', *Science*, **164**, 262–270.

Odum, E. P., 1975, *Ecology*, 2nd edn, Holt, Rhinehart & Winston, New York.

Odum, E. P., and Odum, H. T., 1972, 'Natural areas as necessary components of man's total environment', *Trans. 37th North American Wildlife and Natural Resources Conf.*, pp. 178–189.

Pimentel, D., *et al.*, 1973, 'Food production and the energy crisis', *Science*, **182**, 443–449.

Pimentel, D., *et al.*, 1976, 'Land degradation: effects on food and energy', *Science*, **194**, 149–155.

Pirages, D. C. (Ed.), 1977, *The Sustainable Society*, Praeger, New York.'

Rendel, J., 1975, 'The utilization and conservation of the world's animal genetic resources', *Agriculture and Environment*, **2**, 101–119.

Riddell, R., 1981, *Ecodevelopment*, Gower, Farnborough.

Sachs, I., 1976, 'Environment and styles of development', in Matthews, W. H. (Ed.), *Human Needs*, Dag Hammarskjold Foundation, Uppsala, pp. 41–65.

Simmons, I. G., 1966, 'Wilderness in the mid-twentieth century USA', *Town Planning Rev.*, **36**, 249–256.

Simmons, I. G., 1975, *Rural Recreation in the Industrial World*, Edward Arnold, London.

Simmons, I. G., 1981, *The Ecology of Natural Resources*, Edward Arnold, London.

Sipri, 1977, *Weapons of Mass Destruction and the Environment*, Taylor & Francis, London.

Taghi Farvar, M., and Milton, J. P. (Eds), 1972, *The Careless Technology, Ecology and International Development*, Natural History Press, New York.

Temple, S. A., 1977, 'Plant-animal mutualism: co-evolution with Dodo leads to near extinction of plant', *Science*, **197**, 885–886.

Terborgh, J., 1975, 'Faunal equilibria and the design of wildlife preserves', in Golley, F. B., and Medina, E., *Tropical Ecological Systems Trends in Terrestrial and Aquatic Research*, Ecological Studies, 41, Springer-Verlag, New York, pp. 369–380.

Tinker, J., 1971, 'One flower in ten faces extinction', *New Scientist*, **50**, 408–413.

Tolba, M. K., 1982, *Development without Destruction: Evolving Environmental Perceptions*, Tycooly, Dublin.

Turco, R. P., 1983, 'Nuclear winter: global consequences of multiple nuclear explosions', *Science*, **222**, 1283–1292.

Van Den Bosch, R., and Messinger, P. S., 1973, *Biological Control*, Intertext Books, New York.

Wardle, C., 1977, *Changing Food Habitats in the UK*, Earth Resources Research, London.

Warren, A., 1984, 'The problems of desertification', *Sahara Desert*, in Cloudsley-Thompson, J. (Ed.), *Sahara Desert*, Pergamon, Oxford, pp. 335–342.

WCS (World Conservation Strategy), 1980, *Living Resource Conservation for Sustainable Development*, IUCN/UNEP/WWF, Gland.

Westing, A. H. (Ed.), 1984, *Herbicides in War. The Long-Term Ecological and Human Consequences*, Taylor & Francis, London.

Wilson, E. O., and Wills, E. O., 1975, 'Applied biogeography' in Cody, M. L., and Diamond, J. M. (Eds), *Ecology and Evolution of Communities*, Bel Knapp, Cambridge, Mass. pp. 522–534.

Woodwell, G. M., 1970, 'Effects of pollution on the structure and physiology of ecosystems', *Science*, **168**, 429–433.

Ziswiler, V., 1967, *Extinct and Vanishing Animals*, Springer-Verlag, New York.

PART VI

CONCLUSION

INTRODUCTION

In attempting to provide a review of the impact of human activity on environmental processes, the approach taken in this volume has involved a subdivision of these *processes* according to the major environmental compartments of the atmosphere, hydrophere, geosphere, pedosphere and biosphere. Each of these compartments has been considered in turn and the significance of human activity to its functioning assessed. Other approaches could have been taken, and one alternative might have involved a division of the subject according to major categories of *human activity*. Individual sections could therefore have considered such topics as the effects of forest exploitation and clearance, the impact of agriculture or the effects of mineral extraction. With forests currently covering approximately 3 billion hectares of the earth's surface and with predictions that the tropical forests will shrink by 10–15 per cent by the end of the century, the impact of this branch of human activity is clearly of major significance. A treatment of this impact could focus on such topics as the effects on global climate and the global and regional water balance, the off-site effects on downstream flooding and sediment transport and the more local changes in microclimate, geomorphological processes and soil characteristics. A wider perspective might also have touched upon the increasing concern for the effects of acid rain on forest production stemming from recent reports of widespread destruction of forests in Northern and Central Europe.

Had this approach been followed it would also have been essential to devote a chapter to the effects of urbanization on environmental processes. In 1980 more than 40 per cent of the world's population lived in urban areas and it is predicted that this proportion will increase to nearly 50 per cent in the year 2000. The distinctive nature of the physical environment of these areas has been emphasized by Douglas (1983) in his book *The Urban Environment*. Many of the impacts of urban development are discussed in this volume but the reader must piece together the evidence. Thus, for example, the effects on the precipitation regime are discussed in Chapter 3, Chapter 4 discusses the impact on hydrological processes, Chapters 9 and 11 discuss a number of the geomorphological changes associated with urban development and Chapter 15 reviews species response to urbanization.

Another approach might have included a number of chapters providing case studies of specific environmental *problems* associated with contemporary human impact. Chapter 17 has been included to illustrate this possibility and provides an appraisal of the problem of desertification which has received widespread media coverage in recent years as a consequence of the disastrous famines in the Sahel region of Africa. In this chapter, A. Rapp reviews the causes underlying the phenomenon of desertification and emphasizes the delicate balance between environmental processes and human activity in dryland areas.

A discussion of human impact on environmental processes must inevitably touch upon many broader environmental issues. These largely lie outside the scope of this volume which has adopted an essentially objective view of the human impact, but it is important that the physical geographer should be aware of both the wider implications of this theme and also of his potential contribution to management issues. The final chapter therefore attempts to highlight a selection of topics which, it is hoped, will point the reader towards some of these wider perspectives. These include the growing international awareness of the detrimental effects of human activity on the environment, the need for monitoring strategies and programmes, the development of procedures for environmental impact assessment and the role of modelling and prediction techniques.

REFERENCE

Douglas, I., 1983, *The Urban Environment*, Edward Arnold, London.

Human Activity and Environmental Processes
Edited by K. J. Gregory and D. E. Walling
©1987 John Wiley & Sons Ltd.

17

Desertification

A. RAPP

17.1 INTRODUCTION

The concept of extending environmental concern to the third World—'only one Earth'—made its international breakthrough at the 'UN Conference on the Human Environment', held in Stockholm in 1972. One year later, in 1973, it became known to the world, that a very severe social and environmental catastrophe had struck the Sahel dryland belt south of the Sahara, and had caused the death of about 200 000 people and millions of cattle in a zone stretching across Africa from Senegal on the Atlantic coast to Ethiopia on the Red Sea. The Sahelian catastrophe of starvation and death among people and cattle had been caused by a 5–6 year drought period lasting from 1968 to 1973. (Figs. 17.1 and 17.2). What are the reasons behind this catastrophe and how could it be avoided in the future? These were two of the main themes discussed and analysed at the UN Conference on Desertification (UNCOD), held in Nairobi in 1977. It was the largest international effort launched so far to coordinate and organize the understanding and control of the problem of desertification.

However, most of the countries which had participated in the UNCOD conference did not take the warnings concerning repeated drought catastrophes, desertification and famine seriously, as there was only limited financial and real support for the Plan of Action adopted at UNCOD. In 1984 a follow-up meeting was convened by UNEP, seven years after the conference on desertification. It was concluded that the threat of desertification in 1984 was worse than ever, with a second phase of Sahelian drought in Africa's drylands south of the Sahara having returned in the late 1970s and worsened in severity through 1981–1984 (Figs. 17.1, 17.2). The death toll of people in Ethiopia, Sudan and Chad may this time increase to more than a million victims.

It was concluded at the 1984 UNEP meeting that

desertification is related to four principal groups of factors: natural vulnerability of the ecosystems in drylands and adjoining subhumid

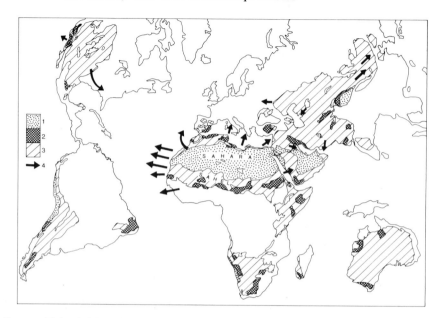

FIGURE 17.1   A Map of deserts and desertification, based on the UNCOD 'World map of desertification' (UNCOD, 1977). Key: 1—Hyperarid zones (desert). 2—Very high risk of desertification. 3—High or moderate risk of desertification. 4—Major plumes of soil dust in the air, observed in satellite images (based on Grigoryev and Kondratyev (1981), and other sources)

territories; population pressures often leading to over-exploitation of resources; economic considerations that hinder the establishment of appropriate land use on a long-term basis . . . ; and political unrest that is not conducive to the long-term actions required for such programmes. (UNEP Governing Council report, 1984)

To what extent are the disasters associated with desertification disasters due to natural causes such as large-scale fluctuations in rainfall, and how important is the impact of human activity? This distinction is an important one to analyse. 'If drought is the cause, we can say there is little man can do to counter a natural phenomenon, leaving us with an excuse to do nothing. If, however, man is the cause, then man has the opportunity to undo the damage he has done or at least to prevent further deterioration' (Dregne, 1983, Preface). We will approach this important issue by trying to answer at first two questions. 'How is desertification defined?' and 'Where does it occur?'.

17.2   DEFINITION AND DISTRIBUTION OF DESERTIFICATION

In this text we use the term desertification to mean the spread of desert-like

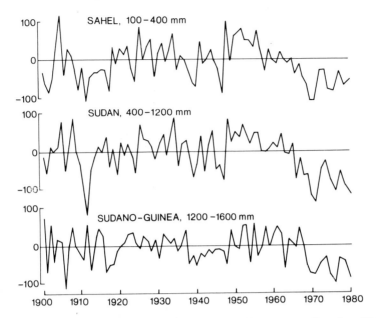

FIGURE 17.2   Annual rainfall variation in three sub-Saharan zones (based on Nicholson, 1983). In these zones stretching across Africa drought recurs periodically as shown by the weather records for the period 1900–1980. Land is exploited to its limits during wet periods, and seriously over-exploited in the subsequent dry spells, leading to land degradation and desertification (based on Hare (1984) and Nicholson (1983)). 100 = one standard deviation

conditions of low biological productivity to drylands outside the previous desert boundaries. Desertification is long-term degradation of drylands, resulting either from over-use by man and his animals, or from natural causes such as climatic fluctuations. It leads to loss of vegetation cover, loss of topsoil by wind or water erosion, or loss of useful plant production as a result of salinization or excessive sedimentation associated with sand dunes, sand sheets or torrents. The effects extend over a longer timescale than the year to year fluctuations of rainfall in these zones of low rainfall and high variability of precipitation but this variability is an important influence. For example the Sahelian zone of thorn-scrub savanna and dry savanna south of the Sahara desert receives an annual rainfall ranging from 100–150 mm at the desert boundary to 600 mm in the south at the Sudanian zone boundary (Fig. 17.1), but the annual variability is very large (Fig. 17.2).

One of the main documents on desertification compiled by the UN agencies is the 'World Map of Desertification' (UNCOD, 1977). A simplifed version is depicted by Fig. 17.1. The map affords a broad overview of the magnitude and extent of the problem, as interpreted by the team of experts responsible for its compilation in 1977. It should be remembered that the map is based on the scanty information available from vast, almost empty areas, and that it can therefore only give a very generalized picture of desertification hazards. Restrictions of scale (Original at 1:25 million) and lack of essential information for global coverage are important limitations. A more appropriate title for this map might be 'A World Map of Desertification Hazards'. One of the recommendations in the Plan of Action of 1977, was to improve existing knowledge of the actual extent and severity of desertification by improved monitoring and mapping, for example by developing remote sensing and ground checks as a means of monitoring and mapping desertification at the regional scale. Fig. 17.9 in this chapter provides an example of such a map of the Nile river basin.

Desertification is either irreversible or reversible degradation. Some of the salinized lands of ancient Babylonia are still unproductive salt flats. However, the wastelands produced by the wind erosion and sand covers of the 'dust bowl' catastrophe of the 1930s in the mid-western States of the USA were restored and recovered as grazing lands a few years later, cured by the US Soil Conservation Service and by investments from a new generation of farmers (Lockeretz, 1978).

17.3 OVERGRAZING

The most widespread land use of drylands such as the Mediterranean steppes or the Sahelian thorn-scrub savanna is animal grazing. Dry ecosystems thrive under moderate use and slight grazing stimulates plant growth. But it is very easy to pass the threshold from optimal grazing to destructive overgrazing followed by land degradation. If too many animals—generally sheep, goats, camels or cows—are allowed to graze and browse for too much of the year, the signs of overgrazing soon develop. Often it is the combination of trampling by many hooves which destroys the structure of the topsoil and the excessive grazing which reduces the protecting vegetation cover which results in the trampled topsoil being blown away by wind erosion or washed away by water erosion (Novikoff, 1983).

Overgrazing results in decrease of palatable and valuable annual and perennial plants. They are gradually replaced by unpalatable species—so-called increasers—which lower the quality of the forage. When the vegetation is badly overgrazed and trampled, wind erosion occurs, causing growing deflation patches with sand ripple marks and sand tails accumulating on lee sides of remaining vegetation tufts or stones exposed from the underlying duricrust (Figs 17.3 and 17.4). The down-wind side of the deflation area is marked by growing sand dunes or sand sheets. The most valuable fine fractions of the topsoil, namely the silt- and clay-size particles and organic matter are transported by the wind over long distances during dust storms. The dust clouds may contribute to loess soil covers if they are deposited on land, or to fine-grained marine sediments in the sea (Fig. 17.1).

FIGURE 17.3   The effects of overgrazing. After five years of overgrazing by sheep and goats this local patch of desert with moving sand ripples in south Tunisia was formed in an area previously occupied by steppe with *Ranterium* shrubs. A topsoil layer of ca. 20 cm thickness was removed by wind erosion. (Photograph: A. Rapp, 1974.)

One particular aspect of overgrazing is the concentration of livestock around villages and watering points. As a result rangeland is frequently destroyed within circles with a radius of 5–6 km around the villages and the wells. This corresponds to the maximum distance that sheep, goats and cows can walk for watering every day in the dry season (Le Houérou, 1976). These overgrazing circles are wider than the inner desert patches which are typical around villages in the Sahelian belt. The inner desert circles around villages were studied by Helldén (1984), using aerial photographs taken in 1962 and Landsat images covering the years 1973 and 1979. The inner circles of completely bare ground around 77 villages in Kordofan, Sudan, exhibited the same radius of a few hundred metres over the period from 1962 to 1979. They are not a useful indicator for monitoring of desertification by satellite imagery, and may have the function of firebreaks or other safety zones (Helldén, 1984).

In thorn-scrub savannas, grazing land can also deteriorate due to *too little* grazing and bush encroachment, as shown by Conant (1982). He described a case from north Kenya, where dense thorn-bush had expanded over former good grazing land, after an area had been abandoned for several years owing to local, internal war.

Desertification induced by overgrazing can be countered by introducing controlled, rotational grazing, increased stall feeding and improvement of fodder

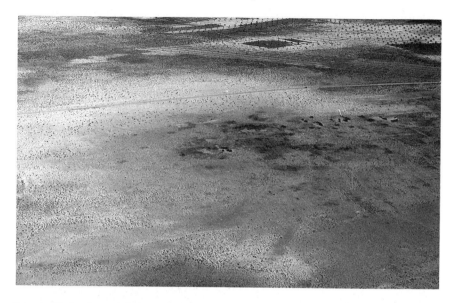

FIGURE 17.4   An aerial view of a local patch of desert developed around the homestead of a newly settled nomad family (house and black huts). Cultivation near the huts, overgrazing, deflation and small sand dunes in the bright semicircle downwind (left) of settlement are visible. This site is in the same area as Fig. 17.3. (Photograph: A. Rapp, 1974.)

for grazing and browsing. (One very important means of counteracting overgrazing is to improve marketing facilities for selling surplus animals and meat. This may involve construction of access roads and slaughterhouses, and the introduction of safe meat price levels and planned spacing of watering points for livestock. Desertification of grazing lands must, however, be seen as a complex process which is also influenced by excessive collection of wood and by overcultivation in rangeland areas. These factors will be considered in the following sections of this chapter.

17.4   OVERCULTIVATION AND EROSION

The population of the arid zones of North Africa and the Middle East has increased approximately six- to seven-fold since the beginning of the present century and the area cultivated for cereals has increased to a similar extent. The productivity of periodically ploughed areas is generally lower than that of similar areas under natural vegetation. Cultivation of steppe lands is also becoming increasingly mechanized. Since the 1960s, cultivation in North Africa's drylands has been increasingly carried out by tractor and disc ploughs. This must be seen as an example of a non-adapted technology in these areas, because

FIGURE 17.5 A large patch of desert south of Medenine, Tunisia. The lower part of this vertical air photograph shows a bright surface with many small sand dunes trending in an east–west direction. A distinct limit is seen to the dark-toned area of steppe vegetation of perennial shrubs covering the upper part of the picture. Dunes and desertification near the lower wadi channel (Oued Fessi) were caused by tractor ploughing, probably in the 1960s. (Air photograph: 1975, T.A.S.)

it destroys the perennial vegetation and leaves the ground barren for several years after harvest, particularly when it is followed by free grazing of the stubble and crop residues. Tens of thousands of hectares of sand dunes have developed in southern Tunisia since the early 1960s according to Le Houérou (1976, p. 133). Mabbutt and Floret (1980) provide a somewhat different view based on the Oglat Merteba study in southern Tunisia. Desertification there is not very spectacular, but takes the form of a continuous reduction in biological productivity (see Fig. 17.5).

Overcultivation of marginal grazing lands has led to severe dryland degradation not only in the Third World, but also in the USA and the Soviet

Union. The so-called 'dust bowl' catastrophe of farmland destruction which struck several states of mid-western USA in the 1930s is an example of local desertification in North America, although it was successfully reversed in the following years. It caused the birth of the US Soil Conservation Service, which became an efficient organization for rehabilitation and recovery of threatened land. However, from time to time land degradation of different kinds still creates severe losses of property and production capacity of land in the USA. One such case was a severe dust storm which occurred in Texas and New Mexico in February 1977 (Fig. 17.6). During this storm, ploughed fields were locally eroded

FIGURE 17.6   Two dust storm plumes registered by the weather satellite GOES-1 on February 23, 1977. One dust plume (A) begins at the Texas–New Mexico border, the other (B) in Colorado. They both have a grey tone and are readily distinguished from the white representing ordinary clouds or snow cover in the Rockies. (From Kessler *et al.*, 1978.)

to depths of more than one metre. Fine sand winnowed from certain vulnerable soils was deposited in lobate sheets that extended several kilometres downwind from the ploughed fields and blowouts. During the storm, dust obscured about 400 000 km^2 of ground surface in the south-central United States. The use of GOES satellite imagery to track the dust represents a technological advance in the study of such storms. In addition aerial reconnaissance and ground studies of one source area in New Mexico were carried out within a week of the storm's dissipation, before the evidence of wind erosion and deposition was removed by reploughing farmland and clearing roads (McCauley *et al.*, 1981) (cf. Fig. 17.6). This case also demonstrates the great contrasts in the approach to dealing with desertification events that may exist between a rich country like the USA and poor developing countries. The documentation of the February 1977 storm effects was very good, with data available from satellite images, aerial photography, weather reports and ground checks. The treatment of the land to aid recovery was quick. One week after the damaging storm the land was reploughed and the roads were cleared, probably at great cost to the community and the taxpayer.

The map presented in Fig. 17.1 also shows some cases of large dust storms observed on satellite images over parts of Africa, the Soviet Union and China (Grigoryev and Kondratyev, 1981). Some of these were in the area affected by the massive dryland degradation which followed the attempts to plough and cultivate former extensive grazing lands on the steppes of the so-called Virgin Lands in Kazakstan, central Asia. From 1954 to 1960 the Soviet government extended cultivation into about 40 million hectares of steppe lands for production of spring wheat. As early as 1956 the yields began to drop and the degradation by wind erosion culminated in the very dry and windy spring of 1963. It has been estimated that 17 million hectares of land were severely damaged by wind erosion in the period 1962–1965 and that 4 million hectares of these were irreversibly destroyed (Eckholm, 1976). This cultivation programme was subsequently abandoned and most of the land was reclaimed by conversion back to controlled grazing. In the drylands of developing countries overcultivation by mechanized farming and by traditional small-scale cultivation are probably causing more land losses than those caused by overgrazing. The so-called Afro-Atlantic dust plumes blow from the Atlantic coast of west Africa to the West Indies in about 5–7 days. In the summer of 1974 a total transport of 200 million tonnes of African soil moved as dust clouds westwards over the Atlantic, according to the observations and calculations of the GARP project (Prospero and Nees, 1977).

17.5 EXCESSIVE WOOD COLLECTION

Wood from trees and bushes is used by the farmers and herders of drylands as firewood for cooking and as material for fences and huts. Trees and bushes grow slowly in dry climates and are therefore increasingly over-exploited by the population of the countryside and towns. In the oil-producing countries people can afford to use oil, gas or kerosene for domestic cooking and heating, but in the countries without oil, a rapidly growing energy crisis is emerging, owing to lack of firewood.

Trees and bushes are of great ecological importance in the dry ecosystems. They shelter the soil from erosion by wind and water. They provide food for people and fodder for animals, and also fuel-wood for domestic cooking, baking and heating. In addition, they provide wood for building of huts and fences. Ibrahim (1980) has studied the pattern of wood consumption by farmers' families in the Sahelian zone of Africa and he found that the average family consumes about 200 trees or large shrubs per year. Of these 50 are used as firewood, 100 for fencing outside the village and 50 for upkeep of fences and huts in the village. Herders also cut trees to feed their animals with leaves, and thus contribute to the wood shortage.

The extent of afforestation in the drylands is rather limited and the areas of planted woodland are not nearly sufficient to meet the enormous demand for wood. Most of the tree plantations are 'green belts' located on the perimeter of towns to protect them against drifting sand and dust storms. A large part of the tree cover in the Sahel belt of Sudan is of local drought-resistant acacia

FIGURE 17.7   Sand dune fixation by tree planting and grazing exclosure (foreground and grey zone on dune in background). El Bashiri oasis, Sudan. (Photograph: A. Rapp.)

trees (*Acacia senegal*). These produce gum arabic which was previously the main cash crop of this zone of Sudan (Ibrahim, 1980, p. 23).

In Morocco's Atlas mountains, the local farmer's cycle of land degradation is reported to destroy about one hectare of land per family per year. Each year the average family needs about one hectare of macchia mountain secondary forest for cooking and baking. After clear-cutting, the land is cultivated without terracing or other conservation measures. Soil erosion by water is rapidly destroying the steep slopes, which after a number of years of cultivation are taken over by grazing goats.

Well-planned and systematic tree-planting to provide shelterbelts, windbreaks, fruit, fodder and firewood is probably one of the top priorities for counteraction of desertification. However, it should preferably be undertaken in relation to clearly defined objectives such as protecting roads, villages and irrigation canals, and not vaguely defined and large-scale objectives such as the establishment of a transnational north Saharan greenbelt. This was conceived as a green barrier of shelterbelt trees extending from the Atlantic along the northern Saharan border zone to the Nile valley, to prevent the desert from spreading northwards. It was based on the misconception that desertification is occurring as an expansion of the desert along a continuous front of moving sand. New patches of desert can, however, form in scattered locations far from the present desert

FIGURE 17.8   Avdat experimental farm for runoff irrigation (foreground) in the Negev desert, Israel. The dark patches behind a series of stone dams on the opposite valley slope are so-called jessour dams used for runoff farming and dating from the Nabatean culture. They are 2000 years old, but still functioning. On the skyline are the ruins of the Roman town of Avdat. (Photograph: A. Rapp.)

boundaries, where local over-exploitation of land, water and vegetation resources occurs (Fig. 17.1).

17.6   SALINIZATION AND OTHER PROBLEMS OF LAND DEGRADATION IN EGYPT

In ancient times Egypt was called 'the gift of the Nile'. The agricultural production of the Nile valley and its delta are entirely dependent on irrigation water from the Nile river. For 5000 years the natural annual flood of the Nile supplied both water and fertile silt to the floodplain and delta agriculture, without any long-term damage or degradation. Since the completion of the construction of the High Dam at Aswan and the damming of Lake Nasser, the flow of the Nile and its irrigation water has been regulated. This occurred in 1970 and made it possible to obtain at least two irrigated crops per year and to produce electricity for the whole country. But there have been several negative consequences for agricultural production in Egypt. In 1982 a meeting of agricultural experts and scientists held a National Conference on the Problems of Land Degradation in Egypt. The three most important forms of land degradation in Egypt were listed as follows (Kishk, 1982; Mensching, 1982).

1.   Salinization and waterlogging has caused losses of considerable areas of fertile agricultural land in the Nile valley and the Nile delta as a result of over-irrigation. Excess application of irrigation water has led to a gradual rise of the groundwater level, bringing dissolved salts up to the root zone of the crops and killing them. With an estimated affected area of 3 million acres in Egypt and an estimated average yield reduction of 20 per cent the total loss of production corresponds to the loss of an area of 600 000 acres. As indicated in Fig. 17.9, areas of Egypt experiencing particularly severe problems of salinization include the lower parts of the Nile delta, the Fayum and other large oases, and also parts of the main Nile valley.

The prevention of salinization and water logging in dryland irrigation involves the digging of deep and efficient drainage ditches, to drain off the salinized water to below the root zone, and decreasing the amount of irrigation. Drainage is, however, a costly and time-consuming operation, which must be well planned and financed. Salinization has come to Egypt along with the new irrigation system of Lake Nasser. The traditional farmers have no earlier experience of it and do not know how to cope with this insidious type of desertification, which sterilizes previously highly productive land and brings disaster to the small-scale farmers.

2.   Deterioration of soil fertility as a result of the loss of Nile alluvial silt now deposited in Lake Nasser and for other reasons. Average yield reduction in the

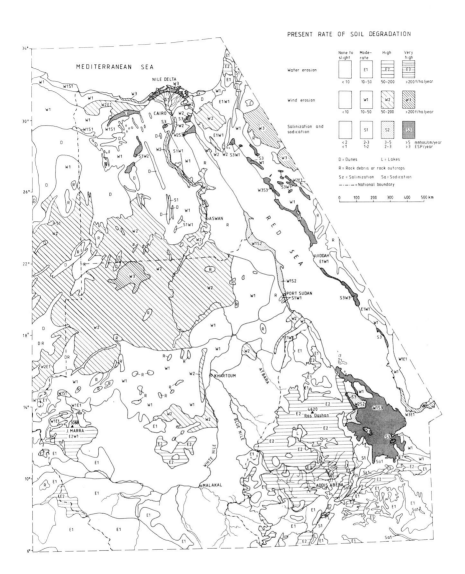

FIGURE 17.9   A map of the present rate of soil degradation in the Nile river basin and adjacent areas. Water erosion (E1 to E3) is marked as predominant in the mountainous areas of Ethiopia and south Sudan. Wind erosion (W1 to W3) dominates on both sides of the Nile Valley from Khartoum to Cairo. Salinization (S1 to S3) strongly affects many irrigated croplands with bad drainage, e.g. the lower parts of the Nile Delta and lowlands near the Red Sea coasts. (Source: FAO, UNEP and UNESCO, 1980.)

whole cultivated area of Egypt was estimated at 10 per cent. This corresponds to a loss of production from 600 000 acres.

3.   Desert encroachment by aeolian sands blown from the desert and deposited as thin sand sheets on the western part of the Nile valley, and as predominantly colluvial wash sediments on the eastern part of the floodplain. The harmful effects of such desert sediments have extended over a zone of cultivated lands at least 3 km wide on both the eastern and western sides of the Nile Valley during the last 20 years (Kishk, 1982). An area of 1 750 000 acres is estimated to be affected with a 20 per cent reduction in crop production, corresponding to loss of production from 350 000 acres.

17.7   DUST STORMS AND DESERTIFICATION

Recent studies of dust storms from deserts and semiarid drylands have shown that large quantities of aeolian soil dust is transported and deposited on land or in the sea, often at long distances from the source areas (see Fig. 16.1). Much information on the erosion, transport and deposition of aeolian sand and dust is presented in publications by Goudie (1978), Morales (1979) and Péwé (1981). The environmental impact of dust storms in drylands is threefold: first, *erosion* and loss of productive topsoil, secondly, *transport* of dust in the air causing air pollution and possible changes in albedo and other climatic variables, and thirdly, *deposition* of wind-borne dust either as loess cover on land or in the sea or lakes, with associated effects on water ecosystems and bottom sediments. Analyses of dust plumes on satellite images and of the content of dust particles in the air has shown that more than 100 million tonnes of African dust is blown westwards each summer over the Atlantic (Prospero and Nees, 1977; Jaenicke, 1979). The importance of this process was observed and discussed by Charles Darwin as early as 1832.

The transport of aeolian dust northwards across the Mediterranean is more irregular and of lesser magnitude than the Afro-Atlantic dust plumes. Studies of loess deposits in North Africa, Israel and southern Europe are providing increasing evidence that such deposits have a long history and were formed during episodes in the Pleistocene and Holocene (Yaalon and Ganor 1973; Rognon, 1984: Rapp, 1983).

Studies of contemporary deposition of Sahelian dust as fallout from the Harmattan winds in Nigeria have been reported by McTainsh (1984). Barth (1982) has described the current serious effects of a combination of water and wind erosion in degrading fossil dune fields and other areas in the Sahelian belt in Mali. The material transported by runoff down the southern slopes of the dunes is deposited and, after drying out, is transported away by wind in sand and dust storms. Moving sand was observed on 28 days during the dry season of 1979 at Mopti airport (Barth, 1982, p. 218).

17.8   CLIMATIC ANOMALIES AND DESERTIFICATION

The recent droughts and famines in the African Sahel and Ethiopia have stimulated much discussion concerning their causes and effects, and also on possible strategies for countermeasures. A basic question that must be considered is whether the recent African drought periods dating from 1968 are due to a climatic trend towards drier conditions south of the Sahara, or if they represent a pattern of fluctuations without any long-term trend. If the climate is changing towards greater aridity in Africa south of the Sahara, then man will face increasing difficulties in a continued use of the Sahelian zone. The persistence of the Sahelian drought has been demonstrated by Lamb (1982), Nicholson (1983) and Olsson (1983). The latter study emphasizes the decrease in number of days with rainfall in excess of 1 mm.

One possible way of assessing whether or not the current droughts mark a new trend towards drier climate is to compare them with records of earlier droughts. Conventionally, climate is représented by the mean and variance of a long series (usually 10 to 30 years) of observation. Short-term variations of meteorological parameters may be viewed as 'weather changes', and as being different from long-term 'climate changes'. An understanding of past climates can be based on weather station records which stretch back to the last century. For the more distant past, historical, archaeological and dendro-chronological evidence for climatic anomalies can be used. Other climatic indicators include specific landforms, sediments and fossils; for example, ancient lakes, sand dunes or river deposits which reflect former environmental conditions.

There is, for instance, growing evidence, that the whole Sahelian belt in Africa south of the Sahara was the scene of at least two periods of desert expansion and sand dune penetration, culminating in the periods 19 000–15 000 BP and 11 000–10 000 BP. The first phase is contemporary with the maximum expansion of the Weichselian ice sheet in northern Europe. The last phase corresponds in time to the final re-advance of the Weichselian ice sheet during the so-called Younger Dryas cold period in Europe. If this interpretation is correct, it means that the two major Late Pleistocene glacial advances in Europe also meant a southward shift of the Saharan desert belt in Africa, a penetration of sand dunes 500–800 km southward into the Sahel, and thus the occurrence of 'natural desertification', since the impact of man and grazing animals in those periods was probably minimal (see Fig. 17.10).

Professor P. Rognon, a leading expert on arid lands and climatic history, has discussed the important hypothesis of parallelism between cold anomalies in Europe/North Atlantic, and dry phases in the Sahelian zone (Rognon, 1984). He defines and classifies climatic fluctuations or 'accidents' which produce a catastrophic impact on the environment into the following types, according to the timescale involved.

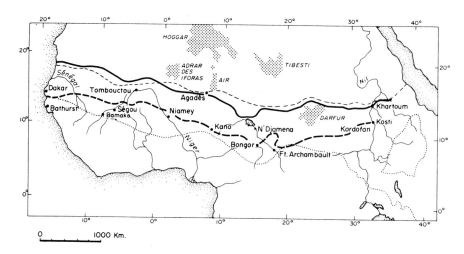

O_____1000 Km.

FIGURE 17.10   The southern limits of active sand dunes (full line) and fossil dunes (heavy dashed line) in the Sahel, indicating the extent of 'natural' desertification during Pleistocene arid phases. The thin dashed line represents the 150 mm annual rainfall isohyet and the thin dotted line is the 1000 mm rainfall isohyet. (*Reproduced, with permission, from Mainguet et al., 1980.*)

(1)   *Weather anomalies* ('anomalies météorologiques') have a duration from some hours to some days. The catastrophic rainfalls of 1969 in Tunisia are an example of this category.

(2)   *Climate anomalies* ('anomalies climatiques') have a duration of one to a few years. The 16-month drought period in Britain and the 10-month drought in western France in 1975–76 are examples of this category.

(3)   *Extended climate anomalies* ('anomalies d'une durée exceptionelle') can continue for more than a decade. The Sahelian drought which began in 1968 and still continues in 1984 can be grouped in category (3) (see Fig. 17.2 and Cross, 1985) and the so-called Great Drought which ruined the agricultural civilization of the Navajo region in North America which was dated by dendrochronology to the period 1273–1285 provides another example.

(4)   *Climate crises* ('crises climatiques') have a duration of a few hundred to a thousand years. The so-called Little Ice Age affected vast areas of the globe for about 300 years (1550–1850) and involved a lowering of the average annual temperatures by 0.5°C to 1.5°C. The beginning of this period seems to have been marked by high rainfall in the Mediterranean and the Sahel, but drought anomalies occurred in the latter area around 1680, during the period 1738–56 and around 1830 (Rognon, 1984, p. 62). Fig. 17.11 illustrates the interesting working hypothesis, proposed by Rognon (1984), that dry climate anomalies in the Sahelian belt may occur

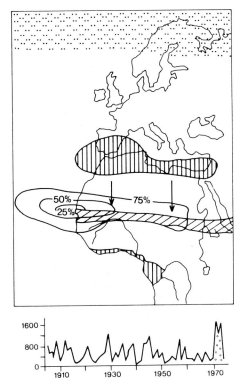

FIGURE 17.11   The climate anomaly of 1968–1972 interpreted as a general southward shift of climate belts in Eurafrica. Cooling in high latitudes (dots) is thought to be contemporary with high rainfall anomalies in North Africa and in areas near the equator and with drought in the Sahel. For comparison, a curve of the frequency of icebergs observed in the Atlantic off Labrador is taken as an indicator of cooling in high latitudes. (Based on Rognon, 1984.) Vertical hatching shows high rainfall anomalies, oblique hatching low rainfall anomalies. Percentage of normal annual precipitation is shown for the Sahelian zone (Reproduced with permission from P. Rognon.)

in phase with cold anomalies in northern Europe and the north Atlantic, as indicated by the years of high frequency of icebergs penetrating south of the 48°N lat. off Newfoundland.

17.9   CONCLUSIONS

The occurrence during the 1970s and 1980s of widespread malnutrition, famine and desertification in the Sahelian belt and other drylands has generated a combination of social and environmental problems with severe effects on man, his animals and the whole environment as the resource base for human survival. The counteraction of dryland degradation is mainly achieved through an improved and ecologically adapted management of soil, water and vegetation. The repeated famines show that traditional approaches to land use in drylands are insufficient to meet the changing situations associated with increased pressure from growing populations and annually fluctuating rainfall. Research into the causes and effects of drought and desertification must continue. Surveys and planning for management aimed at ending the shortages of food, water and firewood must be introduced. Improved land use and alternative employment for part of the threatened populations represent two major strategies to be applied, and must be given a higher political priority than they receive today. Through cooperation between developing and industrialized countries, efficient monitoring of desertification factors could be developed and applied in a number of representative areas. This work should combine collection of information from three levels, namely, ground, air and space. Furthermore, such three-level monitoring systems must be based on meaningful assessments of which indicators of desertification or recovery should be used and recorded. This is a task for continued research and critical evaluation.

REFERENCES

Barth, H. K., 1982, 'Accelerated erosion of fossil dunes in the Gourma region (Mali) as a manifestation of desertification', in Yaalon, D. H. (Ed.), *Aridic soils and geomorphic processes*, Catena Suppl. 1, pp. 211–219.

Conant, F. P., 1982, 'Thorns paired, sharply recurved. Cultural controls and rangeland quality in East Africa', in Spooner, B., and Mann, H. F. (Ed.), *Desertification and Development*, Academic Press, London, pp. 111–122.

Cross, M., 1985, 'Africa's drought may last many years', *New Scientist* 17 January, 9.

Dregne, H., 1983, *Desertification of Arid Lands*, Harwood, Chur.

Eckholm, E., 1976, *Losing Ground*. Norton & Co., New York, 223 pp.

FAO, UNEP and UNESCO, 1980, *Provisional Map of Present Degradation Rate and Present State of Soil, N. Africa & S.W. Asia*, 3 maps, 1:5 mill. scale.

Goudie, A. S., 1978, 'Dust storms and their geomorphological implications', *Journal of Arid Environments*, 1, 291–311.

Grigoryev, A. A., and Kondratyev, K. J., 1981, 'Atmospheric dust observed from space', *WMO Bulletin*, 30(1), 3–9.

Hare, K., 1984, 'Recent climatic experience in the arid and semi-arid lands', *Desertification Control Bulletin*, 10, 15–22 (UNEP, Nairobi).

Helldén, U., 1984, 'Drought impact monitoring', *A Remote Sensing Study of Desertification in Kordofan, Sudan*, Report No. 61. Department of Physical Geography, University of Lund.

Ibrahim, F., 1980, 'Desertification in North-Darfur', *Hamburger Geographische Studien*, 35, 175 (Hamburg).

Jaenicke, R., 1979, 'Monitoring and critical review of the estimated source strength of mineral dust from the Sahara', in Morales, C. (Ed.), *Saharan Dust*, SCOPE report 14, Wiley, Chichester, pp. 233–242.

Kessler, E., Alexander, D. Y., and Rarick, J. F., 1978, 'Duststorms from the U.S. High Plains in late winter 1977', *Proc. Okla. Acad. Sci.*, 58, 116–128.

Kishk, M. A., 1982, 'The first national conference on the problems of land degradation in Egypt', *MAB Egypt, Nat. Comm. Bull., 3, 4,* **1982**, 13–27 (Cairo).

Lamb, P. J., 1982, 'Persistence of Subsaharan drought', *Nature*, **299**, 2, 46–47.

Le Houérou, H. N., 1976, 'Tunisia', in 'Can desert encroachment be stopped?', Rapp, A., Le Houérou, H. N., and Lundholm, B. (Eds), *Ecological Bull. NFR*, **24**, (Stockholm).

Lockeretz, E., 1978, 'The lessons of the Dust Bowl', *Am. Scientist*, **66**, 560–569.

Mabbutt, J., and Floret, C. (Eds), 1980, 'Case studies on desertification', *Nat. res. research*, **XVIII** (UNESCO).

McCauley, J. F., Breed, C. S., Grolier, M. J., and Mackinnon, D. J., 1981, 'The U.S. dust storm of February 1977', in Péwé, T. L. (Ed.), *Desert dust. Origin, Characteristics and Effect on Man*, Geol. Soc. Am. Special Paper 186, pp. 123–146.

McTainsh, G., 1984, 'The nature and origin of the aeolian mantles of central Northern Nigeria', *Geoderma*, **33**, 13–37.

Mainguet, M., Canon, L., and Chemin, M. C., 1980, 'Le Sahara: géomorphologie et paléogéomorphologie éoliennes', Williams, M. A. J., and Faure, H. (Eds), in *The Sahara and the Nile*, A. A. Balkema, Rotterdam, pp. 17–35.

Mensching, H., 1982, *Problems of the Management of Irrigated Land in Areas of Traditional and Modern Cultivation*, IGU Working Group on Resource Management in Drylands, Hamburg.

Morales, C. (Ed.), 1979, *Saharan dust*. SCOPE Report 14, Wiley, Chichester, 297 pp.

Nicholson, S., 1983, 'Sub-Saharan rainfall in the years 1976–80. Evidence of continued drought', *Monthly Weather Rev.*, **111**, 1646–1654.

Novikoff, G., 1983, 'Desertification by overgrazing', *AMBIO*, **12**(2), 102–105 (Stockholm).

Olsson, L., 1983, 'Desertification or climate?' *Lund studies in Geography*, Ser. A. 60, Gleerup, Lund.

Péwé, T. L., 1981, 'Desert dust. Origin, characteristics and effects on Man', *Geol. Soc. Am. Special Paper 186*.

Prospero, J. M., and Nees, R. T., 1977, 'Dust concentration in the atmosphere of the Equatorial North Atlantic', *Science*, **196**, 1196–1198.

Rapp, A., 1983, 'Are Terra rossa soils in Europe eolian deposits from Africa?', *Geolog. Fören. i Stockholm Förhandl.*, **105**, 161–168.

Rapp, A., Le Houérou, H. N., and Lundholm, B. (Eds), 1976, 'Can desert encroachment be stopped?', *Ecological Bull.*, **NFR**, **24**, (Stockholm).

Rognon, P., 1984, 'Les anomalies du climat actuel et les crises climatiques', *Le Courrier du CNRS.*, Suppl. 57, 60–64 (Paris).

UNCOD, 1977, *United Nations Conference on Desertification. World map of desertification*, UNEP, Nairobi.

UNEP, 1984, Governing Council report, UNEP, Nairobi, mimeo.

Yaalon, D., and Ganor, E., 1973, 'The influence of dust on soils during the Quaternary', *Soil Science*, **116**, 146–155.

Human Activity and Environmental Processes
Edited by K. J. Gregory and D. E. Walling
©1987 John Wiley & Sons Ltd.

18

A Perspective

D. E. WALLING and K. J. GREGORY

18.1 INTRODUCTION

The preceding 17 chapters have highlighted the nature and significance of human impact on a variety of environmental processes. Coverage has been selective rather than exhaustive and it should be recognized that the writers have been primarily concerned with processes which have traditionally attracted the interest of the physical geographer. A study of human impact within a broader environmental context would require extension to embrace such topics as air and water pollution, ecotoxicology, changes in global biogeochemical cycles, landscape aesthetics and the general quality of life. Within the confines of this volume it has been emphasized that man must be viewed as a highly significant force in environmental processes and as a potent instrument of change within the contemporary global environment. Many of the effects of human activity described can be classed as direct or intentional but, equally, some of the most significant and wide-ranging changes in environmental processes result from indirect or inadvertent effects. One must not, however, lose sight of the fact that human impact is relevant to only a minute recent portion of the long evolution of this planet and that the earth has been subjected to many other natural changes in environmental processes of great significance in the past. These include continental drift, tectonic activity, changing sea levels, the appearance of living organisms, Pleistocene climatic change and glaciation and other climatic fluctuations. To many, therefore, it might seem that man's activity merely concerns the embroidery on a mantle which has been tailored and restyled through the long period of geological time.

Whatever its relative significance, it is clear that human impact must play an increasingly important role in influencing environmental processes and that this impact must be accepted by the physical geographer as an important influence on contemporary landscape dynamics. Many of the changes occasioned by man's activities are also detrimental in the broader environmental context and the physical geographer must be aware of the wider implications of this topic in the field of environmental management. For example, increased flood magnitude consequent upon forest clearance and land-use change clearly demonstrates the impact of man on hydrological and channel processes, but increased flood stages may in turn cause havoc and devastation where river flood plains are occupied by intensive human activity. Similarly, it has been pointed out in Chapter 16 that desertification with its many attendant problems of human starvation and migration may be the result of albedo changes associated with overgrazing, soil erosion and increases in the dust content of the atmosphere. Situations wherein the intensification of human activity related to increased population and advancing technology gives rise to mounting environmental problems possess many of the features of negative feedback and have stimulated an increased awareness of the interaction between man and his environment. Attention is increasingly being given to these problems at the international level (18.1), new strategies are being developed to monitor changes in the global environment (18.2), procedures are being established to assess the potential impact on the environment of proposed developments in order to avoid future problems (18.3), and models are being developed to predict future trends and to evaluate the impact of alternative development scenarios in order to select that with the least environmental cost (18.4). These themes usefully demonstrate the wider relevance and implications of man's impact on environmental processes.

18.2   INTERNATIONAL AWARENESS

The growth of international awareness and activity in the field of man–environment interactions that characterized the latter years of the 1960s and the following decade has been marked in the literature by the appearance of a considerable number of acronyms for organizations and projects involved. A rapid enumeration suggests that the number involved may approach fifty or more and, while most readers will be familiar with such better-known examples as MAB, UNEP and SCOPE, others such as GEMS and GARP may be less well known.

Both the International Biological Programme (IBP) (1964–74) conceived by the International Council of Scientific Unions (ICSU), and the International Hydrological Decade (IHD) (1965–74) and its successor the International Hydrological Programme (IHP) sponsored by UNESCO, have devoted a considerable portion of their activities to the study of the impact of human activity on ecological and hydrological processes. The Man and the Biosphere (MAB) programme initiated by UNESCO in 1970 was, however, explicitly devoted to the theme of management problems arising from interaction between human activities and the natural system and the work of this intergovernmental body has been complemented by the international interdisciplinary scientific cooperation fostered by the Scientific Committee on Problems of the

Environment (SCOPE). SCOPE was created in 1969 by ISCU as a means of coordinating the relevant activities of its constituent organizations and has as its purpose 'advancing knowledge of the influence of human activities upon the environment and the effects of the resulting changes on human health and welfare, with particular reference to those influences which are global or common to several nations'. The United Nations Conference on the Human Environment which was held in Stockholm in 1972 must also rank as a major landmark in the development of international awareness. This focused public interest on the problems of environmental pollution and gave prominence to the problems associated with the planning and management of human settlements in the context of environmental quality. The associated United Nations Environmental Programme (UNEP) is specifically concerned with implementing a number of the recommendations of the Stockholm Conference and the United Nations Conference on Desertification which was held in Nairobi in 1977 arose from this body.

Other international bodies which have sponsored investigations on mans impact on the physical environment include the Intergovernmental Oceanographic Commission (IOC) of UNESCO which has established a Global Investigation of Pollution in the Marine Environment (GIPME). The Scientific Committee on Water Research (COWAR) was responsible for the conference and workshops held in Alexandria in 1976 on the problems of arid land irrigation (Worthington, 1977) and for the interdisciplinary international conference on man-made lakes which took place in Knoxville in 1971 (Ackermann *et al.*, 1971). The International Geographical Union (IGU) has also contributed to this activity and its Commission on Man and the Environment has produced reports dealing with such topics as the environmental effects of technological developments (Nelson and Scace, 1974) and of complex river development (White, 1977).

Since 1970, the terms 'environment', 'pollution' and 'environmental quality' have been used very emotively in many contexts. The recent increase in international awareness and activity relating to man's impact on the environment must therefore be viewed partly as the result of increasing scientific understanding of the potential problems and partly as a response to popular emotions. Together these two forces are beginning to create a background which should afford new insights into the nature and wider implications of this impact and an improved appreciation of means of regulating future human activity to reduce undesirable effects. The establishment of the Worldwatch Institute which now produces annual 'state of the world' reviews (e.g. Brown, 1984) and the commissioning of the Global 2000 Report to the President (see Barney, 1982) by President Carter in 1977 must both be seen as important milestones along this road. The latter report issues the stern warning:

> If present trends continue, the world in 2000 will be more crowded,
> more polluted, less stable ecologically and more vulnerable to

disruption than the world we live in now. Serious stresses involving population, resources, and environment are clearly visible ahead. Despite greater material output, the world's people will be poorer in many ways than they are today.

18.3   MONITORING ACTIVITY AND PROGRAMMES

One important outcome of the growing awareness of the significance of human impact on landscape processes has been the application of modern technology to monitoring this impact, particularly through the development of national and international monitoring programmes. Recent advances in the use of satellite imagery offer many potential applications in this field, for example, in sensing thermal pollution of water bodies and changes in vegetation cover and surface albedo and in providing potential for long-term sequential observation of specific features of the earth's surface. The use of automatic data collection platforms linked to satellite data retransmission systems similarly offers great potential for real-time data collection, particularly from remote areas.

The development of monitoring programmes designed specifically to monitor human impact has been justified as a means of establishing a baseline against which to evaluate future changes, of providing a synoptic view of contemporary environmental conditions, of detecting current changes, and of affording a basis for predicting future trends. The network of benchmark catchments established in the USA may be viewed in the light of these aims. In this case, the objective is not to monitor change but to identify a baseline of minimum human interference against which data assembled from other stations subject to human activity may be evaluated (Cobb and Biesecker, 1971). Similarly the Harmonized Monitoring Programme operated by the Department of the Environment in the UK aims to provide a countryside survey of water quality trends in major rivers and an evaluation of the flux of material from the land into the oceans (see Simpson, 1980).

At the international level particular attention must be directed to proposals for a united global environmental monitoring programme which was endorsed by the United Nations Conference on the Human Environment in Stockholm in 1972 and which form the basis of the Global Environmental Monitoring System (GEMS) promoted and coordinated by UNEP. A number of strategies have been formulated to further the GEMS objectives (e.g. SCOPE, 1973; Rockefeller Foundation, 1977) but their full implementation faces a number of problems at both the political and operational levels. Considerable progress has, however, been made. For example, the Global Water Quality Monitoring Project (GEMS/WATER) launched by UNEP, UNESCO, the World Health Organization (WHO) and the World Meteorological Organization (WMO) in 1976 aimed to develop monitoring activity on a global basis, to improve the validity and comparability of water quality data at the international level and to provide a long-term assessment of the incidence and trends of water pollution. A global data centre has been established at the Canada Centre for Inland Waters and by 1983 a network of 448 monitoring stations, including 301 river stations, 62 lake and reservoir stations and 85 groundwater stations, had been formally designated in 59 countries and data had been submitted on a routine basis from 344 of these stations. This programme is also closely linked with the SCOPE/UNEP project on Transport of Carbon and Minerals in Major World Rivers (see Degens, 1982).

Developments in global atmospheric monitoring would also seem to have successfully overcome many of the attendant practical difficulties. The Global Atmospheric Research Programme (GARP) organized jointly by ICSU and WMO and established in 1967 has made considerable progress in its attempt to assemble a global data set adequate to provide verifying data for global models of atmospheric behaviour. Similarly the Global Atmospheric Background Monitoring for Selected Environmental Parameters (BAPMON) project promises to provide a valuable global database.

Current monitoring activity can only afford a baseline against which to gauge *future* changes and is unable to provide a definitive assessment of changes that may have already occurred as a result of human impact, perhaps over a period of many centuries. Attention is therefore being given to source of evidence for reconstructing past environmental conditions. Scientists in the USA have already looked to old optical instruments to provide samples of air trapped at the time of their manufacture and deep cores from the polar ice sheets have provided a means of documenting long-term changes in precipitation chemistry. It is, however, the evidence potentially contained in lake sediments which perhaps offers the most fruitful area for future work on environmental reconstruction (see Chapter 7). The physical and chemical properties of lake sediments can provide an invaluable record of environmental conditions within their drainage basins over many thousands of years (see Oldfield, 1977; Pennington, 1981; Haworth and Lund, 1984). Flower and Battarbee (1983) and Davis *et al.* (1983) have, for example, used diatoms recovered from lake sediments to study the recent acidification of Scottish lochs and Davis (1976) has used cores from Frains Lake, Michigan, USA to reconstruct the local rates of erosion over the past 200 years.

18.4 ENVIRONMENTAL IMPACT ASSESSMENT

Recognition of the many detrimental effects of human activity on environmental processes points to the need to assess the potential impact of future activity and to regulate or modify this impact accordingly. The terms environmental impact assessment (EIA) and environmental impact statement (EIS) are now widely used to refer to studies and statements which attempt firstly to produce estimates of future environmental changes attributable to a proposed action,

TABLE 18.1   The Leopold matrix for environmental impact assessment

Part A Environmental 'characteristics' and 'conditions' (vertically in the matrix)	Part B Project actions (horizontally in the matrix)

A   PHYSICAL AND CHEMICAL CHARACTERISTICS

1   EARTH
(a) Mineral resources
(b) Construction material
(c) Soils
(d) Land form
(e) Force fields & background radiation
(f) Unique physical features

2   WATER
(a) Surface
(b) Ocean
(c) Underground
(d) Quality
(e) Temperature
(f) Recharge
(g) Snow, ice & permafrost

3   ATMOSPHERE
(a) Quality (gases, particulates)
(b) Climate (micro, macro)
(c) Temperature

4   PROCESSES
(a) Floods
(b) Erosion
(c) Deposition (sedimentation, precipitation)
(d) Solution
(e) Sorption (ion exchange, complexing)
(f) Compaction and settling
(g) Stability (slides, slumps)
(h) Stress-strain (earthquake)
(i) Air movements

B   BIOLOGICAL CONDITIONS

I   FLORA
(a) Trees
(b) Shrubs
(c) Grass
(d) Crops
(e) Microflora
(f) Aquatic plants
(g) Endangered species
(h) Barriers
(i) Corridors

A   MODIFICATION OF REGIME
(a) Exotic flora or fauna introduction
(b) Biological controls
(c) Modification of habitat
(d) Alteration of ground cover
(e) Alteration of groundwater hydrology
(f) Alteration of drainage
(g) River control and flow modification
(h) Canalization
(i) Irrigation
(j) Weather modifications
(k) Burning
(l) Surface or paving
(m) Noise and vibration

B   LAND TRANSFORMATION AND CONSTRUCTION
(a) Urbanization
(b) Industrial sites and buildings
(c) Airports
(d) Highways and bridges
(e) Roads and trails
(f) Railroads
(g) Cables and lifts
(h) Transmission lines, pipelines and corridors
(i) Barriers including fencing
(j) Channel dredging and straightening
(k) Channel revetments
(l) Canals
(m) Dams and impoundments
(n) Piers, seawalls, marinas & sea terminals
(o) Offshore structures
(p) Recreational structures
(q) Blasting and drilling
(r) Cut and fill
(s) Tunnels and underground structures

C   RESOURCE EXTRACTION
(a) Blasting and driling
(b) Surface excavation
(c) Subsurface excavation and retorting
(d) Well drilling and fluid removal
(e) Dredging
(f) Clear cutting and other lumbering
(g) Commercial fishing and hunting

TABLE 18.1 *(continued)*

Part A Environmental 'characteristics' and 'conditions' (vertically in the matrix)	Part B Project actions (horizontally in the matrix)

2 FAUNA
(a) Birds
(b) Land animals including reptiles
(c) Fish & shellfish
(d) Benthic organisms
(e) Insects
(f) Microfauna
(g) Endangered species
(h) Barriers
(i) Corridors

C CULTURAL FACTORS
1 LAND USE
(a) Wilderness & open spaces
(b) Wetlands
(c) Forestry
(d) Grazing
(e) Agriculture
(f) Residential
(g) Commercial
(h) Industrial
(i) Mining and quarrying

2 RECREATION
(a) Hunting
(b) Fishing
(c) Boating
(d) Swimming
(e) Camping & hiking
(f) Picnicking
(g) Resorts

3 AESTHETIC & HUMAN INTEREST
(a) Scenic views and vistas
(b) Wilderness qualities
(c) Open space qualities
(d) Landscape design
(e) Unique physical features
(f) Parks & reserves
(g) Monuments
(h) Rare & unique species or ecosystems
(i) Historical or archaeological sites and objects
(j) Presence of misfits

D PROCESSING
(a) Farming
(b) Ranching and grazing
(c) Feed lots
(d) Dairying
(e) Energy generation
(f) Mineral processing
(g) Metallurgical industry
(h) Chemical industry
(i) Textile industry
(j) Automobile and aircraft
(l) Oil refining
(l) Food
(m) Lumbering
(n) Pulp and paper
(o) Product storage

E LAND ALTERATION
(a) Erosion control and terracing
(b) Mine sealing and waste control
(c) Strip mining rehabilitation
(d) Landscaping
(e) Harbour dredging
(f) Marsh fill and drainage

F RESOURCE RENEWAL
(a) Reforestation
(b) Wildlife stocking and management
(c) Ground water recharge
(d) Fertilization application
(e) Waste recycling

G CHANGES IN TRAFFIC
(a) Railway
(b) Automobile
(c) Trucking
(d) Shipping
(e) Aircraft
(g) Pleasure boating
(h) Trails
(i) Cables and lifts
(j) Communication
(k) Pipeline

H WASTE EMPLACEMENT AND TREATMENT
(a) Ocean dumping
(b) Landfill

TABLE 18.1   *(opposite)*

Part A Environmental 'characteristics' and 'conditions' (vertically in the matrix)	Part B Project actions (horizontally in the matrix)
(a) Cultural patterns (life style) (b) Health and safety (c) Employment (d) Population density	(c) Emplacement of tailings, spoil and overburden (d) Underground storage (e) Junk disposal (f) Oil well flooding (g) Deep well emplacement (h) Cooling water discharge (i) Municipal waste discharge including spray irrigation (j) Liquid effluent discharge (k) Stabilization and oxidation ponds (l) Septic tanks, commercial & domestic (m) Stack and exhaust emission (n) Spent lubricants
5 MAN-MADE FACILITIES AND ACTIVITIES (a) Structures (b) Transportation network (movement, access) (c) Utility networks (d) Waste disposal (e) Barriers (f) Corridors	
D ECOLOGICAL RELATIONSHIPS SUCH AS: (a) Salinization of water resources (b) Eutrophication (c) Disease insect vectors (d) Food chains (e) Salinization of surficial material (f) Brush encroachment (g) Other	**I CHEMICAL TREATMENT** (a) Fertilization (b) Chemical deicing of highways, etc. (c) Chemical stabilization of soil (d) Weed control (e) Insect control (pesticides)
OTHERS	**J ACCIDENTS** (a) Explosions (b) Spills and leaks (c) Operational failure
	OTHERS

Source: Leopold *et al.* (1971).

and secondly attempt to suggest the likely impact of these changes on man's future wellbeing. In the USA, the National Environmental Policy Act (NEPA) of 1969 marked an important turning-point in this context, because the act required Federal Agencies to produce environmental impact assessments for all major actions, or in legislative terminology to 'identify and develop methods and procedures which will ensure that presently unqualified environmental amenities and values are given appropriate consideration in decision making along with economic and technical considerations'. By the end of 1975 statements on nearly 7000 actions had been filed with the United States Council on Environmental Control. Subsequently a number of US states and cities have adopted similar requirements and the EIA system has become firmly established as a means of providing a measure of environmental protection. This situation has produced similar formalised procedures in many other countries including,

TABLE 18.2   The reduced data matrix for a phosphate mining environmental impact statement

Project Actions

Environmental 'characteristics' and 'conditions'	II B.b. Industrial sites and buildings	II B.d. Highways and bridges	II B.h. Transmission lines	II C.a. Blasting and drilling	II C.b. Surface excavation	II D.f. Mineral processing	II G.c. Trucking	II H.c. Emplacement of tailings	II J.b. Spills and leaks
I A.2.d. Water quality					2/2	1/1		2/2	1/4
I A.3.a. Atmospheric quality						2/3			
I A.4.b. Erosion	2/2				1/1			2/2	
I A.4.c. Deposition, sedimentation	2/2				2/2			2/2	
I B.1.b. Shrubs					1/1				
I B.1.c. Grasses					1/1				
I B.1.f. Aquatic plants					2/2			2/3	1/4
I B.2.c. Fish					2/2			2/2	1/4
I C.2.e. Camping and hiking					2/4				
I C.3.a. Scenic views and vistas	2/3	2/1	2/3		3/3			2/1	3
I C.3.b. Wilderness qualities	4/4	4/4	2/2	1/1	3/3	2/5	3/5	3/5	
I C.3.h. Rare and unique species	2/5			5/10	2/4	5/10	5/10		
I C.4.b. Health and safety							3/3		

Source: Leopold *et al.* (1971)

for example, Australia, Canada, Japan, West Germany, France, Denmark and Eire.

An environmental impact statement commonly requires analysis of the physical, biological and social conditions existing prior to the proposed action, of the expected environmental impacts on those existing conditions, of any adverse effects which cannot be avoided should the proposal be implemented, of viable alternatives to the proposed action, of the relationship between local short-term uses of the environment and the maintenance of long-term productivity and stability, and of any irreversible and irretrievable commitments of resources which would be involved in the proposed action. It is clear that any attempt to 'cost' environmental changes and damage associated with a proposed development and to thereby evaluate alternative management or development strategies will prove difficult and at best *highly* subjective. Furthermore, discussion of impact in terms of human wellbeing and health, amenity recreation, agricultural productivity and similar criteria is beyond the scope of this text, but the EIA system must be viewed as a mechanism whereby many of the effects of man on environmental processes discussed in the preceding chapters can be given formal consideration.

A large number of different methodologies and procedures have been developed for formulating an environmental impact statement (e.g. Canter, 1977; SCOPE, 1979; PADC Environmental Impact Assessment and Planning Unit, 1983) and, as a single example, that produced by Professor Luna Leopold and his colleagues of the US Geological Survey (Leopold *et al.*, 1971) may be briefly reviewed. The Leopold system involves an open-cell matrix listing 100 project actions along the horizontal axis and 88 environmental characteristics and conditions which might be affected by these actions displayed along the vertical axis. These are listed in Table 18.1. A ranking system scaled from 1 (least severe) to 10 (most severe) is used to designate the *magnitude* and *importance* of each possible impact. This system was amongst the first to be developed and has been subsequently criticized because of its bias towards the physical–biological environment, its lack of objectivity and the difficulty of identifying interactions (see SCOPE, 1979), but neither these limitations nor the details of the ranking system need concern us here. Table 18.1 provides a classic example of how much of the material contained in this book could be synthesized and how it could be used to ensure that its implications are not neglected when large-scale developments are planned.

Table 18.2 illustrates the reduced matrix presented by Leopold *et al.* (1971) to demonstrate the most significant facets of an environmental impact assessment for a proposed phosphate mining development in the Los Padres National Forest, California. Clearly, the success of this type of analysis depends heavily upon the extent and adequacy of knowledge concerning the interaction of environmental processes and attributes with human activity. It must be accepted that in certain cases this knowledge will be lacking and that improved understanding of the effects of man on environmental processes will in turn improve the capability for producing meaningful environmental impact assessments. Many of the environmental problems associated with the development of the Alaskan oilfields might have been foreseen and reduced if a fuller understanding of the dynamics of permafrost areas (see Chapter 10) had been available. As knowledge accumulates, environmental impact statements will no longer be restricted to *identification* of the major areas of impact and a general measure of their severity but will provide *quantitative* estimates of the extent of the changes involved. An essential prerequisite for such progress is the development of process-based modelling strategies which permit the effects of particular human activities to be predicted and the impact of a number of alternative development scenarios to be simulated. Thus, whereas the main emphasis of this book has been in identifying the nature and magnitude of past and present human impact on environmental processes, successful development of predictive models will permit future impact to be predicted.

18.5 PREDICTION AND MODELLING STRATEGIES

A variety of predictive modelling techniques have been employed in investigating and assessing the impact of man on environmental processes and these range in complexity from simple empirical and statistical relationships to complex mathematical simulation models. Simple predictive procedures could involve little more than the recognition of basic process interrelationships and causal mechanisms and may rely heavily upon the experience and intuition of the personnel involved. Thus a forester may have developed a detailed appreciation of the links between forest clearance, logging practice, runoff processes, erosion, nutrient cycling and stream response and may be able to predict the impact of a particular forest management programme on river water quality. Similarly, a civil engineer or soil scientist could predict the likelihood of slope failure during the construction of road cuttings from a knowledge of the linkages between material properties, moisture regime and slope angle. This type of model may rely upon simple design equations embracing relatively few variables.

The University Soil Loss Equation (USLE) described in Chapter 13 provides a further example of how it is possible to produce quantitative estimates of soil loss under different land use scenarios from a knowledge of the statistical relationship between soil erosion and topography, soil type, rainfall characteristics and land management practice. Likewise, functional relationships between river channel dimensions and measures of runoff and sediment transport could be used to evaluate the potential impact of reservoir construction, river regulation or inter-basin transfers on river channel morphology.

Simulation models move beyond the use of statistically derived functional relationships and involve attempts to simulate the dynamics of the processes involved, often on a continuous basis. Whereas many statistical models are essentially empirical in nature and relate to the range of conditions from which they were derived, simulation models possess the potential for modelling environmental changes from a knowledge of process dynamics and are essentially deterministic. Perhaps the greatest advances in this sphere of modelling are associated with atmospheric and hydrological processes. Thus Thuillier (1976) describes the application of regional atmospheric quality models to the San Francisco Bay area which permit the simulation of spatial variation in the concentrations of ozone, nitrogen oxides, hydrocarbons and carbon monoxide at a regional resolution of 25 km^2. Koch *et al.* (1976) outlines the use of a multiple source urban diffusion model which has been used to analyse the effectiveness of various SO_2 control strategies in San Francisco and Boston; Wetherald and Manabe (1979) discuss the use of a simplified three-dimensional general circulation model (GCM) to predict global temperature changes resulting from an increase in the present atmospheric CO_2 concentration; and Williams (1979) reports the application of energy balance, radiative-convective and GCM models to predict the impact of large-scale energy conversion systems on global climate.

Examples of hydrological simulation models used to predict future environmental changes consequent upon human impact include that established by the Federal Institute of Hydrology in West Germany to predict the effects of hydropower development, domestic and industrial water abstraction, increasing demand for agricultural irrigation and long-term changes in rainfall regime on the low flow discharges of the River Rhine (Wildenhahn, 1984), and the Susa project model developed by the Danish Committee for Hydrology (Hansen and Dyhr-Nielsen, 1983). This latter model provides a means of simulating the effects of water resource development, and particularly increased groundwater abstraction, in the 750 km^2 Susa catchment on low flows, the capacity for sewage disposal, water quality and the recreational value of the streams. Golubev (1980) also reports a simple global model of nitrate leaching which he uses to identify those areas of the globe most susceptible to increased nitrate leaching as a result of increasing rates of fertilizer application.

Simulation models have also been developed for other suites of environmental processes, including cloud-seeding projects, coastal stability, soil profile development and soil fertility and land subsidence. In the latter case, Gambolati and his co-workers (1973, 1974) describe a mathematical simulation of the subsidence of Venice, and Final and Farouq Ali (1975) have demonstrated the possibility of simulating ground deformation associated with oil field development on the Bolivar coast of Western Venezuela. The Erosion Productivity Impact Calculator (EPIC) model currently being developed by the US Department of Agriculture also provides an interesting example of a nationally applicable model developed in the USA which permits long term forecasting of the effects of various land use and crop management scenarios on soil erosion, soil fertility and crop yields. The significance of a predictive model of this nature is clearly emphasized by Larson *et al.* (1983) who suggest that the continuation

TABLE 18.3   Environmental projections developed for 'The Global 2000 Report to the President'

Factor	Projected change	
World population	55 per cent increase	1975–2000
Number of cattle	32 per cent increase	1976–2000
Number of sheep and goats	20 per cent increase	1976–2000
Global CO_2 emissions	10–17 per cent increase	1985–1990
Global particulate emissions	11–17 per cent increase	1985–1990
CO_2 content of atmosphere	Double pre-industrial level by 2025–2050	
Global mean air temperature	Increase of $>6°C$ by 2150–2200	
Global desert area	62 per cent increase	1975–2000
Global closed forest area	17 per cent decrease	1978–2000
Global arable land area	4 per cent increase	1975–2000
Global fertilizer consumption	275 per cent increase	1972–2000
Global irrigated area	22 per cent increase	1975–1990
Global water use	274 per cent increase	1967–2000
Land area utilized for global mineral production	206 per cent increase	1976–2000
Solid waste generated by global extraction of stone, clay, iron ore, phosphate rock, copper and uranium	252 per cent increase	1976–2000

Based on Barney (1982).

of 1977 rates of soil erosion in the USA for 100 years would reduce the yields of the nation's cropland by a minimum of 5–10 per cent, assuming that technology and other factors remain constant.

A number of examples of the practical application of models to assist in environmental impact assessment are provided by studies of proposals for large-scale development of the shale-oil industry and coal mining in Colorado, Wyoming and Montana, USA. Guy (1977) outlines how various statistical models were used to furnish sediment data relating to the potential environmental impact of an extension of open-cast coal mining near Decker, Montana. Information was assembled concerning erosion and sediment yield under natural conditions and after reclamation of the mined area, sedimentation of adjacent reservoirs during the mining period, and problems associated with diverting six ephemeral streams away from the disturbed area during the period of mining activity and returning these streams to a man-made valley across the reclaimed spoil upon completion of mining. More sophisticated simulation techniques were employed to evaluate the effects of increased coal exploitation and associated economic development on the regional water resources of the Yampa River basin in Colorado and Wyoming (Steele, 1978). Projections indicated that coal production in this area could triple within the next 15 years and the population was expected to double or triple over the same period. Much of the coal mined was to be converted to electricity or gas locally. Seven coal-resource development alternatives were identified for this region and models were used to evaluate hydrological problems associated with wastewater discharges, with the construction of proposed reservoirs, with increased sediment yield from surface mining and construction activities, and with disruption of groundwater recharge areas by mining and degradation of groundwater quality by disposal of waste from the coal-conversion processing plants. For example, maps were produced indicating the likely increases in the dissolved solids concentration in groundwater around two waste disposal areas after 20, 60 and 200 years.

The scope and value of predictive techniques is also well demonstrated by the environmental projections contained in *The Global 2000 Report to the President. Entering the Twenty-First Century* (Barney, 1982). This report, commissioned by President Carter and produced by a team of US scientists, provides a global view of the interdependence of population, resources and environment and of the growing number of environmental problems facing the contemporary world. It is clear from Table 18.3 that the 55 per cent increase in global population forecasted for the period 1975–2000 and the rapid pace of technological development is likely to generate a major impact on the global environment.

18.6   THE PROSPECT

Written more than 120 years after George Perkin Marsh's *Man and Nature* and following after a century of further work, a decade of papers, and a decade of readings (Chapter 1), this text cannot claim a novel message or revelative conclusions. Nevertheless, it is hoped that the emphasis on processes and the attempt to highlight and elucidate the role of man in influencing the dynamics of the physical environment provides its justification. Its chapters emphasize that man must be viewed as a potent influence on contemporary environmental processes. Just as Nir (1983) argues for anthropic geomorphology as a separate branch of geomorphology specifically concerned with the effects of human activity, so similar justifications could be advanced for other fields of process study. Recognition of the significance of human activity should also generate an awareness of the many environmental problems that may stem from these influences and encourage the physical geographer to play a role in developing improved understanding of potential problems involved. Incorporation of knowledge of the various interactions between human activity and environmental processes into models should provide a means of extending the scope of environmental impact assessments beyond enumeration of general areas of impact towards quantitative estimates of the nature and extent of that change.

The contemporary world provides evidence of considerable change within the physical environment as a consequence of the impact of human activity. One can only speculate on the extent to which the pace of change will quicken as a result of technological advancement and increasing population pressure. It could be argued that man's power to disturb is even now running ahead of available methods for assessing and responding to environmental impact and the theme of this book will doubtless continue to represent a profitable avenue for future study and research.

REFERENCES

Ackermann, W. C., White, G. F., and Worthington, E. B. (Eds.), 1971, *International Symposium on Man-made Lakes*, Amer. Union, Geophys. Monog., 17.

Barney, G. P., 1982, *The Global 2000 Report to the President. Entering the Twenty-First Century*, Penguin, Harmondsworth.

Brown, L. R., (Ed.), 1984, *State of the World 1984: A Worldwatch Institute Report on Progress toward a Sustainable Society*, Norton, New York.

Canter, L. W., 1977, Environmental Impact Assessment, McGraw-Hill, New York.

Cobb, E. D., and Biesecker, J. E., 1971, 'The national hydrologic bench-mark network', *US Geol. Survey Circular, 460-D*.

Davis, M. B., 1976, 'Erosion rates and land use history in Southern Michigan', *Environmental Conservation*, **3**, 139–148.

Davis, R. B., Norton, S. A., Hess, C. T., and Brakke, D. F., 1983, 'Paleolimnological reconstruction of the effects of atmospheric deposition of acids and heavy metals on the chemistry and biology of lakes in New England and Norway', in Merilainen, J., Huttunen, P., and Battarbee, R. W. (Eds), *Paleolimnology*, Junk, The Hague, pp. 113-123

Degens, E. (Ed.), 1982, 'Transport of carbon and minerals in major world rivers', *Mitt. Geol. Palaont Inst. Hamburg*, **52**.

Final, A., and Farouq Ali, S. M., 1975, 'Numerical simulation of oil production with simultaneous ground subsidence', *Jnl. Soc. Petroleum Eng.*, **15**, 411–424.

Flower, R. J., and Battarbee, R. W., 1983, 'Diatom evidence for recent acidification of two Scottish lochs', *Nature*, **305**, 130–133.

Gambolati, G., and Freeze, R. A., 1973, 'Mathematical simulation of the subsidence of Venice, 1: Theory', *Water Resources Res.*, **9**, 721–733.

Gambolati, G., Gatto, P., and Freeze, R. A., 1974, 'Mathematical simulation of the subsidence of Venice, 2: Results', *Water Resources Res.*, **10**, 563–577.

Golubev, G. N., 1980, 'Nitrate leaching hazards: A look at the potential global situation', *IIASA Working Paper WP-80-89*.

Guy, H. P., 1977, 'Sediment information for an environmental impact statement regarding a surface coal mine, western United States', *Erosion and Solid Matter Transport in Inland Waters*, IAHS Publ., 122, pp. 98–108.

Hansen, E., and Dyhr-Nielsen, M., 1983, 'The Susa project: Modelling for water-resources management', *Nature and Resources*, **19**, 10–18.

Haworth, E. Y., and Lund, J. W. G. (Eds), 1984, *Lake Sediments and Environmental History*, Leicester University Press, Leicester.

Koch, R. C., Felton, D. J., and Hwang, P. H., 1976, 'Sampled chronological input model (SCIM) applied to air quality planning in two large metropolitan areas', *Environmental Modelling and Simulation, US Env. Protection Agency Publ., EPA-600/9-76-016*, pp. 92–96.

Larson, W. E., Pierce, F. J., and Dowdy, R. H., 1983, 'The threat of soil erosion to long-term crop production', *Science*, **219**, 458–465.

Leopold, L. B., Clark, F. E., Hanshaw, B. B., and Balsley, S. R., 1971, 'A procedure for evaluating environmental impact', *US Geol. Surv. Circular, 645*.

National Academy of Sciences, 1976, 'Implementation of the global environment monitoring system', *Int. Env. Programs Committee Rept.*

Nelson, J. G., and Scace, R. C. (Eds), 1974, *Impact of Technology on Environment: Some Global Examples*, University of Calgary, Calgary.

Nir, D., 1983, *Man, A Geomorphological Agent: A Introduction to Anthropic Geomorphology*, Reidel, Dordrecht.

Oldfield, F., 1977, 'Lakes and their drainage basins as units of sediment-based ecological study', *Progress in Physical Geography*, **1**, 460–504.

PADC Environmental Impact Assessment and Planning Unit, (Ed), 1983, *Environmental Impact Assessment*, Martinus Nijhoff, The Hague.

Pennington, W., 1981, 'Records of a lake's life in time: the sediments', *Hydrobiologia*, **79**, 197–219.

Rockefeller Foundation, 1977, *International Problems in Environmental Monitoring*, Rockefeller Foundation, New York.

SCOPE (Scientific Committee on Problems of the Environment), 1973, *Global Environmental Monitoring System: Action Plan for Phase 1*, SCOPE, Paris.

SCOPE, 1979, *Environmental Impact Assessment*, Wiley, Chichester.

Simpson, E. A., 1980, 'The harmonization of the monitoring of the quality of rivers in the United Kingdom', *Hydrol. Sci. Bull.*, **25**, 13–23.

Steele, T. D., 1978, 'Assessment techniques for modelling water quality in a river basin impacted by coal-resource development', *Modelling the Water Quality of the Hydrological Cycle*, IAHS Publ. 125, pp. 322–332.

Thuillier, R. H., 1976, 'Air quality modelling—a user's viewpoint', *Environmental Modelling and Simulation, US Env. Protection Agency Publ., EPA-600/9-76-016*, pp. 35–39.

Wetherald, R. T., and Manabe, S., 1979, 'Sensitivity studies of climate involving changes in CO_2 concentrations', in Bach, W., Pankrath, J., and Kellogg, W. (Eds), *Man's Impact on Climate*, Elsevier, Amsterdam, pp. 57–64.

White, G. F. (Ed.), 1977, *Environmental Effects of Complex River Developments*, Westview Press, Boulder.

Wildenhahn, E., 1984, 'Changes of low water conditions in the Rhine River basin', *Problems of Regional Hydrology*, Institut fur Physische Geographie, Freiburg, pp. 36–37.

Worthington, E. B. (Ed.), 1977, *Arid Irrigation in Developing Countries: Environmental Problems and Effects*, Pergamon, Oxford.

Williams, J., 1979, 'Modelling the impact of large-scale energy conversion systems on global climate', in Bach, W., Pankrath, J., and Kellogg, W. (Eds), *Man's Impact on Climate*, Elsevier, Amsterdam, pp. 253–267.

Index

Subjects indexed below are restricted to major references. Locations and authors are not included. References to illustrations and tables are indicated by numbers in italics. Mr C. T. Hill kindly compiled the index.